Cycles and Bridges in Graphs

Mathematics and Its Applications *(East European Series)*



Volume 49

# Cycles and Bridges in Graphs

*by*


Heinz-Jürgen Voss
*College of Education Dresden, Germany*

KLUWER ACADEMIC PUBLISHERS
DORDRECHT / BOSTON / LONDON

DEUTSCHER VERLAG DER WISSENSCHAFTEN
BERLIN

Library of Congress Cataloging-in-Publication Data

Voss, Heinz-Jürgen.
Cycles and bridges in graphs / Heinz-Jürgen Voss.
284 p. 24 cm. – (Mathematics and its applications (East European series) ; v. 49)
Includes bibliographical references and indexes.
ISBN 0-7923-0899-9 (acid free paper)
1. Paths and cycles (Graph theory) 2. Bridges (Graph theory) I. Title. II. Series: Mathematics and its applications (Kluwer Academic Publishers). East European series ; v. 49.
QA166.22.V67 1991
511′.5–dc20 90-43689

ISBN 0-7923-0899-9

*Language editing:* E. Primrose, Leicester, U. K.

Kluwer Academic Publishers incorporates
the publishing programmes of
D. Reidel, Martinus Nijhoff, Dr W. Junk and MTP Press.

Distributors

*for the United States and Canada:*
Kluwer Academic Publishers, 101 Philip Drive, Norwell, MA 02061, USA

*for Albania, Bulgaria, China, Cuba, Czechoslovakia, Germany, Hungary, Democratic People's Republic of Korea, Mongolia, Poland, Roumania, U.S.S.R., Vietnam, and Yugoslavia:*
Deutscher Verlag der Wissenschaften GmbH, P. B. 1216, O-1080 Berlin, Germany, ISBN 3-326-00368-4 / Mathematische Monographien, Band 23, ISSN 0543-1042

*for all other countries:*
Kluwer Academic Publishers Group, P. O. Box 322, 3300 AH Dordrecht, The Netherlands

Joint edition published by
Kluwer Academic Publishers, Dordrecht, The Netherlands and
Deutscher Verlag der Wissenschaften GmbH, Berlin, Germany.



PRINTED IN GERMANY

# Contents

# Series Editor's Preface

'Et moi, ..., si j'avait su comment en revenir, je n'y serais point allé.'

Jules Verne

The series is divergent; therefore we may be able to do something with it.

O. Heaviside

One service mathematics has rendered the human race. It has put common sense back where it belongs, on the topmost shelf next to the dusty canister labelled 'discarded nonsense'.

Eric T. Bell

Mathematics is a tool for thought. A highly necessary tool in a world where both feedback and nonlinearities abound. Similarly, all kinds of parts of mathematics serve as tools for other parts and for other sciences.

Applying a simple rewriting rule to the quote on the right above one finds such statements as: 'One service topology has rendered mathematical physics ...'; 'One service logic has rendered computer science ...'; 'One service category theory has rendered mathematics ...'. All arguably true. And all statements obtainable this way form part of the raison d'être of this series.

This series, *Mathematics and Its Applications,* started in 1977. Now that over one hundred volumes have appeared it seems opportune to reexamine its scope. At the time I wrote

"Growing specialization and diversification have brought a host of monographs and textbooks on increasingly specialized topics. However, the 'tree' of knowledge of mathematics and related fields does not grow only by putting forth new branches. It also happens, quite often in fact, that branches which were thought to be completely disparate are suddenly seen to be related. Further, the kind and level of sophistication of mathematics applied in various sciences has changed drastically in recent years: measure theory is used (non-trivially) in regional and theoretical economics; algebraic geometry interacts with physics; the Minkowsky lemma, coding theory and the structure of water meet one another in packing and covering theory; quantum fields, crystal defects and mathematical programming profit from homotopy theory; Lie algebras are relevant to filtering; and prediction and electrical engineering can use Stein spaces. And in addition to this there are such new emerging subdisciplines as 'experimental mathematics', 'CFD', 'completely integrable systems', 'chaos, synergetics and large-scale order', which are almost impossible to fit into the existing classification schemes. They draw upon widely different sections of mathematics."

By and large, all this still applies today. It is still true that at first sight mathematics seems rather fragmented and that to find, see, and exploit the deeper underlying interrelations more effort is needed and so are books that can help mathematicians and scientists do so. Accordingly MIA will continue to try to make such books available.

If anything, the description I gave in 1977 is now an understatement. To the examples of interaction areas one should add string theory where Riemann surfaces, algebraic geometry, modular functions, knots, quantum field theory, Kac-Moody algebras, monstrous moonshine (and more) all come together. And to the examples of things which can be usefully applied let me add the topic 'finite geometry'; a combination of words which sounds like it might not even exist, let alone be applicable. And yet it is being applied: to statistics via designs, to radar/sonar detection arrays (via finite projective planes), and to bus connections of VLSI chips (via difference sets). There seems to be no part of (so-called pure) mathematics that is not in immediate danger of being applied. And, accordingly, the applied mathematician needs to be aware of much more. Besides analysis and numerics, the traditional workhorses, he may need all kinds of combinatorics, algebra, probability, and so on.

In addition, the applied scientist needs to cope increasingly with the nonlinear world and the extra mathematical sophistication that this requires. For that is where the rewards are. Linear models are honest and a bit sad and depressing: proportional efforts and results. It is in the nonlinear world that infinitesimal inputs may result in macroscopic outputs (or vice versa). To appreciate what I am hinting at: if electronics were linear we would have no fun with transistors and computers; we would have no TV; in fact you would not be reading these lines.

There is also no safety in ignoring such outlandish things as nonstandard analysis, superspace and anticommuting integration, $p$-adic and ultrametric space. All three have applications in both electrical engineering and physics. Once, complex numbers were equally outlandish, but they frequently proved the shortest path between 'real' results. Similarly, the first two topics named have already provided a number of 'wormhole' paths. There is no telling where all this is leading – fortunately.

Thus the original scope of the series, which for various (sound) reasons now comprises five subseries: white (Japan), yellow (China), red (USSR), blue (Eastern Europe), and green (everything else), still applies. It has been enlarged a bit to include books treating of the tools from one subdiscipline which are used in others. Thus the series still aims at books dealing with:

- a central concept which plays an important role in several different mathematical and/or scientific specialization areas;
- new applications of the results and ideas from one area of scientific endeavour into another;
- influences which the results, problems and concepts of one field of enquiry have, and have had, on the development of another.

This is a systematic book on bridges and cycles in graphs. The simplest kind of bridge is an edge (together with its adjacent vertices) whose removal increases the number of components of the graph. The general notion of a bridge (relative to a subgraph) is somewhat more sophisticated and it is also a relatively new concept – dating from 1966. It has turned out to be a quite powerful one, suitable for the analysis of a large number of problems in graph theory, particularly problems concerning cycles. The very well-known author of this substantial volume has contributed greatly to the topic of bridges and cycles in graphs and a great deal included here is related to his own work.

It is my own personal impression that bridges in graphs constitute one of those nonsurface concepts which appear in a subject when it has reached a certain maturity and sophistication and which then turn out to be influential for a long time. That seems to me a good reason to welcome this book, the first systematic one on these topics, in this series.

The shortest path between two truths in the real domain passes through the complex domain.

J. Hadamard

La physique ne nous donne pas seulement l'occasion de résoudre des problèmes ... elle nous fait pressentir la solution.

H. Poincaré

Never lend books, for no one ever returns them; the only books I have in my library are books that other folk have lent me.

Anatole France

The function of an expert is not to be more right than other people, but to be wrong for more sophisticated reasons.

David Butler

Bussum, July 1990

Michiel Hazewinkel

# Preface

This book is based on lectures given to students and colleagues at the Ilmenau Institute of Technology (GDR), at the Lenin University Kishinev (USSR) and at the College of Education Dresden (GDR). A series of lectures was given at the 30th Semester on Combinatorics and Graph Theory at the Stefan Banach International Mathematical Center in Warsaw (Poland) in 1987. The book is devoted to cycles and bridges in graphs, particularly, to bridges of cycles, where throughout the book a graph is understood to be a finite, undirected graph without loops and multiple edges.

The emphasis of the book is on the concept of a "bridge", particularly, on the concept of a "bridge of a cycle". The concept of a bridge was first introduced by W. T. Tutte, 1966, and O. Ore, 1967a. It is widely used in graph theory, where the bridges of a subgraph $J$ of a graph $G$ are, roughly speaking, the minimal subgraphs that are attached to the remainder of the graph at vertices of $J$, the so-called vertices of attachment of the corresponding bridge. Thus a bridge is either a "diagonal" of $J$ or it is obtained from a component $K$ of $G-J$ by adding all edges incident with $K$ and all vertices of $J$ adjacent to $K$. In most cases $J$ ist a cycle.

Nowadays, the concept of bridge can be found in introductory text-books on graph theory, for example, it is used in the second volume of the book "Einführung in die Theorie der endlichen Graphen" by H. Sachs, 1972, and in the introductory book "Graph Theory with Applications" by J. A. Bondy and U. S. R. Murty, 1976.

Different authors use different names for bridges: a bridge is called $J$-component, bridge, $J$-fragment by W. T. Tutte, 1966, O. Ore, 1967a, and C. Thomassen, 1981b; piece by C. Berge, 1970; Gespinst by H. Sachs, 1972; Gespinst and Spinnwebe by H. Walther and H.-J. Voss, 1974; Querstücke and Relativkomponente by R. Bodendiek and K. Wagner, 1985; and Brücke by R. Halin, 1981.

Bridges were first intensively used in investigating cycles of planar graphs. The first and most famous example is the proof of Tutte's well-known theorem that every planar 4-connected graph has a Hamiltonian cycle. The proof is accomplished by showing that a non-Hamiltonian 2-connected planar graph always contains a cycle $C$ such that each of its bridges $B$ has at most three vertices of attachment, i.e., $B$ and $C$ have at most three common vertices. Thus the most famous paper on this subject is Tutte's article "Bridges and Hamiltonian circuits in planar graphs" (W. T. Tutte, 1977).

The concept of bridge has also been successfully used in investigating properties of graphs related to the connectivity of a graph. Here we mention Tutte's book "Connectivity in graphs" (W. T. Tutte, 1966) and papers of Mader, Halin and others.

Concerning these two subjects we recommend to the reader Tutte's book "Graph theory" (W. T. Tutte, 1984). A good survey on these topics can also be found in Hobbs' article "$J$-components, Bridges, and $J$-fragments" (A. M. Hobbs, 1979). H. Sachs initiated the concept "bridge" to be used not only in investigating planar graphs but also in studying nonplanar graphs.

Results have been obtained in the following directions:

1) Overlap graphs (bridge graphs). Applications of the results to planar graphs.
2) Non-separating cycles.
3) The "length" of bridges of longest cycles in 2-connected graphs.
4) The circumference of bridges of longest cycles in 2-connected graphs.
5) The class of all 2-connected graphs with given circumference. Applications of the results to critical graphs.
6) The length of cycles and the number of diagonals of cycles in graphs with given minimum degree and/or with given girth.
7) Bridges of longest cycles in graphs with given minimum degree. Extensions of theorems of G. A. Dirac and O. Ore.
8) Cycles with prescribed bridges.
9) 2-connected graphs with given maximum degree and given circumference having maximal order.

In the Introduction of this book a detailed survey on the problems and results dealt with in this book is presented.

If one considers the subgraph $G-H$ of a graph $G$ with subgraph $H$, then one automatically investigates properties of $H$-bridges. Hence many problems and results are related to bridges in graphs. As a consequence of this we restrict our attention to the nine subjects above.

We can consider two directions of applying bridges in graph theory:

A) Bridges can be subjects in graph theory, e.g., in the papers of J. A. Bondy, 1980b, and H. J. Veldman, 1983b, who present sufficient conditions that a cycle or path has bridges with prescribed properties. So Veldman defined a so-called $D_\lambda$-cycle, which is a cycle $C$ such that $G-C$ has only components of order less than $\lambda$.

B) Theorems of graph theory can be proved by a "bridge method", i.e., results of graph theory can be obtained by studying bridges of certain subgraphs. For example, W. T. Tutte's famous theorem on Hamiltonian cycles in planar 4-connected graphs has been proved by finding a cycle with bridges of a certain type.

Both directions A) and B) will be presented in this book. To a great extent the book is based on the author's own research work.

Many people have contributed, either directly or indirectly, to this book. First of all, I wish to express my gratitude to Professors G. A. Dirac, T. Gallai and H. Sachs, through whom I have come to love graph theory, and from whom in principle I have learnt almost everything about research work in graph theory.

I am particularly indebted to Professors R. Bodendiek, I. A. Bondy, C. Hoede, H. Sachs, G. Schaar, D. M. Stiebitz, B. Toft and H. Walther for valuable suggestions, Professors H. A. Jung, M. Simonovits, B. Toft, T. Zamfirescu and Dr. W. Wessel for reading some chapters. Particularly I wish to express my thanks to Dr. K. Gräf who introduced me to the English language, twenty years ago, and now carefully checked the English text. I would like to thank Deutscher Verlag der Wissenschaften and especially Mrs. B. Mai for their efficiency and great skill in producing this book. But above all I have to thank my wife Waltraud; her unbroken encouragement and moral support have been the firm basis of my work.

Heinz-Jürgen Voss

Dresden, April 1990

# Introduction

The material of the book is presented in eleven chapters. Each chapter is introduced by a section 0. This contains a detailed survey of the problems and results which are dealt with in the chapter or are closely related to the subject of the chapter. In particular, the latter problems and results are discussed in detail.

In this Introduction we give a short survey of the material presented in this book. If anyone is interested in a more detailed survey of the contents of any chapter, then we recommend him to read the introductory Section 0 of that chapter.

**Chapter 1** presents the fundamentals of bridges in graph theory, in particular, on bridges of cycles. For this purpose we introduce the concept of overlap graph (bridge graph by Tutte). The vertices of this graph are the bridges of a given cycle, where two vertices are joined by an edge if the corresponding bridges overlap.

Properties of overlap graphs are presented. We include Tutte's results that

1) a 2-connected graph $G$ is planar if and only if for each cycle $C$ of every subgraph $H$ of $G$ the overlap graph $O(H:C)$ is bipartite, and

2) two vertices $X_1$, $X_2$ of a planar graph $G$ are on no common face in any embedding of $G$ if and only if the overlap graph of some cycle of $G$ has two adjacent bridges $B_1$, $B_2$ such that $X_i$ is an inner vertex of $B_i$, $i = 1, 2$.

From Tutte's characterization of planar 3-connected graphs a proof of K. Kuratowski's theorem is derived.

In **Chapter 2** separating and peripheral cycles are treated. A cycle is separating if it has at least two regular bridges. An induced cycle is called peripheral if it has at most one bridge.

Thomassen's characterization of all graphs having no separating cycles and Toft's and Thomassen's characterization of all graphs without peripheral cycles are presented. We include the following theorems that

1) each edge of any 3-connected graph is in at least two peripheral cycles,

2) a 3-connected graph is non-planar if and only if one edge is contained in at most three peripheral cycles, and

3) the peripheral cycles of a 3-connected graph generate the cycle space.

These results are due to Kelmans, McLane and Tutte.

**Chapter 3** is devoted to the so-called length of an $H$-bridge $B$, where $H$ is a certain proper subgraph of $G$. If an $H$-bridge $B$ has $\alpha \geqq 2$ vertices of attachment, then, obviously, $B$ contains a tree whose $\alpha$ end vertices are the $\alpha$ vertices of attachment of $B$. The number of edges of a largest such tree is called the length of $B$. The length of bridges is a useful tool in investigating bridges of longest cycles, odd and even cycles and smallest cycles. We include Hauschild's theorem that the circumference of a bridge of a longest cycle is smaller than the circumference of the whole graph. Extensions and variations of this theorem will be given.

Properties of vertices of longest cycles which are neighbours of vertices of attachment of some bridge are studied; Woodall's well-known hopping lemma is included as one of the deepest results in the theory of longest cycles. The intersection pattern of two or more longest cycles is presented, in particular, results of Grötschel and Zamfirescu are included.

In **Chapter 4** the following two classes of graphs are considered. The class $\mathfrak{Z}_l$ of all 2-connected graphs with circumference $l$ and the class $\mathfrak{Y}_\lambda$ of all connected graphs with maximum path length $\lambda$.

The graphs of $\mathfrak{Z}_l$ and $\mathfrak{Y}_\lambda$ of large order have the following property: they contain many isomorphic bridges, where two $H$-bridges $B$, $B'$ are called isomorphic if they have the same set of vertices of attachment and there is an isomorphism $\varphi$ of $B$ onto $B'$ such that the vertices of attachment are fixed under the action of $\varphi$.

The results on $\mathfrak{Z}_l$ and $\mathfrak{Y}_\lambda$ are due to Voss and Simonovits, respectively. Reductions in $\mathfrak{Z}_l$ and $\mathfrak{Y}_\lambda$ are introduced.

**Chapter 5** presents a small contribution to the reconstruction of graphs. The investigation of properties of isomorphic bridges is continued.

Two non-adjacent vertices are said to be symmetric if they are kernels of isomorphic bridges, i.e., these two vertices have the same neighbourhood. Dörfler and Imrich proved that if $G$ contains at least one class with at least $k$ symmetric vertices, $k \geqq 2$, and no class with precisely $k-1$ symmetric vertices, then $G$ is reconstructable. The result of Dörfler and Imrich will be extended to graphs having large classes of isomorphic bridges with certain properties.

In **Chapter 6** the investigations of isomorphic bridges are continued further. The results of Chapter 4 are extended to various classes of graphs and applied to both to critically $k$-chromatic graphs and to graphs which are edge-critical with respect to the independence number. We succeed in showing that if these graphs have a large order and contain a subgraph of a given structure, then they also contain a "large" subgraph of the same structure.

As a byproduct a result of Voss is presented that in every 2-connected non-bipartite graph the length of a largest odd cycle is not greater than twice the length of an even cycle diminished by 2.

In **Chapter 7** we investigate valency conditions for the existence of cycles of given lengths. An example of such a problem is the well-known result of Dirac that each 2-connected graph with minimum degree $\delta(\geqq 2)$ has a Hamiltonian cycle or a cycle of length at least $2\delta$. A second example generalizing Dirac's theorem is a result of Bondy, Bermond and Linial, which gives Ore-type conditions on the valencies ensuring the existence of a long cycle. They proved for 2-connected graphs that if the degree sum of any pair of non-adjacent vertices is at least $m$, then the graph $G$ has a Hamiltonian cycle or a cycle of length at least $m$.

These theorems are extended to assertions on cycles through given vertices and edges, on odd and even cycles, and on cycles in bipartite and regular graphs. In particular, the graphs of these theorems are characterized on which the given lower bounds are attained.

Jackson's well-known result is also included; this asserts that every 2-connected $r$-regular graph on at most $3r$ vertices has a Hamiltonian cycle. The chapter will be concluded with valency conditions on the existence of cycles having only bridges of a given type. An example is the investigation of cycles $C$ which have only bridges with fewer than $\lambda$ inner vertices ($\lambda$ an integer). If $\lambda = 2$, then $C$ is said to be a dominating cycle.

**Chapter 8** ist devoted to valency conditions for the existence of cycles with a given number of diagonals.

Many problems are known in which diagonals of cycles play an important role. Such problems are the strong perfect graph conjecture, Hadwiger's conjecture and Kelmans' conjecture. The solution of Kelmans' conjecture (independently by Kelmans and Thomassen) states that every 3-connected non-planar graph on at least six vertices has a cycle with three pairwise crossing chords (a refinement of Kuratowski's theorem for 3-connected graphs). The strong perfect graph conjecture asserts that $G$ is perfect if and only if in both $G$ and its complement $\overline{G}$ every odd cycle of length at least 5 has a diagonal. Chordal graphs, parity graphs and other graphs are also defined by conditions involving the number of diagnonals in certain cycles.

One of the main results of the chapter is the proof of the assertion that, except for three graphs, every non-bipartite 2-connected graph with minimum degree at least 3 contains both an odd and an even cycle with two or more chords. This result implies that each 4-chromatic graph $G$, $G \ncong K_4$, contains an odd cycle with two diagonals, as Erdős has conjectured.

In **Chapter 9** the investigations of Chapters 7 and 8 are continued. We deal with conditions for the existence of long cycles and of cycles with many diagonals. Partially, graphs of high minimum degree or/and large girth are to be studied. Sometimes the connectivity of the graphs is involved. A typical result presented is that every non-bipartite 2-connected graph $G$ with minimum degree at least 3 and girth at least $t \geqq 4$ contains both an odd and an even cycle of length at least $2^{t/4-7}$ and both an odd and an even cycle with at least $2^{t/12-8}$ diagonals.

In **Chapter 10** we investigate conditions on the number of edges of graphs (their size) ensuring the existence of a "long" (odd or even) cycle or of an (odd or even) cycle with "many" diagonals. Such problems belong to the so-called extremal graph theory.

Turán's theorem on the size of graphs of order $n$ ensuring the existence of a complete $k$-graph is the most famous theorem in this field.

A common feature of all extremal problems is that they can be carried over from questions involving the valencies of the vertices into questions involving the number of edges. Therefore, many theorems of Chapters 7-9 can be carried over to extremal graph theorems. An example is Dirac's theorem that any 2-connected graph with minimum degree $\delta\,(\geqq 2)$ has a Hamiltonian cycle or a cycle of length at least $2\delta$. This is used in the proof of the well-known theorem of Erdős and Gallai that each graph of order $n$ and at least $\frac{l}{2}(n-1)$ edges contains a cycle of length at least $l$.

We dealt with graphs of given *minimum* degree in Chapters 7-9. In **Chapter 11** the class of all 2-connected graphs of order $n$ with given *maximum* degree $\Delta$ is considered. By a result of Bondy and Entringer the smallest circumference $c_\Delta(n)$ in this class satisfies the inequalities

$$4 \log_{\Delta-1} n - 4 \log_{\Delta-1} \log_{\Delta-1} n - 20 < c_\Delta(n) < 4 \log_{\Delta-1} n + 4.$$

The following subclasses are of special interest: the subclass of all cubic graphs and the subclass of all planar cubic graphs. In both subclasses the smallest circumference, denoted by $\bar{c}_3(n)$ and $\tilde{c}_3(n)$, respectively, is investigated. For all even $n$ I have succeeded in determining all graphs in these subclasses at which the circumference attains the value $\bar{c}_3(n)$ or $\tilde{c}_3(n)$, respectively.

A survey of results on subclasses of higher connectivity is given.

With respect to the notation of the theorems, propositions and lemmas we remark that they are denoted by $i.j.k$, where $i$, $j$, $k$ are three integers; $i$ denotes the chapter, $j$ the section and $k$ the number of the theorem, or proposition, or lemma in section $i.j$.

## Basic concepts

In this book only undirected finite graphs will be considered that are simple in general, i.e., they have no loops and no multiple edges.

We believe that the reader will be acquainted with the fundamentals of graph theory. But to speak a common language we present the basic concepts and their notation here.

We remark that the concept of bridge will be presented in detail in section 1.1.

The concepts are from J.A. Bondy and U.S.R. Murty, 1976, and B. Bollobás, 1978a. The notation is slightly different and will be presented here. This thorough presentation of the concepts of graph theory is somewhat similar to that of B. Bollobás in his book "Extremal Graph Theory" (B. Bollobás, 1978a). Unless otherwise stated every set is finite. The number of elements of a set $M$ is denoted by $|M|$, and $M \subset N$ means $M \subseteqq N$ *and* $M \neq N$. The set of all 2-subsets of $M$ is denoted by $M^{(2)}$.

A graph $G$ is an ordered pair of disjoint sets $(V, E)$ such that $E \subseteqq V^{(2)}$ and $V \neq \emptyset$. The set $V$ is the set of vertices of $G$ and $E$ is the set of edges. An edge $\{X, Y\}$ is said to *join* or *link the vertex X to the vertex Y* and is denoted by $(X, Y)$. We also say that $X$ and $Y$ are *adjacent vertices* and the vertex $X$ is *incident with the edge* $(X, Y)$. Two distinct edges with a common end vertex are *adjacent.* Two graphs are *isomorphic* if there exists a 1-1 correspondence between their vertex sets that preserves adjacency in both directions. Usually we make no distinction between isomorphic graphs, unless we want to specify the vertices and edges. If $G$ and $H$ are isomorphic graphs, we write $G \cong H$. The *vertex set of a graph G* is denoted by $V(G)$ and the *edge set* by $E(G)$. If the letter $G$ occurs without explanation, then it stands for an arbitrary graph. Vertices of $G$ are denoted by latin capitals, sometimes with subscripts. Small latin letters (sometimes with subscripts) denote edges of $G$. Instead of $X \in V(G)$ and $e \in E(G)$ we usually write $X \in G$ and $e \in G$, to denote that $X$ is a vertex of $G$ and $e$ is an edge of $G$, respectively. The number of vertices of $G$, called the *order* of $G$, is denoted by $n(G)$. The number of edges of $G$, called the *size* of $G$, is denoted by $e(G)$.

The graph $G' = (V', E')$ is a *subgraph* of a graph $G = (V, E)$ if $V' \subseteqq V$ and $E' \subseteqq E$. In this case we write $G' \subseteqq G$. If $G'$ and $G$ have no common vertices, then $G' \subseteqq G$ means that $G$ contains a subgraph $H$ such that $G' \cong H$. If $V' = V$, then $G'$ is a *spanning subgraph* of $G$. If $W \subseteqq V$, then the graph $(W, E \cap W^{(2)})$ is said to be the *subgraph induced by W*, and is denoted by $G\langle W \rangle$. We say that $H$ is an *induced subgraph* of $G$ if $H \subseteqq G$ and $H = G\langle V(H)\rangle$.

The set of vertices adjacent to a vertex $X \in G$ is denoted by $N(X)$, and $v(G{:}X) = |N(X)|$ is said to be the *valency* or *degree* of $X$ in $G$. If it is clear which the underlying graph is, we write $v(X)$. The *minimum degree* of the vertices of $G$ is denoted by $\delta(G)$ and the *maximum degree*, by $\Delta(G)$. If $\delta(G) = \Delta(G) = r$, i.e., each vertex of $G$ has valency $r$, then $G$ is said to be *regular of degree r* or *r-regular*. A 3-regular graph is said to be *cubic*. If $E' \subseteqq E(G)$, then

$G - E'$ denotes the graph resulting from $G$ if we delete the edges belonging to $E'$, i.e., $G - E' = (V(G), E(G) - E')$. Similarly, if $W \subseteqq V(G)$, then $G - W$ is the graph obtained from $G$ by the removal of the vertices belonging to $W$. Of course, if a vertex $X \in W$ is deleted, then each edge incident with $X$ is also deleted, i.e. if $G = (V, E)$, then $G - W = G\langle V - W\rangle = (V - W, E \cap (V - W)^{(2)})$. If $W = \{X\}$, we usually write $G - X$ instead of $G - \{X\}$, similarly we write $G - e$ instead of $G - \{e\}$ if $E' = \{e\} \subseteqq E$. If $H \subseteqq G$, then we may write $G - H$ instead of $G - V(H)$. If $e \in V^{(2)} - E$, then the graph obtained from $G$ by adding $e$ is $G \cup \{e\}$ or $G \cup e$.

Let $X$ and $Y$ be two not necessarily distinct vertices of $G$. By an *X,Y-walk w* we mean an alternating sequence of vertices and edges, say $X_1, e_1, X_2, e_2, \ldots, X_l, e_l, X_{l+1}$, such that $X_1 = X$, $X_{l+1} = Y$ and $e_i = (X_i, X_{i+1}) \in E(G)$, $1 \leqq i \leqq l$. We usually put $w = X_1X_2\ldots X_{l+1}$ or $w = [X_1, X_2, \ldots, X_{l+1}]$, because from this form it is clear which the edges in the sequence are. If we consider $w$ as a sequence of edges only, we write $e_1, e_2, \ldots, e_l$ or $(X_1, X_2, \ldots, X_{l+1})$. The *length $\lambda(w)$ of the walk w* is $\lambda(w) =_{\text{def}} l = e(w)$, its number of edges. The walk above is a *path* or *$X_1,X_{l+1}$-path* (also *$X_1X_{l+1}$-path*) if all its vertices are distinct. We also introduce an *$X_1,X_1$-path* $(X_1)$ which has length 0 and consists only of the vertex $X_1$. If $l \geqq 3$ and $X_1 = X_{l+1}$, but the other vertices are distinct from each other and $X_1$, then we call the walk a *cycle* or *circuit*.

Let $H$, $H_1$, $H_2$ be subgraphs of $G$. A path $p$ of $G$ with end vertices $X$, $Y$ is an *H,H-path* if $V(p) \cap V(H) = \{X, Y\}$ and $E(p) \cap E(H) = \emptyset$. If $V(p) \cap V(H) = \{X\}$, then $p$ is an *H-path*. The only edge of an *H,H-path* of length 1 is called a *diagonal* of $H$. Correspondingly, *$H_1,H_2$-paths* are defined.

A path $p$ and a cycle $C$ are identified with the graphs $(V(p), E(p))$ and $(V(C), E(C))$, respectively. In particular, $X_1X_2\ldots X_lX_{l+1}$ and $X_{l+1}X_l\ldots X_2X_1$ denote the same path, so an $X,Y$-path is also a $Y,X$-path. Similarly,

$$X_1X_2\ldots X_lX_1, \quad X_2X_3\ldots X_lX_1X_2 \quad \text{and} \quad X_lX_{l-1}\ldots X_2X_1X_l$$

denote the same cycle. An edge of the form $(X_1, X_j)$ $(3 \leqq j \leqq l - 1)$ is a *diagonal* of this cycle. We denote a cycle of length $l$ by $C_l$. A cycle is *odd (even)* if its length is odd (even).

If $p = X_1X_2\ldots X_l$ is a path, $P = X_i$, $Q = X_j$ and $1 \leqq i \leqq j \leqq l + 1$, then the *P,Q-segment* of $p$ is the $P,Q$-path $X_iX_{i+1}\ldots X_{j-1}X_j$. We refer to it as the *partial path of p with ends P, Q* and denote it by $p[P, Q]$ and $PpQ$. If $p$ is an $X,Y$-path and $q$ is a $Y,Z$-path, then $XpYqZ = p[X, Y] \cup q[Y, Z]$ is the *X,Z-walk obtained by stringing* the two paths together. Similarly we may string together segments of paths to obtain a walk or a path with the self-explanatory notation

$$X_1p_1X_2p_2\ldots p_lX_{l+1} = p_1[X_1, X_2] \cup p_2[X_2, X_3] \cup \ldots \cup p_l[X_l, X_{l+1}].$$

A graph is *connected* if every pair of vertices is joined by a path. Otherwise the graph is said to be *disconnected*. A maximal connected subgraph is said to be a *component* of the graph. A connected graph not containing cycles is a *tree*, and a graph without cycles is a *forest*. Clearly a forest is a graph each component of which is a tree. A tree of order $n$ has $n - 1$ edges and a forest of order $n$ with $c$ components has $n - c$ edges.

If the graph $G$ has a cycle, then the *girth* $g(G)$ and the *circumference* $c(G)$ are the lengths of the smallest and the longest cycles, respectively.

It might be appropriate to remark here that, following recent custom, we use the words "maximum" and "maximal" with different meanings. "Maximal" refers to a maximal element of an ordered set in which, unless otherwise stated, the ordering is given by inclusion. "Maximum" refers to an element of maximal size. Thus $p$ is a maximal path of a graph $G$ if it is not properly contained in any other path, and $q$ is a maximum path of $G$ if $G$ does not contain a path $w$ longer than $q$ (i.e., $\lambda(w) \leqq \lambda(q)$ for every path $w$ in $G$). The words "minimum" and "minimal" are similarly used with different meanings.

The *complement* $\bar{G}$ of a graph $G = (V, E)$ has the same vertex set, where two distinct vertices are adjacent if and only if they are non-adjacent in $G$. Hence $\bar{G} = (V, V^{(2)} - E)$. $I_n$ denotes the *edgeless graph on n vertices*. $K_n = \bar{I}_n$ is said to be the *complete n-graph*. Thus $K_n$ has $n$ vertices and each pair of distinct vertices is joined by an edge. $K_n^-$ is obtained from $K_n$ by deleting one edge. $K_{p,q}$ denotes the graph obtained from two disjoint graphs $I_p$, $I_q$ by joining each vertex of $I_p$ to each vertex $I_q$ by an edge. Each subgraph of some $K_{p,q}$ is called *bipartite*. A bipartite graph has only even cycles. A graph is *non-bipartite* if and only if it has at least one odd cycle.

The *Petersen graph* is obtained from two disjoint 5-cycles $X_1X_2X_3X_4X_5X_1$ and $Y_1Y_2Y_3Y_4Y_5Y_1$ by adding the edges $(X_1, Y_1)$, $(X_2, Y_3)$, $(X_3, Y_5)$, $(X_4, Y_2)$, $(X_5, Y_4)$.

A graph $G$ of order at least $k + 1 \geqq 1$ is called *k-connected* if the deletion of any set $S \subseteqq V(G)$, $|S| \leqq k - 1$, results in a connected graph. If for a vertex set $S$ of a connected graph $G$ the subgraph $G - S$ is disconnected, then $S$ is said to be a *separating set*. A set $S$ of cardinality $k$ which separates a $k$-connected graph $G$ is called a *cut set* of $G$. If $k = 1$, then the unique vertex of a cut set is said to be a *cut vertex*. The largest $k$ for which a connected non-complete graph is $k$-connected, $k \geqq 1$, is called its *connectivity* $\varkappa(G)$, where $\varkappa(K_n) =_{\text{def}} n - 1$, $n \geqq 2$.

We shall use the following two consequences of Menger's famous theorem:

1) *Any two distinct vertices X, Y of a k-connected graph are linked by k internally disjoint X,Y-paths.*

2) *Any two disjoint vertex sets A, B of a k-connected graph G of cardinalities* $|A| \geqq k$, $|B| \geqq k$ *are linked by k disjoint paths.*

Similar concepts can be introduced with respect to edge sets.

Let $H_1$, $H_2$ be two subgraphs of $G$ with at least one edge. Then $H_1 \triangle H_2$ denotes the subgraph of $G$ consisting of all edges of the symmetric difference $E(H_1) \triangle E(H_2)$ and all vertices of $G$ incident with an edge of $E(H_1) \triangle E(H_2)$.

Let $(P, Q) \in E(G)$ and $p$ be a $P,Q$-path with $V(p) \cap V(G) = \{P, Q\}$. If $G$ is replaced by $(G - (P, Q)) \cup p$, then $(P, Q)$ is said to be *subdivided* by vertices of valency 2. A graph $H$ is said to be a *subdivision* of $G$ if $H$ can be obtained by subdividing some edges of $G$.

In addition we make the following definition. Let $\alpha$ be a real number. $\lfloor \alpha \rfloor$ denotes the largest integer not greater than $\alpha$, and $\lceil \alpha \rceil$ is the smallest integer not smaller than $\alpha$.

# Chapter 1

# Overlap properties of bridges of cycles and overlap graphs

## 1.0. Introduction

The concept of bridge is mostly used in investigating properties of cycles of graphs, in particular of planar graphs. For instance many proofs of the famous theorem of K. Kuratowski, 1930, on planar graphs are based on the study of the properties of $C$-bridges, where $C$ is a cycle with certain properties.

There is an immense literature on graph embeddings: in accord with the preface we restrict our investigations to problems which are not related to embeddings of graphs in the plane or in higher surfaces. We recommend the reader to the survey article on this subject of C. Thomassen, 1988, to the book of A. T. White, 1973, the article of A. T. White and L. W. Beineke, 1978, and of J. Širan, R. Bodendiek and K. Wagner, 1989.

Here we only treat the overlap graphs and their applications to the crossing number of graphs, Kuratowski's theorem, Hamiltonian cycles and paths in planar graphs, and the separation of vertices by cycles in planar graphs.

### 1° Overlap graphs and circle graphs

In the first chapter we shall investigate the overlap properties of bridges of cycles.

We shall characterize the graphs containing a cycle with two or three overlapping bridges by a subgraph condition. In order do characterize the overlap properties of bridges of a cycle we introduce and study the so-called overlap graph (overlap graph by C. Thomassen, 1980, 1983b, and bridge graph by W. T. Tutte, 1984).

If all bridges are *singular*, i.e., they are edges, then this concept corresponds to the concept of the circle graphs, investigated by several authors (see, e.g., M. C. Golumbic, 1980, and W. Wessel and R. Pöschel, 1985). We note that different from our notation in M. C. Golumbic, 1980, an overlap graph is a circle graph, only.

### 2° The crossing number of graphs

It is well known that every graph $G = (V, E)$ can be represented in the plane as follows: 1) to each vertex $X \in V$ a point $\varphi(X)$ of the plane is assigned so that distinct points of the plane are assigned to distinct vertices, 2) to each $X,Y$-edge $e$ a Jordan curve $\varphi(e)$ of the plane with end points $\varphi(X)$, $\varphi(Y)$ is assigned so that no inner point of $\varphi(e)$ is the image of a vertex of $G$, two distinct Jordan curves $\varphi(e)$, $\varphi(e')$, $e \neq e'$, $e, e' \in E$, have at most one common point, and three distinct Jordan curves $\varphi(e)$, $\varphi(e')$, $\varphi(e'')$, $e \neq e' \neq e'' \neq e$, $e, e', e'' \in E$, do not intersect in a common inner point. A common inner point of two Jordan curves is said to be a *cross vertex* of $G$. The image of $G$ so obtained is called a *representation* of $G$ in the plane.

If $G$ has a representation in the plane without cross vertices, then $G$ is said to be a *planar graph* and the representation is referred to as a *plane representation*, or a *plane graph*, or an *embedding* of $G$ in the plane.

Obviously, every planar graph is isomorphic to its plane representations.

If $G$ is not planar, then there is a representation of $G$ in the plane with the minimum number $\gamma(G)$ of cross vertices. The number $\gamma(G)$ is said to be the *crossing number* of $G$ (introduced by F. Harary and A. Hill, 1962-3). The crossing number of graphs has been investigated by F. Harary and A. Hill, 1962-3, R. K. Guy, 1967, 1972, and others. We refer the reader to the survey article on this subject by R. K. Guy, 1971. Interesting results closely related to the crossing number are due to H. Harborth.

In this section we shall present a relation between the number of cross vertices of a certain representation of $G$ in the plane and the number of edges of the corresponding circle graph.

### 3° Planar graphs

A nice theorem of W. T. Tutte, 1958, 1984, says (see also C. Thomassen, 1981b):

*A graph $G$ is planar if and only if for each cycle $C$ the overlap graph $O(G:C)$ is bipartite.*

Hence the overlap graphs are closely related to the planar graphs, and the bridges of cycles play an important role in embedding graphs in the plane.

From Tutte's theorem the well-known characterization of planar graphs by K. Kuratowski, 1930, can be derived:

*A graph $G$ is planar if and only if $G$ does not contain neither a subdivision of $K_5$ nor a subdivision of $K_{3,3}$.*

(For the definition of a subdivision of a graph see p. 7.)

The theorem of W. T. Tutte, 1984, mentioned above indicates that in the theory of embedding graphs in the plane the concept of overlapping bridges plays an important role. Let $G$ be a plane graph, i.e., a graph embedded in the plane without crossings of edges. Let $G$ contain a cycle $C$ with two overlapping $C$-bridges $B_1$, $B_2$. Then the bridges $B_1$, $B_2$ cannot be drawn without crossings either inside $C$ or outside $C$. Then precisely one bridge lies inside C and the other bridge outside $C$. Consequently, in $G$ an inner vertex of $B_1$ and an inner vertex of $B_2$ are separated by the cycle $C$. For such pairs of vertices W. T. Tutte, 1960, 1975, has proved the following result:

*Let $X$, $Y$ be vertices of a planar graph $G$. Then $G$ has a plane representation such that $X$, $Y$ are adjacent to a common face unless $G$ contains a separating cycle $C$ such that the $C$-bridges of $G$ containing $X$ and $Y$, respectively, as inner vertices are overlapping.*

Hence two vertices $X$, $Y$ of a planar graph are called *far apart* if and only if there is a cycle $C$ of $G-\{X, Y\}$ such that $X$, $Y$ are in different overlapping $C$-bridges of $G$.

A consequence of the considerations above is that no cycle in a planar graph has three pairwise overlapping bridges. The first combinatorial proof of the famous theorem of K. Kuratowski, 1930, by G. A. Dirac and S. Schuster, 1954, is based on this idea. Elegant proofs of this theorem, due to M. Burštein, 1978, W. T. Tutte, 1960, 1984, and C. Thomassen, 1980, are based on the observation that the overlap graph of each cycle in a plane graph has no odd cycle. With the help of our characterization theorems we shall give such a proof in section 1.5.

A very good overview on embedding graphs in the plane can be found in the papers of C. Thomassen, 1980, 1981b, 1988. Bridges of cycles have been intensively used by W.T. Tutte, 1956, 1977, 1984, in the proof of his famous theorem:

*Every vertex 4-connected planar graph has a Hamiltonian cycle.*

The key of the proof of this theorem is the following lemma:

*Every 2-connected non-Hamiltonian planar graph contains a cycle with a bridge having at most three vertices of attachment.*

C. Thomassen, 1983b, has proved this lemma in a sharper version implying a proof of M. D. Plummer's conjecture (M. D. Plummer, 1973).

*Every vertex 4-connected planar graph is Hamiltonian connected,* i.e., *each pair of vertices is joined by a Hamiltonian path.*

A new proof of Tutte's theorem is due to N. Chiba and T. Nishizeki, 1989, providing a linear-time algorithm for finding Hamiltonian cycles in planar 4-connected graphs.

These theorems on 4-connected planar graphs contrast with results relating to 4-connected non-planar graphs. For example, J. Širáň, 1983, showed that each edge of any 4-connected non-planar graph of order at least 6 is in a subdivision of $K_{3,3}$.

The concept of a $C$-bridge of a certain cycle $C$ is used not only in graph theoretical proofs of Kuratowski's theorem but also in algorithms for deciding whether a graph is planar or not and for finding an embedding of a planar graph in the plane. Such algorithms are due to G. Demoucron, Y. Malgrange and R. Pertuiset, 1964 (see J. A. Bondy and M. S. R. Murty, 1976), W. Tutte, 1984, and others.

As C. Thomassen, 1981b, writes, almost all known proofs of Kuratowski's theorem can be turned into polynomially bounded algorithms for testing planarity of graphs. The asymptotically fastest (but not conceptually simplest) algorithm for this was found by J. E. Hopcroft and R. E. Tarjan, 1974. For the algorithmic aspects of Kuratowski's theorem the reader is referred to the survey paper of S. G. Williamson, 1980.

We conclude this introduction with a conjecture of D. König, 1936 (p. 199), that a Kuratowski-like theorem is true for all orientable and non-orientable surfaces. This was proved by N. Robertson and Ph. Seymour, 1984 and 1990; shorter independent proofs are due to R. Bodendiek and K. Wagner, 1989, for the orientable surfaces and to D. Archdeacon and Ph. Huneke, 1989, for the non-orientable surfaces. In these shorter proofs bridges are again used.

## 1.1. Bridges of a subgraph $H$

At the beginning we repeat the definition of a bridge in a graph $G$ with respect to a subgraph $H$.

Let $H$ be a proper subgraph of a graph $G$. Let $S$ be the set of all edges of $G$ not in $H$. On $S$ the relation $\sim$ will be defined as follows: for two edges $e_1$, $e_2$ of $S$ the relation $e_1 \sim e_2$ holds if and only if there is a path $p$ of $G$ with end edges $e_1$ and $e_2$ and no inner vertex of $p$ belongs to $H$ (it is possible that $p$ has length 1). It can easily be seen that $\sim$ is an equivalence relation. Then an equivalence class $M$ of $S$ together with all vertices of $G$ incident with edges of this class is called a *bridge $B$ of $H$ in $G$* or a *bridge in $G$ with respect to $H$* or an *$H$-bridge* in $G$. An isolated vertex not in $H$ is also referred to as an $H$-bridge.

If $\| M \| = 1$, then $B$ has precisely one edge. If both end vertices of this edge belong to $H$, then $B$ is called a *singular bridge* or a *diagonal* of $H$. If the bridge $B$ is not a singular one, then it is called a *regular* (or *non-degenerate*) *bridge*. The $H$-bridges $B_1$, $B_2$, $B_3$, $B_5$ of Fig. 1.1.1 are regular, and the $H$-bridge $B_4$ is singular, where $H$ is a cycle.

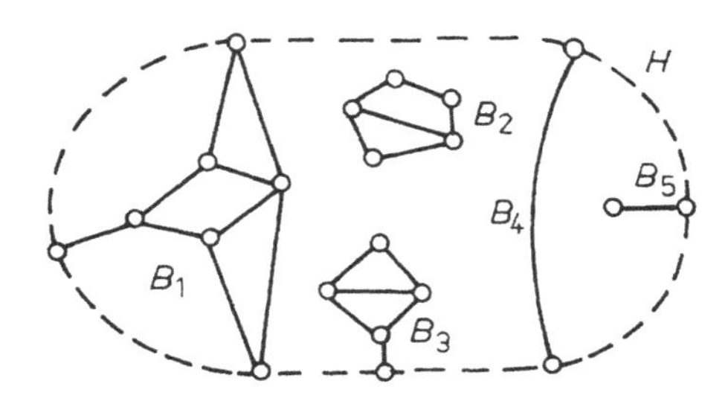


Fig. 1.1.1

The vertices of $B \cap H$ are called *attaching vertices* of $B$ (*vertices of attachment* of $B$), and the vertices of $B \setminus H$ are referred to as the inner vertices of $B$. The $H$-bridges $B_1$, $B_2$, $B_3$, $B_4$, $B_5$ of Fig. 1.1.1 have 3, 0, 1, 2, 1 vertices of attachment, respectively.

The *kernel of a regular bridge $B$* is the subgraph induced by the inner vertices of $B$. The *kernel of a singular bridge* is the empty graph. The following lemma can easily be proved.

**Lemma 1.1.1.** *The kernel of a bridge is connected.* □

With the help of this lemma we can give a new definition of the notion of $H$-bridge in $G$: each component $K$ of $G - H$, together with all vertices of $G$ adjacent to $K$ and all edges of $G$ incident with $K$, forms a *regular $H$-bridge* in $G$. An edge not in $H$ linking two vertices of $H$ is a *singular $H$-bridge*.

## 1.2. Overlap graphs and circle graphs

Here and in the following sections we consider bridges of cycles only.

Let $B$ be a bridge of a cycle $C$ in a graph $G$ having at least two vertices of attachment $A_1, A_2, \ldots, A_s$, $s \geqq 2$, which are on $C$ in this cyclic order. Then the $s$ arcs $C[A_i, A_{i+1}]$, $C[A_s, A_1]$, $1 \leqq i \leqq s-1$, are called *segments* of $C$ with respect to $B$.

Let $B$ and $B'$ be two bridges of a cycle $C$ in a graph $G$. We say $B$ and $B'$ *avoid each other* if and only if there are two vertices $X$ and $Y$ on $C$ such that all vertices of attachment of $B$ are on at most one of the arcs $C[X, Y]$ of $C$ with ends $X$ and $Y$ and all vertices of attachment of $B'$, if any, are on the other such arc $C[Y, X]$.

The bridges $B_1$, $B_3$ of Fig. 1.2.1 avoid each other.

Obviously, if $B$ or $B'$ has at most one vertex of attachment, then $B$ and $B'$ avoid each other.

Next let $B$ have at least two vertices of attachment. The vertices of attachment of $B$ subdivide $C$ into segments with respect to $B$. Then $B$ and $B'$ avoid each other if and only if either $B'$ has no vertex of attachment or there is a segment of $C$ with respect to $B$ containing all vertices of attachment of $B'$.

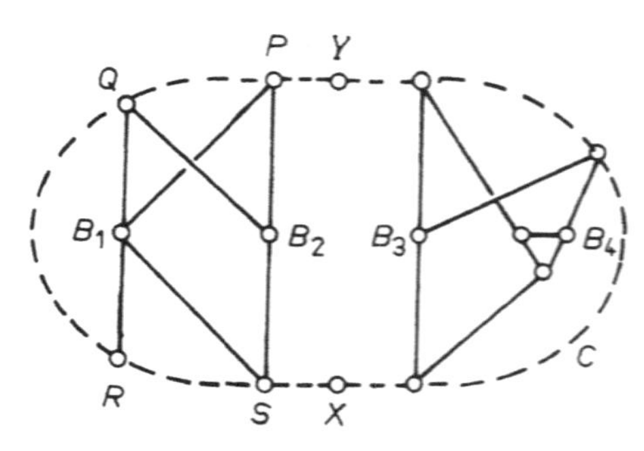


Fig. 1.2.1

Two bridges $B$ and $B'$ of a cycle $C$ are said to be *overlapping* if and only if they do not avoid each other (the bridges $B_1$, $B_2$ and also $B_3$, $B_4$ of Fig. 1.2.1 overlap each other). Furthermore, $B$ and $B'$ are said to be *skew* if and only if there are four vertices $P, Q, R, S$ in this cyclic order on the cycle $C$ such that $P$ and $R$ are in $B$ and $Q$ and $S$ are in $B'$ (see the two skew $C$-bridges $B_1$, $B_2$ of Fig. 1.2.1).

The $C$-bridges $B_3$ and $B_4$ of Fig. 1.2.1 overlap each other, but they are not skew. Thus we may ask the question: which overlapping bridges of a cycle are skew? The answer is given in the lemma of W. T. Tutte, 1977.

**Lemma 1.2.1.** *Let $C$ be a cycle of a graph $G$. If $B$, $B'$ are overlapping $C$-bridges, then either they are skew or both $B$ and $B'$ have precisely three vertices of attachment which coincide.*

In Fig. 1.2.1 the $C$-bridges $B_1$, $B_2$ and $B_3$, $B_4$ overlap each other. The bridges $B_1$, $B_2$ are skew and the bridges $B_3$, $B_4$ are not skew and have precisely three vertices of attachment which coincide.

*Proof of Lemma 1.2.1.* Since $B$ overlaps $B'$, both $C$-bridges have at least two vertices of attachment. We consider two cases:

Case 1. Assume that $B'$ contains an inner vertex of a segment $S$ of $C$ with respect to $B$. Since $B'$ overlaps $B$, there is a vertex of attachment of $B'$ in $C - S$. Therefore, $B$ and $B'$ are skew. The same result is obtained if $B$ contains an inner vertex of a segment $S'$ of $C$ with respect to $B'$.

Case 2. Assume that $B$ and $B'$ have the same set of vertices of attachment. Let this set have cardinality $s$. If $s = 2$, then $B'$ does not overlap $B$. This contradicts the hypothesis of the lemma. Hence $s \geqq 3$. If $s = 3$, then we have the statement of the lemma. If $s \geqq 4$, then $B$ and $B'$ are skew. □

In order to investigate the overlap properties of three or more bridges of a cycle $C$ in a graph $G$ we introduce the overlap graph or bridge graph of $G$ with respect to $C$ in the sense of C. Thomassen, 1980, 1983 b, and W. T. Tutte, 1984, respectively.

**Definition 1.2.2.** Let $C$ be a cycle of a graph $G$. Let the $C$-bridges be the vertices of a new graph $O(G:C)$—the so-called *overlap graph* of $G$ with respect to $C$. Two vertices $B$, $B'$ of $O(G:C)$ are *joined* by an edge if and only if the $C$-bridges $B$, $B'$ overlap. The spanning subgraph of $O(G:C)$ in which $B$, $B'$ are adjacent if and only if $B$ and $B'$ are skew is called the *skew-overlap graph* of $G$ with respect to $C$.

If $C$ is an Hamiltonian cycle, then all bridges are *diagonals* of $C$; in this case the overlap graph is called a *circle graph*.

Fig. 1.2.2 depicts an overlap graph and Fig. 1.2.3 a circle graph.

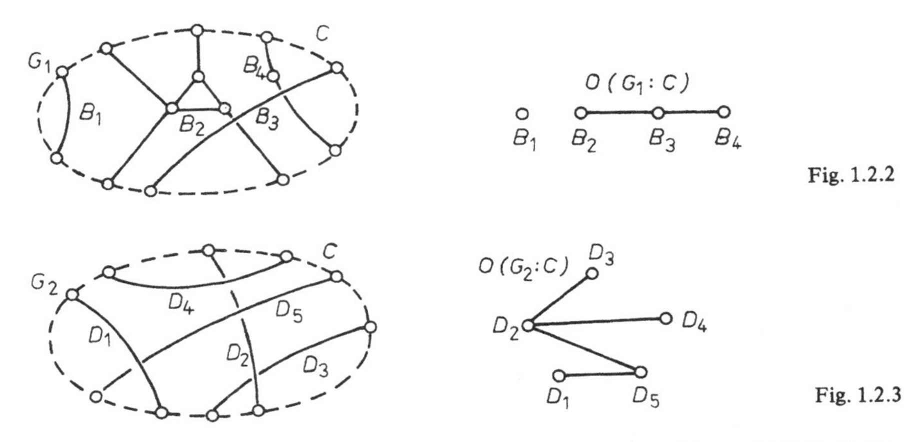


Fig. 1.2.2

Fig. 1.2.3

The overlap graphs have been used by C. Thomassen, 1980, 1983b, and W. T. Tutte, 1984.

A graph is said to be an *overlap graph* (a *circle graph*) if it is isomorphic to $O(G:C)$ for some graph $G$ and cycle $C$.

Circle graphs have been characterized by S. Even and A. Itai, 1971, and J. C. Fournier, 1978. C. P. Gabor, W.-L. Hsu and K. J. Supowit, 1985, settled the recognition problem for them by giving a polynomial-time algorithm. W. Wessel and R. Pöschel, 1985, presented families of circle graphs. Further investigations are due to M. C. Golumbic, 1980, and W. Wessel, 1985, 1987. We note that different from our notation in M. C. Golumbic, 1980, an overlap graph is a circle graph, only.

Here we pose the problem of investigating the properties of the overlap graph $O(G:C)$. We present first results in the sections 1.3–1.6.

In section 1.5, I shall show that there are graphs which are not overlap graphs and that there are also overlap graphs which are not circle graphs.

Let $G'$ be a representation of $G$ in the plane such that $H$ is represented by a simple closed Jordan curve $H'$ and all diagonals of $H$ are represented by simple Jordan curves lying in the finite region of $H'$. The number of cross vertices equals the number of edges in the corresponding circle graph $O(G:H)$.

If $H'$ is chosen as a geometric circle and the diagonals are represented by straight line segments, then we arrive at a second definition of a circle graph: the vertices of a circle graph are the chords of the geometric circle and two chords are joined by an edge if and only if they intersect in an inner point (see, e.g., W. Wessel and R. Pöschel, 1985, and M. C. Golumbic, 1980).

## 1.3. Subgraphs of overlap graphs

In this section we deal with subgraphs of the overlap graph $O(G:C)$ of a graph $G$ with respect to a cycle $C$. We ask: what can be said about the structure of a graph $G$ if some properties of $O(G:C)$ are known, for instance if it is known that $O(G:C)$ contains a cycle or a path of a certain length? We shall especially investigate the question: what structure do the bridges of a cycle $C$ in the graph $G$ have if the overlap graph $O(G:C)$ has an induced cycle or path of a certain length?

We start with the case when $O(G:C)$ contains a triangle. First in Lemma 1.3.1 we shall consider the stronger hypothesis that $C$ has three pairwise skew bridges, i.e., they form a triangle in the skew-overlap graph.

Let $F_i$, $1 \leqq i \leqq 9$, denote the graphs depicted in Fig. 1.3.1, where $L$ always denotes the outer cycle. The vertices of attachment of the bridges of $L$ are represented by small rectangles.

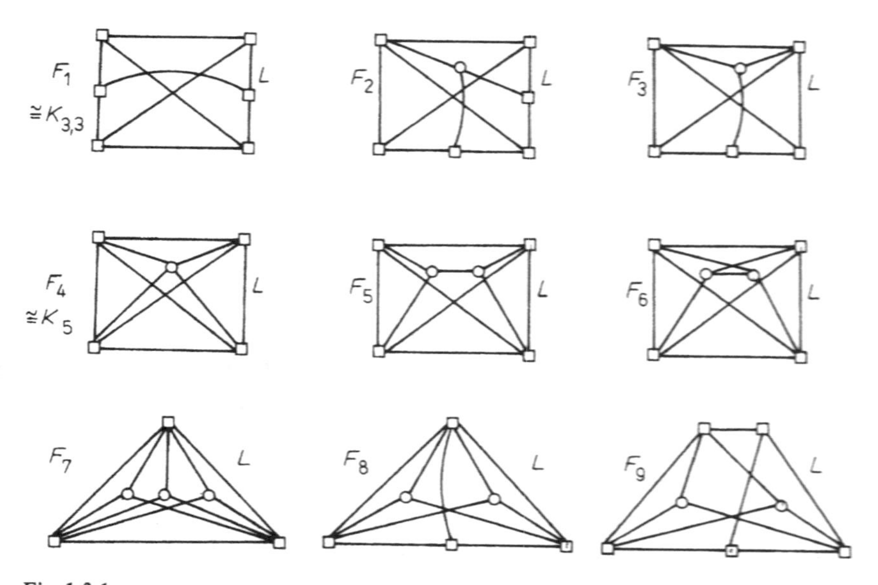


Fig. 1.3.1

**Lemma 1.3.1.** *Let $B_1$, $B_2$, $B_3$ be three pairwise skew bridges of a cycle C. Then $C \cup B_1 \cup B_2 \cup B_3$ contains a subdivision of at least one of the $F_i$'s, $1 \leqq i \leqq 6$, so that C is a subdivision of L* (for the graphs $F_i$ see Fig. 1.3.1).

For convenience we say that a *graph* with a cycle $C$ is a $C$-subdivision of $F_i$ if it is obtained by subdividing $F_i$ so that $L$ becomes the cycle $C$.

*Proof.* Assume that $C \cup B_1 \cup B_2 \cup B_3$ contains no $C$-subdivision of $F_i$ for all $i$, $1 \leqq i \leqq 6$. Let the indices of $B_1$, $B_2$, $B_3$ be chosen so that $B_3$ has the largest number of vertices of attachment.

Since $B_1$, $B_2$ are skew, there are four different vertices $P_1$, $P_2$, $Q_1$, $Q_2$ on $C$ in this cyclic order and such that $P_1, Q_1 \in B_1$ and $P_2, Q_2 \in B_2$ (see Fig. 1.3.2 a). Then these four vertices split $C$ into four segments $C[P_1, P_2]$, $C[P_2, Q_1]$, ... If $B_3$ has inner vertices of at least two "non-neighbouring" segments of $C$, then $C \cup B_1 \cup B_2 \cup B_3$ contains a $C$-subdivision of $F_1$. Hence two "neighbouring" segments of $C$ have no inner vertex belonging to $B_3$.

Without loss of generality, let $C[P_1, P_2]$ and $C[P_2, Q_1]$ be these segments. Then all vertices of attachment of $B_3$ lie on $C[Q_1, P_1] \cup \{P_2\}$.

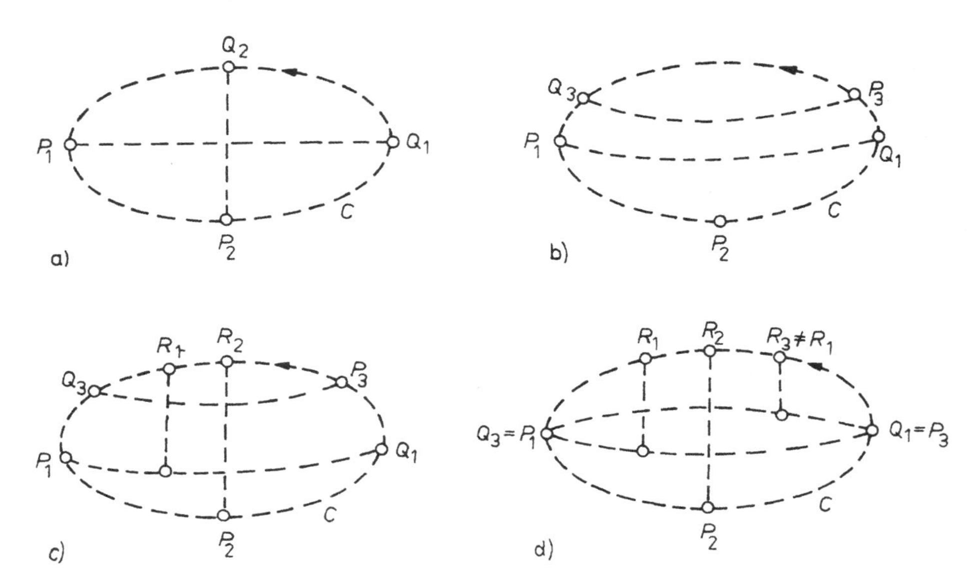


Fig. 1.3.2

(1) *The bridge $B_3$ has at least three vertices of attachment.*

*Proof of (1).* Assume that $B_3$ has precisely two vertices of attachment. Then $B_1$, $B_2$ also have precisely two vertices of attachment. Since $B_i$, $B_j$ are skew for all $1 \leqq i < j \leqq 3$, $C \cup B_1 \cup B_2 \cup B_3$ contains a $C$-subdivision of $F_1$. Contradiction! □

We consider two cases.

Case 1. All vertices of attachment of $B_3$ lie on $C[Q_1, P_1]$, i.e., $P_2 \notin B_3$. Let $C[P_3, Q_3] \subseteqq C[Q_1, P_1]$ be the smallest arc of $C$ containing all vertices of attachment of $B_3$ (see Fig. 1.3.2 b). Since both pairs $B_3$, $B_1$ and $B_3$, $B_2$ are skew, there are inner vertices $R_1$, $R_2$ of $C[P_3, Q_3]$ such that $R_i \in B_i$ $(i = 1, 2)$.

Case 1.1. Assume that $P_1 \neq Q_3$ or $Q_1 \neq P_3$. Without loss of generality let $P_1 \neq Q_3$. This situation is depicted in Fig. 1.3.2 c, where it may be either $Q_1 \neq P_3$, $R_1 \neq R_2$ or $Q_1 \neq P_3$, $R_1 = R_2$ or $Q_1 = P_3$, $R_1 \neq R_2$ or $Q_1 = P_3$, $R_1 = R_2$. In all these cases $C \cup B_1 \cup B_2 \cup B_3$ contains a $C$-subdivision of $F_i$ for some $i$, $1 \leqq i \leqq 3$.

Case 1.2. Assume that $P_1 = Q_3$ and $Q_1 = P_3$. By (1) the bridge $B_3$ has a vertex of attachment $R_3 \in C[P_3, Q_3] - \{P_3, Q_3\}$. If $R_3 = R_1$, then $B_1$ and $B_3$ have three common vertices of attachment. Since $B_3$, $B_1$ are skew and $B_3$ has the largest number of vertices of attachment, $B_3$ has a fourth vertex of attachment. Hence $R_3$ can be chosen so that $R_3 \neq R_1$ (see Fig. 1.3.2 d). Then $C \cup B_1 \cup B_2 \cup B_3$ contains a $C$-subdivision of $F_1$ or $F_2$ or $F_3$ depending on the position of $R_1$, $R_2$, $R_3$ on $C$.

Case 2. Assume that $P_2 \in B_3$. By (1) at least two vertices of $B_3$ lie on $C[Q_1, P_1]$. Let $C[P_3, Q_3] \subseteqq C[Q_1, P_1]$ denote the smallest arc of $C$ containing all vertices of attachment of $B_3 - \{P_2\}$.

Case 2.1. Assume that $P_1 \neq Q_3$ or $Q_1 \neq P_3$. Without loss of generality, let $P_1 \neq Q_3$.

Case 2.1.1. Assume that an inner vertex of $C[P_3, Q_3]$ belongs to $B_2$ (see Fig. 1.3.3 a). Without loss of generality let $Q_2$ be this vertex. Then $C \cup B_1 \cup B_2 \cup B_3$ contains a $C$-subdivision of $F_2$ or $F_3$ if $P_3 \neq Q_1$ and $P_3 = Q_1$, respectively.

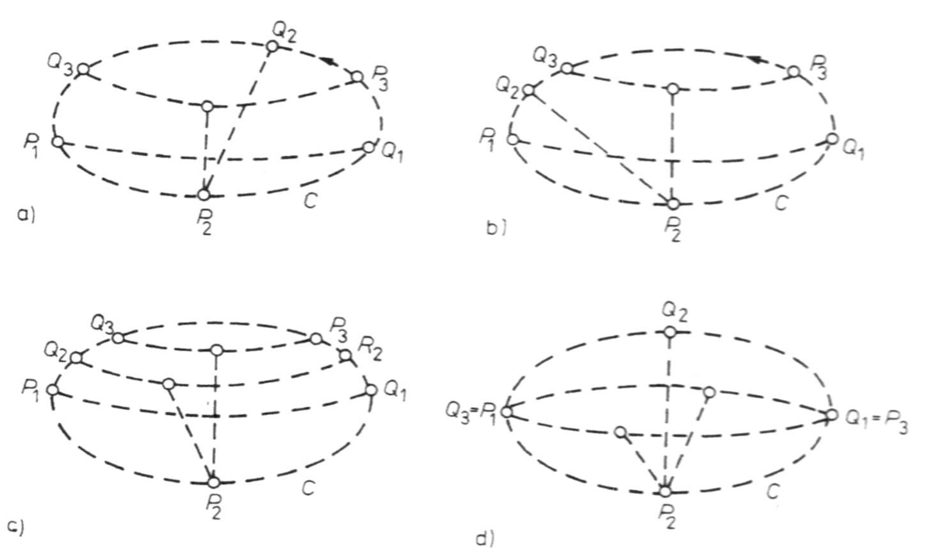


Fig. 1.3.3

Case 2.1.2. Assume that no inner vertex of $C[P_3, Q_3]$ belongs to $B_2$. If $Q_1 = P_3$, then $Q_2 \in C[Q_3, P_1] - \{P_1\}$ and if $Q_1 \neq P_3$, then $Q_2 \in C[Q_3, P_1] - \{P_1\}$ or $Q_2 \in C[Q_1, P_3] - \{Q_1\}$. By symmetry (note: by hypothesis $P_1 \neq Q_3$) we may assume that $Q_2 \in C[Q_3, P_1] - \{P_1\}$ (see Fig. 1.3.3 b). Since $B_2$ contains no inner vertex of $C[P_3, Q_3]$ and $B_2$, $B_3$ are skew, $B_2$ contains a vertex $R_2$ of $C[P_2, P_3] - \{P_2\}$ (see Fig. 1.3.3 c).

Next we have to consider the two possible cases "$P_3 = Q_1$" and "$P_3 \neq Q_1$" and to combine them with the two possible positions of $Q_2$, namely "$Q_2 \neq Q_3$" or "$Q_2 = Q_3$", and with the four possible positions of $R_2$, namely "$R_2 = P_3$" or "$R_2 \in C[Q_1, P_3] - \{Q_1, P_3\}$" or "$R_2 = Q_1$" or "$R_2 \in C[P_2, Q_1] - \{P_2, Q_1\}$".

In all cases where $Q_2 \neq Q_3$ or $R_2 \neq P_3$ the subgraph $C \cup B_1 \cup B_2 \cup B_3$ of $G$ contains a $C$-subdivision of one $F_i$ $(1 \leq i \leq 3)$.

In all cases where $Q_2 = Q_3$ and $R_2 = P_3$ the bridges $B_2$ and $B_3$ have three common vertices of attachment, namely $P_2$, $Q_2$, $R_2$. Since $B_3$, $B_2$ are skew and, by hypothesis, $B_3$ has the largest number of vertices of attachment, the bridge $B_3$ has a fourth vertex of attachment not in $\{P_2, Q_2, P_2\}$. This vertex is on $C[P_3, Q_3] - \{P_3, Q_3\}$. Then $C \cup B_1 \cup B_2 \cup B_3$ contains a $C$-subdivision of $F_2$ or $F_3$.

Case 2.2. Assume that $P_1 = Q_3$ and $Q_1 = P_3$ (see Fig. 1.3.3 d). Since $B_1$, $B_3$ are skew and $B_3$ is the bridge with the largest number of vertices of attachment, the arc $C[P_3, Q_3] - \{P_3, Q_3\}$ contains a fourth vertex $R_3$ of $B_3$. If $Q_2 \neq R_3$, then $C \cup B_1 \cup B_2 \cup B_3$ contains a subdivision of $F_3$, and if $Q_2 = R_3$, then $C \cup B_1 \cup B_2 \cup B_3$ contains a subdivision of $F_4$ or $F_5$ or $F_6$. □

Lemma 1.3.1 implies the following somewhat extended result:

**Theorem 1.3.2.** *Let $B_1$, $B_2$, $B_3$ be three pairwise overlapping bridges of a cycle C. Then $C \cup B_1 \cup B_2 \cup B_3$ contains a subdivision of at least one of the $F_i$'s, $1 \leqq i \leqq 9$, so that C is a subdivision of L* (for the graphs $F_i$ see Fig. 1.3.1). □

Next we investigate the case when the three bridges form an induced path in $O(G:C)$.

**Lemma 1.3.3.** *Let $(B_1, B_2, B_3)$ be an induced path in $O(G:C)$. Then $B_1$, $B_2$ and $B_2$, $B_3$ are skew.*

*Proof.* Assume that $B_1$, $B_2$ are not skew. By Lemma 1.2.1 the bridges $B_1$ and $B_2$ have the same set of three attaching vertices. Consequently, $B_2$, $B_3$ overlap if and only if $B_1$, $B_3$ overlap. Contradiction! □

**Proposition 1.3.4.** *Let C be a cycle in G. Let the overlap graph $O(G:C)$ have an induced cycle $(B_1, B_2, \ldots, B_s, B_1)$ of length $s \geqq 4$. Then there exist 2s vertices $X_1$, $Y_1$, $X_2$, $Y_2$, ..., $X_s$, $Y_s$ on C in this cyclic order, $X_i \neq Y_i$ for all $1 \leqq i \leqq s$, such that:*

(i) *$X_i$, $Y_{i+1}$ are vertices of attachment of $B_i$ with the property that $C[X_i, Y_{i+1}]$ contains all vertices of attachment of $B_i$ for all $1 \leqq i \leqq s$* (indices modulo s),

(ii) *there exist s internally disjoint C,C-paths $p_1, \ldots, p_s$ so that $p_i$ joins $X_i$ and $Y_{i+1}$, where $p_i \subseteqq B_i$* (see Fig. 1.3.4).

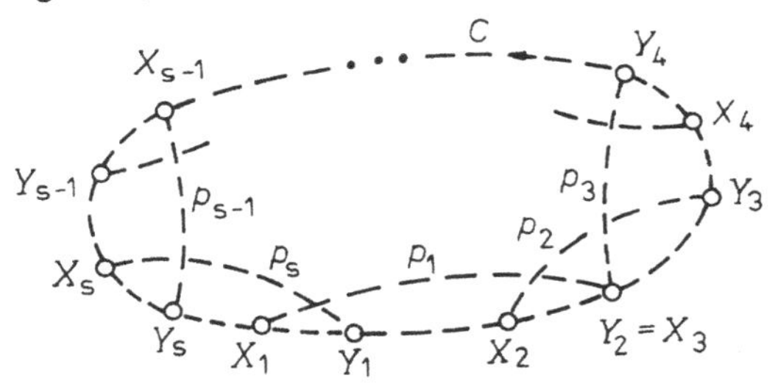


Fig. 1.3.4

**Remark 1.3.5.** It may happen that $X_{i+1} = Y_i$ for some indices *i*.

*Proof of Proposition 1.3.4.* Since $B_i$ is a C-bridge with attaching vertices $X_i$, $Y_{i+1}$, there is a C,C-path $p_i$ of $B_i$ with ends $X_i$, $Y_{i+1}$. Hence (ii) follows from (i). Thus we have only to prove assertion (i). We define:

$$M_i =_{\text{def}} \{B_j \mid j \leqq i-2 \text{ or } j \geqq i+2\}.$$

(1) *The attaching vertices of all bridges of $M_i$ lie on one segment $S_i$ of $B_i$.*

*Proof of (1).* Since no bridge of $M_i$ overlaps $B_i$, all attaching vertices of $B_j$, $j \leqq i-2$ or $j \geqq i+2$, are on one segment of C with respect to $B_i$. This is the assertion of Proposition 1.3.4 in the case $s = 4$.

Now let $s \geqq 5$. If there are two different segments S and S′ of $B_i$ containing all attachment vertices of $B_{i+2}$ and $B_{i+3}$, respectively, then $B_{i+2}$ and $B_{i+3}$ do not overlap. This contradicts the hypothesis that $B_{i+2}$ and $B_{i+3}$ overlap. Therefore, the attaching vertices of both $B_{i+2}$ and $B_{i+3}$ are in the same segment $S_i$ of C with respect to $B_i$. By induction all vertices of attachment of $B_{i+2}$, $B_{i+3}$, ..., $B_{i-2}$ are in $S_i$. □

We introduce a direction on C. Let $X_i$, $Y_{i+1}$ denote the end vertices of $S_i$ such that $Y_{i+1}$ precedes $X_i$ on $S_i$. Let $\bar{S}_i$ be the arc of C with ends $X_i$ and $Y_{i+1}$ such that $S_i \cup \bar{S}_i = C$ (and $X_i$ precedes $Y_{i+1}$ on $\bar{S}_i$). Hence $\bar{S}_i = C[X_i, Y_{i+1}]$.

(2) *$\bar{S}_i$ contains all vertices of attachment of $B_i$ and $S_i$ contains all vertices of attachment of $B_{i+2}$, $B_{i+3}$, …, $B_{i-2}$.* □

(3) *$\bar{S}_i$ and $\bar{S}_{i+1}$ have a "proper" intersection,* i.e., *$\bar{S}_i \cap \bar{S}_{i+1}$ is a path of length $\geqq 1$ and* $\bar{S}_i \neq (\bar{S}_i \cap \bar{S}_{i+1}) \neq \bar{S}_{i+1}$ (see Fig. 1.3.5).

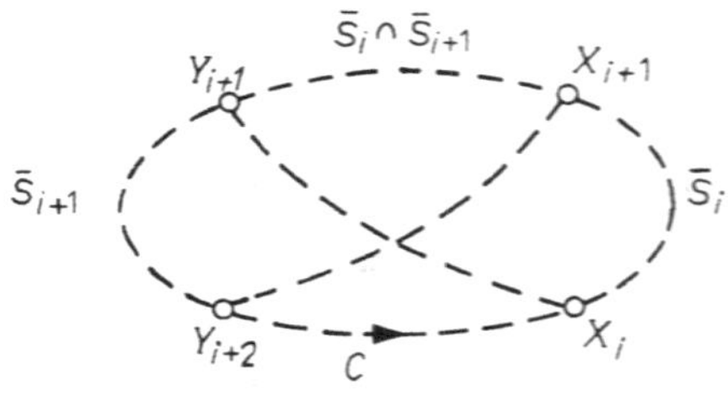


Fig. 1.3.5

*Proof of (3).* Since $B_i$ and $B_{i+1}$ overlap, by Lemma 1.3.3 they are skew. Hence the intersection $\bar{S}_i \cap \bar{S}_{i+1}$ is a path of length $\geqq 1$.

In order to complete the proof it is sufficient to show that $\bar{S}_i \cap \bar{S}_{i+1} \neq \bar{S}_{i+1}$. By (1) no vertex of attachment of $B_{i+2}$ is in $\bar{S}_i - \{X_i, Y_{i+1}\}$. Since $B_{i+2}$, $B_{i+1}$ are skew, there is an attaching vertex $A_{i+2}$ of $B_{i+2}$ in $\bar{S}_{i+1} - \{X_{i+1}, Y_{i+2}\}$, which is not a vertex of $\bar{S}_i - \{X_i, Y_{i+1}\}$. Therefore, $A_{i+2} \in \bar{S}_{i+1} - \{X_{i+1}, Y_{i+2}\}$ and $A_{i+2} \notin \bar{S}_i - \{X_i, Y_{i+1}\}$. Consequently, $\bar{S}_i \cap \bar{S}_i \neq \bar{S}_{i+1}$. □

Assertion (1) implies that

(4) *for $1 \leqq i < j \leqq s$ the intersections $\bar{S}_i \cap \bar{S}_{i+1}$ and $\bar{S}_j \cap \bar{S}_{j+1}$ are internally disjoint.* □

Moreover, applying (1), (2), (3) and (4) we can derive by induction:

(5) *$\bar{S}_1, \bar{S}_2, \ldots, \bar{S}_s$ lie on $C$ so that the pairwise internally disjoint paths $\bar{S}_1 \cap \bar{S}_2$, $\bar{S}_2 \cap \bar{S}_3, \ldots, \bar{S}_{s-1} \cap \bar{S}_s$, $\bar{S}_s \cap \bar{S}_1$ of length $\geqq 1$ are on $C$ in this cyclic order.* □

Obviously, $\bar{S}_i \cap \bar{S}_{i+1} = C[X_{i+1}, Y_{i+1}]$ for all $1 \leqq i \leqq s$ (indices taken modulo $s$). Assertion (5) immediately implies the validity of assertion (i) of Proposition 1.3.4. □

Now we derive some consequences from Proposition 1.3.4.

**Proposition 1.3.6.** *Let $C$ be a cycle of a graph $G$. Let $O(G:C)$ contain an induced odd cycle $(B_1, B_2, \ldots, B_s, B_1)$, $s \geqq 5$. Then there is a cycle $\bar{C}$ of $C \cup \bigcup_{i=1}^{s} B_i$ with vertices $Q_1, Q_2, \ldots, Q_{2s}$ in cyclic order joined by $s$ internally disjoint $\bar{C},\bar{C}$-paths $q_1, q_2, \ldots, q_s$ so that $q_i$ joins $Q_i$ and $Q_{i+s}$ $(1 \leqq i \leqq s)$, where $Q_{2j-1} \neq Q_{2j}$ for all $1 \leqq j \leqq s$* (see Fig. 1.3.6 a, b).

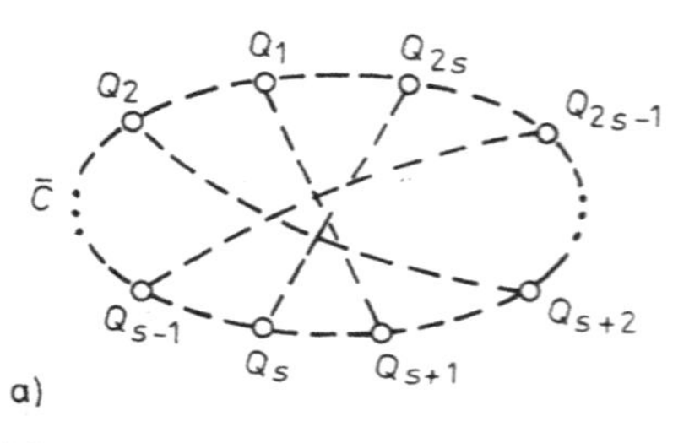


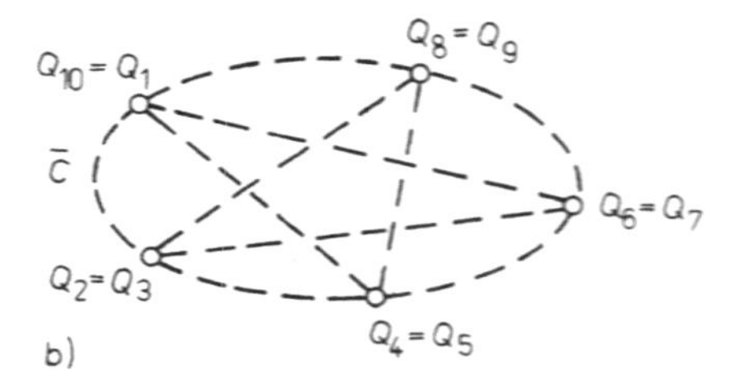


Fig. 1.3.6

*Proof.* Let $s \geqq 5$ be an odd integer. Let $C$ and $B_i$ be defined as in Proposition 1.3.4. The validity of Proposition 1.3.6 follows immediately from Proposition 1.3.4 by defining

$$\bar{C} =_{\text{def}} X_1p_1Y_2CX_3p_3Y_4CX_5p_5Y_6 \ldots X_{s-2}p_{s-2}Y_{s-1}CX_sp_sY_1CX_2p_2Y_3CX_4 \ldots X_{s-1}p_{s-1}Y_sCX_1,$$

where $p_i$ is a $C,C$-path of $B_i$ with ends $X_i$, $Y_{i+1}$, and $q_i =_{\text{def}} C[X_i, Y_i]$ (indices modulo $s$). □

I remark that for even $s$ this construction results in a "prism graph".

**Proposition 1.3.7.** *Let $s \geqq 5$ be an odd integer. Let $C$ and $B_i$ satisfy the hypotheses of Proposition 1.3.4. Then there is a cycle $C^*$ of $C \cup \bigcup_{i=1}^{s} B_i$ and of $G$ such that $O(G:C^*)$ has a triangle and the three bridges of the triangle are pairwise skew.*

*Proof.* Proposition 1.3.7 follows immediately from Proposition 1.3.4 by defining the circuit

$$C^* =_{\text{def}} Y_{s-1}CX_3p_3Y_4CX_5p_5Y_6CX_7 \ldots X_{s-4}p_{s-4}Y_{s-3}CX_{s-2}p_{s-2}Y_{s-1}$$

of $C \cup \bigcup_{i=1}^{s} B_i$, where $p_i$ is a $C,C$-path of $B_i$ with ends $X_i$, $Y_{i+1}$ (indices modulo $s$). Since the vertices of attachment of the $C$-bridges $B_s$, $B_1$ are on $C[X_s, Y_2]$ and $C[X_s, Y_2] = C^*[X_s, Y_2]$, the subgraphs $B_s$ and $B_1$ are also $C^*$-bridges.

Let $B$ denote a third $C^*$-bridge containing the $C^*,C^*$-path

$$X_2p_2Y_3CX_4p_4Y_5CX_6p_6Y_7 \ldots X_{s-3}p_{s-3}Y_{s-2}CX_{s-1}p_{s-1}Y_s.$$

Since $X_2$ and $Y_s$ are vertices of attachment of $B$, the subgraphs $B_s$, $B_1$ and $B$ are three pairwise skew $C^*$-bridges in $G$. This completes the proof of Proposition 1.3.7. □

By the same method for even $s$ a similar result can be proved:

**Proposition 1.3.8.** *Let $s \geqq 4$ be an even integer. Let $C$ and $B_i$ satisfy the hypotheses of Proposition 1.3.4. Then there is a cycle $C^*$ of $C \cup \bigcup_{i=1}^{s} B_i$ such that $O(G:C^*)$ has a quadrangle.* □

## 1.4. Connected overlap graphs

Let $C$ be a cycle of a graph $G$. At page 12 we stated that every $C$-bridge with at most one vertex of attachment avoids every other bridge. This implies that each such bridge is an isolated vertex of $O(G:C)$. Consequently, if $O(G:C)$ is a connected graph with at least two vertices, then every $C$-bridge has at least two vertices of attachment.

We ask: how can we characterize all graphs $G$ with a cycle $C$ having a connected overlap graph $O(G:C)$?

First we prove some lemmas.

**Lemma 1.4.1.** *Let $C$ be a cycle of a graph $G$, $K$ a connected subgraph of $O(G:C)$ and $X$, $Y$ two vertices of $C$. Let $K$ have a vertex of attachment on both arcs $C[X, Y] - \{X, Y\}$ and $C[Y, X] - \{Y, X\}$. Then one bridge $B$ of $K$ has a vertex of attachment on both arcs.*

*Proof.* Let $A'$ and $A''$ denote vertices of $K$ on $C[X, Y] - \{X, Y\}$ and $C[Y, X] - \{Y, X\}$, respectively. Let $A'$, $A''$ be vertices of $B'$ and $B''$ ($B', B'' \in K$), respectively. Since $K$ is connected, there is an induced $B',B''$-path $B' = B_1, B_2, \ldots, B_q = B''$ of $O(G:C)$. Since $B''$ has a vertex in $C[Y, X] - \{Y, X\}$, there is a smallest index $i$ such that $B_i$ has a vertex in $C[Y, X] - \{Y, X\}$. If $i = 1$, then with $B =_{\text{def}} B'$ the lemma is true. If $i > 1$, then $B_{i-1}$ has no vertex of attachment in $C[Y, X] - \{Y, X\}$, and $B_{i-1}$ overlaps $B_i$. Hence $B_i$ has a vertex of attachment in $C[X, Y] - \{X, Y\}$ also. With $B =_{\text{def}} B_i$ the lemma is true. □

For sets $K$ of $C$-bridges (or subgraphs of $O(G:C)$) we again define $K$-segments of $C$. Let $K$ be a set of $C$-bridges having at least two vertices of attachment $A_1, A_2, \ldots, A_s$, $s \geqq 2$, which are on $C$ in this cyclic order. Then the $s$ edge-disjoint arcs $C[A_i, A_{i+1}]$, $C[A_s, A_1]$, $1 \leqq i \leqq s-1$, are called *segments of C with respect to K*. For two disjoint sets $K$, $K'$ of $C$-bridges we again define the terms: $K$ avoids $K'$, $K$ overlaps $K'$, $K$ and $K'$ are skew. $K$ and $K'$ *avoid each other* if and only if either one of them has at most one vertex of attachment or both have at least two vertices of attachment and a segment of $C$ with respect to $K$ contains all vertices of attachment of $K'$. The sets $K$ and $K'$ *overlap* if and only if they do not avoid each other. Furthermore, $K$ and $K'$ are said to be *skew* if and only if there are four vertices $P$, $Q$, $R$, $S$ in this cyclic order on $C$ such that $P$ and $R$ are in $K$ and $Q$ and $S$ are in $K'$.

If $|K| = |K'| = 1$, then we arrive again at the corresponding concepts concerning $C$-bridges (see page 13).

According to Lemma 1.2.1 we have the following result:

**Lemma 1.4.2.** *Let C be a cycle of a graph G. If K and K′ are disjoint overlapping sets of C-bridges, then either they are skew or both K and K′ have precisely three vertices of attachment which coincide.*

The proof is just the same as for Lemma 1.2.1. □

For two connected subgraphs of $O(G:C)$ we shall prove the following result:

**Theorem 1.4.3.** *Let C be a cycle of a graph G. Let K and K′ be two connected subgraphs of $O(G:C)$, joined by no edge. Then K avoids K′.*

*Proof* (by contradiction). Assume that $K$ overlaps $K'$. By Lemma 1.4.2 we have two cases: 1) $K$ and $K'$ are skew and 2) $K$ and $K'$ have precisely three vertices of attachment which coincide.

Case 1. Let $K$ and $K'$ be skew. Then there are four vertices $P$, $Q$, $R$, $S$ in this cyclic order on $C$ such that $P, R \in K$ and $Q, S \in K'$. By Lemma 1.4.1 there is a bridge $B \in K$ having a vertex $P'$ on $C[S, Q] - \{S, Q\}$ and a vertex $R'$ on $C[Q, S] - \{Q, S\}$. Hence $K'$ has the vertex $Q$ on $C[P', R'] - \{P', R'\}$ and the vertex $S$ on $C[R', P'] - \{R', P'\}$. By Lemma 1.4.1 there is a bridge $B' \in K'$ having a vertex $Q'$ on $C[P', R'] - \{P', R'\}$ and a vertex $S'$ on $C[R', P'] - \{R', P'\}$. Therefore $P'$, $Q'$, $R'$, $S'$ are four vertices in this cyclic order on $C$, where $P', R' \in B$ and $Q', S' \in B'$. Hence $B$ and $B'$ are skew and $(B, B')$ is a $K,K'$-edge of $O(G:C)$.

Case 2. Let $K$ and $K'$ have the same set of three vertices of attachment $X$, $Y$, $Z$. If $K$ is an isolated vertex $B$ of $O(G:C)$, then $B$ contains $X$, $Y$, $Z$. Let $K$ have $q \geqq 2$ $C$-bridges $B_1, \ldots, B_q$. Since $K$ is connected, each bridge $B \in K$ overlaps a different bridge $B^* \in K$. Since $B$ and $B^*$ cannot be skew, both $B$ and $B^*$ contain $X$, $Y$, $Z$. Consequently, all bridges of

$K \cup K'$ contain $X$, $Y$, $Z$ and $K \cup K'$ induces a complete graph of $O(G:C)$. Hence $K$ and $K'$ are joined by an edge of $O(G:C)$.

In both cases we arrive at a contradiction. This proves the theorem. □

Next we characterize the connected overlap graphs $O(G:C)$. For this purpose we make the following definition: a 2-separation $(G_1, G_2)$ of a graph $G$ consists of two subgraphs $G_1$, $G_2$, both with at least three vertices such that $G = G_1 \cup G_2$ and $G_1 \cap G_2$ consists of two vertices only. With this concept we can formulate the following result:

**Theorem 1.4.4.** *Let $C$ be a cycle of a graph $G$. Then the overlap graph $O(G:C)$ is disconnected if and only if there is a two-separation $(G_1, G_2)$ of $G$ such that*

(1) $\{X, Y\} =_{\text{def}} V(G_1) \cap V(G_2) \subseteqq V(C)$,

(2) $C[X, Y] \subseteqq G_1$ *and* $C[Y, X] \subseteqq G_2$ *and*

(3) *neither $G_1$ nor $G_2$ is an arc of $C$.*

*Proof.* 1) Let $G$ have a 2-separation $(G_1, G_2)$, satisfying (1), (2), (3). By (3) the subgraph $G_i$ ist not a $C$-arc. Hence $G_i$ contains a $C$-bridge $B_i$.

If $B_1$ and $B_2$ belong to the same component $K$ of $O(G:C)$, then by Lemma 1.4.1 the component $K$ contains a $C$-bridge having at least one vertex on both arcs $C[X, Y] - \{X, Y\}$ and $C[Y, X] - \{Y, X\}$. This contradicts the definition of a 2-separation with (1), (2), (3). Consequently, $B_1$, $B_2$ are in different components of $O(G:C)$, and $O(G:C)$ is disconnected.

2) Let $O(G:C)$ be disconnected. We consider two cases.

Case 1. Let $G$ contain a $C$-bridge $B$ with at most one vertex of attachment $A$. Then $B$ is a component of $O(G:C)$, namely, an isolated vertex. Choose an edge $[X, Y]$ of $C$ such that $X = A$ if $A \in B$. Let $G_1 =_{\text{def}} B \cup [X, Y]$ and $G_2 =_{\text{def}} G - (X, Y) - (B - \{X\})$. Then $(G_1, G_2)$ is a 2-separation of $G$ satisfying (1), (2), (3).

Case 2. Let $G$ contain $C$-bridges with at least two vertices of attachment. Since $O(G:C)$ is disconnected, it contains two components $K$, $K'$. Theorem 1.4.3 implies that $K$ avoids $K'$. Hence all vertices of attachment of $K'$ are on a segment $S$ of $C$ with respect to $K$. If $X$, $Y$ are the end vertices of $S$, then by Theorem 1.4.3 the graph $G$ has a 2-separation $(G_1, G_2)$ with $V(G_1) \cap V(G_2) = \{X, Y\}$ and without loss of generality $C[X, Y] \cup K \subseteqq G_1$ and $C[Y, X] \cup K' \subseteqq G_2$. Hence $(G_1, G_2)$ satisfies (1), (2), (3). □

Theorem 1.4.4 immediately implies a theorem of W. T. Tutte, 1984:

**Theorem 1.4.5.** *Let $C$ be a cycle of a 3-connected graph $G$. Then $O(G:C)$ is connected.* □

We remark that a higher connectivity of $G$ does not imply a higher connectivity of $O(G:C)$. We shall construct an example graph.

Let $C =_{\text{def}} [X_1, X_2, \ldots, X_{2p}, X_1]$, $p \geqq 3$, be a cycle of the graph $\Gamma_p =_{\text{def}} C \cup [X_1, X_3, \ldots, X_{2p-1}, X_1]$. The graph $G_{p,q}$ is obtained from $\Gamma_p$ by adding a $K_q$, $q \geqq 1$, and joining each vertex of $K_q$ to each vertex of $\Gamma_p$. The graph $G_{p,q}$ has connectivity $q+2$ and its $C$-bridges are $B_0 =_{\text{def}} G_{p,q} - E(\Gamma_p)$, $B_j =_{\text{def}} [X_{2j-1}, X_{2j+1}]$, $B_p =_{\text{def}} [X_{2p-1}, X_1]$ for all $1 \leqq j \leqq p-1$, where $E(\Gamma_p)$ denotes the edge set of $\Gamma_p$. These bridges form a star in $O(G:C)$, where $B_0$ has valency $p$ and $B_j$ has valency 1 for all $1 \leqq j \leqq p$. Then the graph $G$ is a $(q+2)$-connected graph, $q \geqq 1$, and $O(G:C)$ has connectivity 1.

## 1.5. Further properties of overlap graphs and circle graphs

In the preceding §§ 1.3-1.4 properties of bridges of cycles have been described by their overlap graphs. Here we shall deal with the question: what graphs are overlap graphs? With the help of Proposition 1.3.4 we shall show that there are graphs which are not overlap graphs.

**Theorem 1.5.1.** *The graphs $H_1$, $H_2$, $H_3$ of Fig. 1.5.1 are not overlap graphs.*

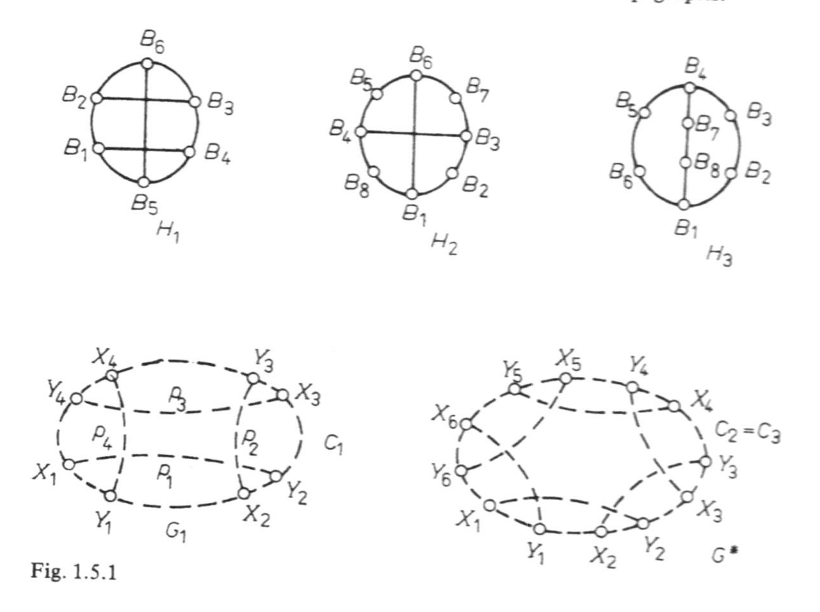


Fig. 1.5.1

*Proof.* Assume that there is a graph $G_i$ with circuit $C_i$ such that $O(G_i : C_i) \cong H_i$.

Since $H_i$ has no triangle by Lemma 1.3.3, two adjacent bridges of $O(G_i : C_i)$ are skew in $G_i$.

Case $i = 1$. $(B_1, B_2, B_3, B_4, B_1)$ is an induced 4-cycle of $O(G_1 : C_1)$. By Proposition 1.3.4 there exist eight vertices $X_1, Y_1, X_2, \ldots, X_4, Y_4$ on $C$ in this cyclic order, $X_i \neq Y_i$, such that $X_i, Y_{i+1}$ are vertices of attachment of $B_i$ with the property that $C[X_i, Y_{i+1}]$ contains *all* vertices of attachment of $B_i$, $1 \leqq i \leqq 4$, where the indices are taken mod 4 (compare with Fig. 1.5.1, where $B_i$ is represented by a $C,C$-path $p_i$ of $B_i$ with ends $X_i$ and $Y_{i+1}$).

Obviously,

(1) *if $B_j$, $5 \leqq j \leqq 6$, overlaps $B_i$, $1 \leqq i \leqq 4$, then $B_j$ has a vertex on $C[X_i, Y_{i+1}] - \{X_i, Y_{i+1}\}$.* □

Since $B_5$ overlaps $B_4$, $B_1$ and avoids $B_2$, $B_3$ and $B_6$ overlaps $B_2$, $B_3$ and avoids $B_4$, $B_1$, the vertices of attachment of $B_5$ or $B_6$ are in $C[Y_4, X_2]$ or $C[Y_2, X_4]$, respectively. Consequently, $B_5$ avoids $B_6$. Contradiction!

In the cases $i = 2$ or 3 we only outline the proof.

Case $i = 2$. $(B_1, B_2, \ldots, B_6, B_1)$ is an induced 6-cycle of $O(G_2 : C_2)$. Proposition 1.3.4 has the same consequences as in the case $i = 1$, we have only to replace the eight vertices $X_1, \ldots, Y_4$ by twelve vertices $X_1, \ldots, Y_6$. The resulting subgraph $G^*$ of $G_2$ is depicted in

Fig. 1.5.1. Since $B_7$ overlaps precisely $B_3$, $B_6$ and $B_8$ overlaps precisely $B_1$, $B_4$, the vertices of attachment of $B_7$ and $B_8$ are in $C[Y_3, X_4] \cup C[Y_6, X_1]$ and $C[Y_1, X_2] \cup C[Y_4, X_5]$, respectively. Consequently, $B_7$ overlaps $B_8$. Contradiction!

Case $i = 3$. $\mathfrak{K} =_{\text{def}} (B_1, B_2, \ldots, B_6, B_1)$ is an induced 6-cycle of $O(G_3 : C_3)$. As in the case $i = 2$, applying Proposition 1.3.4 we arrive at a subgraph $G^*$ of $G_3$ depicted in Fig. 1.5.1. Since $B_7$ overlaps $B_4$ and avoids $\mathfrak{K} - \{B_4\}$, and $B_8$ overlaps $B_1$ and avoids $\mathfrak{K} - \{B_1\}$, the vertices of attachment of $B_7$ and $B_8$ are in $C[Y_3, X_6]$ and $C[Y_6, X_3]$, respectively. Consequently, $B_7$ avoids $B_8$. Contradiction! □

We have defined a circle graph as an overlap graph of a Hamiltonian graph with respect to a Hamiltonian cycle $H$. Obviously, all bridges of $H$ are diagonals of $H$.

Let $H$ have at least two diagonals. If $s \geqq 2$ diagonals of $H$ are incident with the same vertex $X$ of $H$, then we can replace $X$ by a path $p$ of length $s$ so that in the new circuit no two of these diagonals overlap or have a common vertex on $p$ (see Fig. 1.5.2).

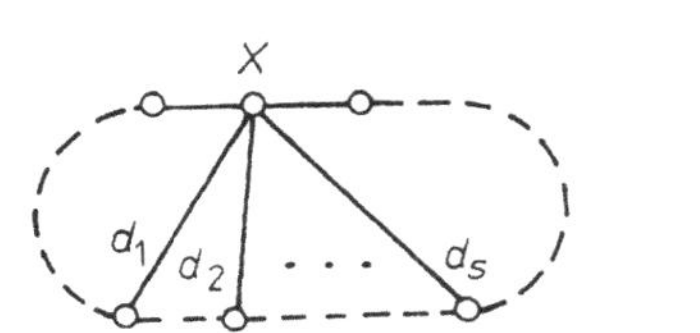


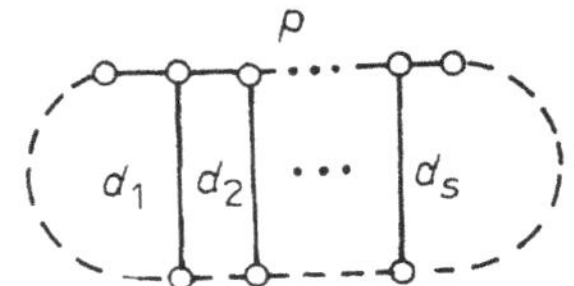


Fig. 1.5.2

If the vertex $X$ of $H$ is incident with no diagonal, then we replace $X$ and the edges incident with $X$ by the edge $(X_1, X_2)$, where $X_1$, $X_2$ are the two neighbours of $X$ on $H$.

In both cases let the new graph and the new Hamiltonian cycle be denoted by $G'$ and $H'$, respectively. Then $O(G : H) = O(G' : H')$. Repeatedly applying these two operations to all vertices of $H$ we obtain a cubic Hamiltonian graph $G^*$ (not necessarily simple) with a Hamiltonian cycle $H^*$ so that $O(G : H) = O(G^* : H^*)$. This is sufficient for considering only circle graphs of a cubic Hamiltonian graph $G$ with respect to a Hamiltonian cycle $H$. The cycle $H$ can be drawn in the plane as a geometric circle $C$ and the diagonals as geometric chords of this circle. Now the circle graph can be defined as the overlap graph or intersection graph of the chords of this geometric configuration. This is the usual way of introducing the concept of circle graph.

By Theorem 1.5.1 we know that there are graphs which are not overlap graphs.

Next we present overlap graphs which are not circle graphs.

**Theorem 1.5.2.** *The graphs $H_4$, $H_5$, $H_6$ of Fig. 1.5.3 are overlap graphs but not circle graphs.*

*Proof.* Obviously, $H_i$ is the overlap graph of $G_i$ of Fig. 1.5.3 with respect to the circuit $C_i$. The proof that $H_i$ is not a circle graph is similar to the proof of Theorem 1.5.1. □

Obviously, all trees, all complete graphs, and all cycles are circle graphs. If $G^*$ is a circle graph, then a cubic Hamiltonian graph $G$ and a Hamiltonian cycle $H$ of $G$ are said to be a *representation of* $G^*$ if and only if $O(G : H) \cong G^*$.

Let $G$, $H$ be a representation of a circle graph. By the following operation (O) a new representation of a circle graph can be obtained:

**Operation (O).** If $(X_1, Y_1)$ and $(X_2, Y_2)$ are edges of $H$ and $X_1, Y_1, X_2, Y_2$ are on $H$ in this cyclic order, then

$$G^* =_{\text{def}} G - \{(X_1, Y_1), (X_2, Y_2)\} \cup \{(X_1, X_2), (Y_1, Y_2)\}$$

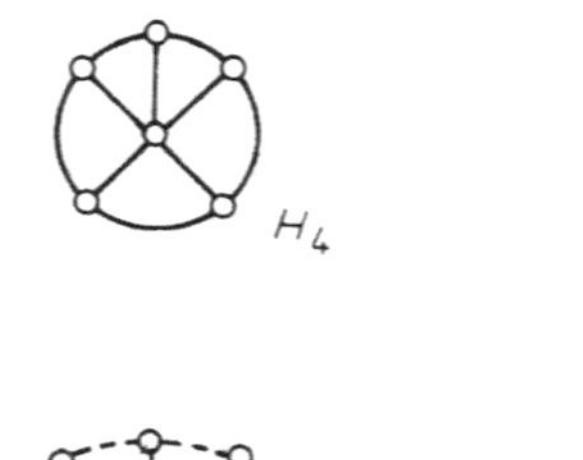


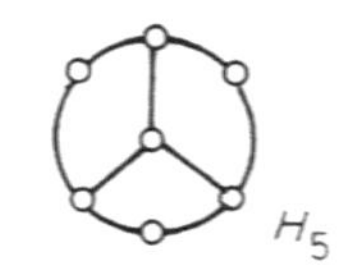

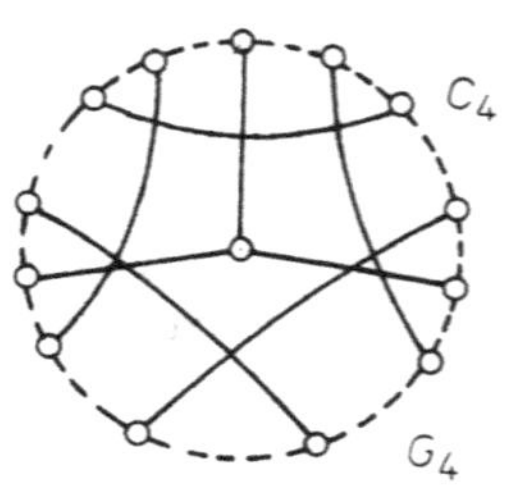


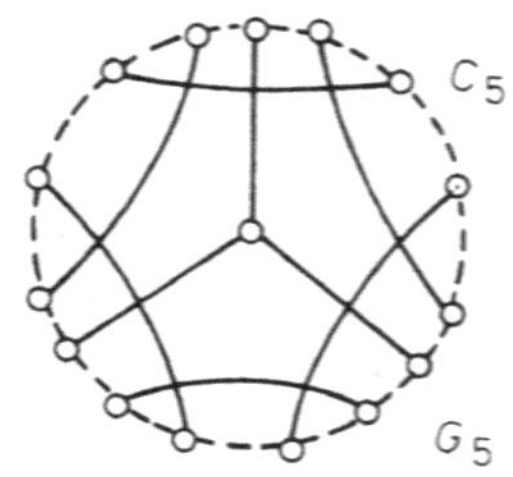


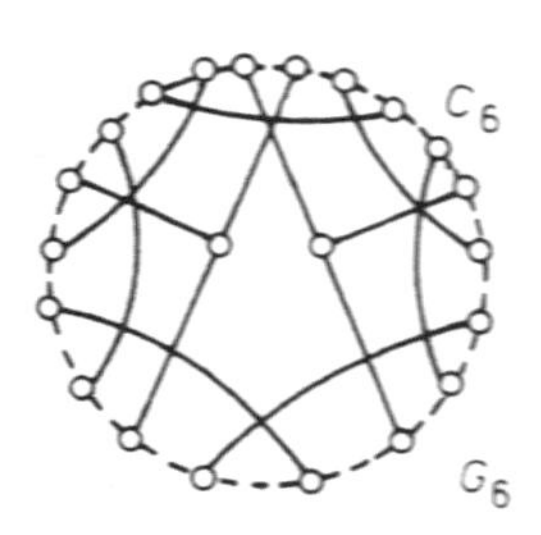


Fig. 1.5.3

is a new cubic Hamiltonian graph with the new Hamiltonian cycle

$$H^* =_{\text{def}} H - \{(X_1, Y_1), (X_2, Y_2)\} \cup \{(X_1, X_2), (Y_1, Y_2)\}.$$

Then $O(G^*: H^*)$ is a new circle graph.

Let $d_1, \ldots, d_n$ be the diagonals of $H$ in a representation $G$, $H$ of a circle graph. We label the two end vertices of $d_i$ by $i$ $(1 \leqq i \leqq n)$. Then the graph $O(G: H)$ is completely characterized by the corresponding cyclic ordered sequence of $i$'s $(1 \leqq i \leqq n)$. Each possible such sequence can be obtained by exchanging two neighbouring symbols; this corresponds to the application of the operation (O) to $H$, where $X_1$, $Y_1$, $X_2$, $Y_2$ form a 3-path, i.e., $(Y_1, X_2)$ is an edge of $H$. A consequence of this is the following result.

**Proposition 1.5.3.** *From a representation $G$, $H$ of a circle graph of order $n$ the representations of all circle graphs of order $n$ can be obtained by repeated application of the operation* (O). □

Proposition 1.5.3 describes a method of constructing new circle graphs, which depends on a representation of a given circle graph. A "representation-free" method is obtained by applying (O), where $(X_1, Y_2)$ is assumed to be a diagonal. This result due to W. Wessel and R. Pöschel, 1985, is the following

**Theorem 1.5.4.** *Let $\underline{G}$ be a circle graph and $N(P)$ the neighbourhood of a vertex $P$ of $\underline{G}$. A new circle graph is obtained by replacing the induced subgraph $G[N(P)]$ by its complement.* □

The graphs of Fig. 1.5.4 can be obtained from a 3-path by recursive application of the operation described in Theorem 1.5.4.

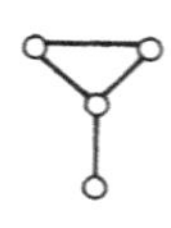

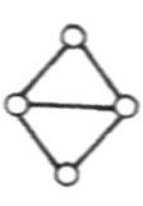

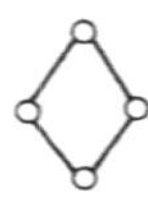

Fig. 1.5.4

There is a large literature on circle graphs. We refer the reader to W. Wessel and R. Pöschel, 1985.

We suggest also the question: which graphs are overlap graphs? Characterize these graphs!

Recently M. Koebe, 1990, presented a characterization of overlap graphs (spider graphs by M. Koebe, 1990a and b, which results in a polynomial-time recognition algorithm of overlap graphs.

## 1.6. Overlap graphs and the planarity criteria of Tutte and Kuratowski

The first proof of the famous theorem of K. Kuratowski, 1930, using bridges of a certain cycle, is due to G. A. Dirac and S. Schuster, 1954, which is included in several text books on graph theory. We follow a proof of M. Burštein, 1978 (the main idea is implicitly contained in a paper of W. T. Tutte, 1965, 1984).

The famous theorem of Kuratowski (and related results) were proved in a very nice paper of C. Thomassen, 1981b. For a more detailed account on this topic we refer the reader to this paper. Here we only give a proof of Kuratowski's theorem based on the results on bridges of the preceding sections and on the following nice criterion of W. T. Tutte, 1984, that a 2-connected graph $G$ is non-planar if and only if it has a subgraph $H$ in which there is a cycle $C$ such that $O(H:C)$ is non-bipartite.

In the literature there are often statements on plane graphs that are said to be intuitively clear, but to have complicated formal proofs. C. Thomassen, 1981b, pointed out that this need not be so. Since our aim here is only to show the use of bridges in the theory of planar graphs, we prefer the intuitive point of view.

Our first aim is to prove Tutte's criterion. For this purpose we prove some propositions and lemmas.

We consider only 2-connected plane graphs. If the vertices and edges of a plane graph are deleted from the plane, then there remain connected maximal parts which are homeomorphic to a disc. These connected parts are said to be the *faces* of the plane graph. Each face of a 2-connected plane graph is bounded by a cycle, called a *facial cycle.*

**Proposition 1.6.1.** *Let $C$ be a cycle of a graph $G$ such that $O(G:C)$ has no edge and $C \cup B$ is planar for each $C$-bridge $B$. Then $G$ has an embedding in the plane such that $C$ bounds the exterior face.*

*Proof.* The proof will be accomplished by induction on the number of $C$-bridges.

If $O(G:C)$ has precisely one vertex $B$, then $G = C \cup B$ is planar. Suppose that our proposition is true if $C$ has $k$ bridges $(1 \leqq k \leqq m)$. Next, let $C$ be a cycle of $G$ having $m+1$ $C$-bridges such that $C \cup B$ is planar for each $C$-bridge $B$.

Since $O(G:C)$ is disconnected, by Theorem 1.4.4 there is a 2-separation $(G_1, G_2)$ of $G$ such that

(i) $\{X, Y\} =_{\text{def}} V(G_1) \cap V(G_2) \subsetneqq V(C)$,

(ii) $C[X, Y] \subsetneqq G_1$ and $C[Y, X] \subsetneqq G_2$,

(iii) at least one $C$-bridge has its vertices of attachment in $C[X, Y]$ and at least one $C$-bridge has its vertices of attachment in $C[Y, X]$.

Hence $C_1 =_{\text{def}} C[X, Y] \cup (X, Y)$ is a cycle of $G_1 \cup (X, Y)$ and $C_2 =_{\text{def}} C[Y, X] \cup (Y, X)$ is a cycle of $G_2 \cup (X, Y)$.

Each $C$-bridge in $G$ is either a $C_1$-bridge in $G_1 \cup (X, Y)$ or a $C_2$-bridge in $G_2 \cup (X, Y)$.

By (iii) the cycle $C_i$ has at least one $C_i$-bridge. Hence the cycle $C_i$ has at most $m$ bridges, $i = 1, 2$.

If $B$ is a $C$-bridge that is also a $C_i$-bridge, then $C_i \cup B$ is planar. The induction hypothesis implies that $G_i \cup (X, Y)$ has an embedding $\underline{G}_i$ in the plane such that $C_i$ is the bound of the outer face of $\underline{G}_i$, $i = 1, 2$.

In order to obtain an embedding of $G$ in the plane put $\underline{G}_2$ in the outer face of $\underline{G}_1$, delete $(X, Y)$ in both $\underline{G}_1$ and $\underline{G}_2$ and identify the vertices $X$ of $\underline{G}_1$ and $\underline{G}_2$ and the vertices $Y$ of $\underline{G}_1$ and $\underline{G}_2$. □

**Proposition 1.6.2.** *Let $C$ be a cycle of a graph $G$ such that $O(G:C)$ is bipartite and $C \cup B$ is planar for each $C$-bridge $B$. Then $G$ is a planar graph.*

*Proof.* Let $G_i$ consist of the cycle $C$ and the bridges of one set $M_i$ of the bipartition $M_1$, $M_2$ of $O(G:C)$, $i = 1, 2$, of $\underline{G}_1$ and the inner face of $\underline{G}_2$.

By Proposition 1.6.1 the graph $G_i$ has a planar embedding $\underline{G}_i$ such that $C$ bounds the outer face.

An embedding of the graph $G$ is obtained by identifying the cycle $C$ of $\underline{G}_1$ with the cycle $C$ of $\underline{G}_2$. □

The next result is intuitively clear.

**Proposition 1.6.3.** *Let $\underline{G}$ be a plane graph with cycle $\underline{C}$, and let $\underline{B}_1$, $\underline{B}_2$ be two $\underline{C}$-bridges. If $\underline{B}_1$ overlaps $\underline{B}_2$, then $\underline{B}_1$ and $\underline{B}_2$ cannot be embedded in the same region of $\underline{C}$.* □

We shall need the following result later.

**Lemma 1.6.4.** *Let $G$ be any 2-connected graph. Then $G$ is a cycle $C$, or a cycle $C$ with precisely one additional $C,C$-path, or $G$ has a cycle $C$ having at least two bridges in $G$.*

*Proof.* Since $G$ is 2-connected, $G$ has a cycle $C$. If $G = C$, the lemma holds. If $C$ has at least two bridges, then the proof is complete. Hence let $C$ have precisely one bridge $B$.

Let $S$ be a segment of $C$ with respect to $B$ with end vertices $X$ and $Y$ such that $S = C[X, Y]$.

The bridge $B$ contains a $C,C$-path $p$ with ends $X$, $Y$. Then $C^* =_{\text{def}} C[Y, X] \cup p$ is a cycle with the $C^*$-bridge $C[X, Y]$. If $C^*$ has a second bridge, the lemma is true. If $C^*$ has only the bridge $C[X, Y]$, then the graph $G$ consists of the cycle $C^*$ with precisely one additional $C^*,C^*$-path. □

We start to prove Tutte's planarity criterion. This will be done by proving two propositions. After that the criterion will be formulated.

**Proposition 1.6.5.** *Let $G$ be a 2-connected graph. If $G$ has a subgraph $H$ such that $O(H:C)$ is non-bipartite, then $G$ is non-planar.*

*Proof.* Suppose there is a plane embedding of $G$ in the plane. Consequently, $H$ is also embedded in the plane. Since $O(H:C)$ is non-bipartite, the $C$-bridges inside $C$ or outside $C$ are joined by an edge of $O(H:C)$. Hence two overlapping $C$-bridges are embedded in the same region of $C$. This contradicts Proposition 1.6.3! □

**Proposition 1.6.6.** *Let $G$ be a 2-connected graph. If for each cycle $C$ of every subgraph $H$ of $G$ the overlap graph $O(H:C)$ is bipartite, then $G$ is planar.*

*Proof.* The proof will be accomplished by induction on the number of edges.

The 2-connected graphs with at most five edges are planar graphs. Suppose that each graph $G$ with at most $m$ edges satisfying the hypotheses of Proposition 1.6.6 is planar.

Next, let $G$ be a graph with $m+1$ edges satisfying the hypotheses of Proposition 1.6.6.

If each cycle of the 2-connected graph $G$ has at most one bridge, then by Lemma 1.6.4 the graph $G$ is a cycle or $G$ is a cycle $C$ with precisely one additional $C,C$-path. In both cases $G$ is planar.

Let $G$ have a cycle $C$ with at least two $C$-bridges. By the hypotheses of our proposition the overlap graph $O(G:C)$ is bipartite. For each $C$-bridge $B$ the subgraph $C \cup B$ has at most $m$ edges and satisfies the hypotheses of Proposition 1.6.6. By the induction hypothesis $C \cup B$ is planar. Then Proposition 1.6.2 implies that $G$ is planar. □

We summarize Propositions 1.6.5–6 and formulate the planarity criterion of W.T.Tutte, 1965, 1984, M.Burštein, 1978, and C.Thomassen, 1981b.

**Theorem 1.6.7.** *A 2-connected graph $G$ is non-planar if and only if it has a subgraph $H$ in which there is a cycle $C$ such that $O(H:C)$ is non-bipartite.* □

A graph $H$ is said to be a *subdivision* of a graph $G$ if the graph $H$ can be obtained from $G$ by repeatedly replacing an edge by a path $p$ such that $p$ has only its ends in the old graph, which are also the ends of the deleted edge.

With the help of Theorem 1.6.7 we shall prove the theorem of K.Kuratowski, 1930.

**Theorem 1.6.8.** *A graph $G$ is non-planar if and only if it contains a subdivision of $K_5$ or $K_{3,3}$.*

*Proof.* It is sufficient to prove Theorem 1.6.8 for 2-connected graphs only.

1) Let $G$ contain a subdivision $D$ of $K_5$ or $K_{3,3}$. Let $C$ be a cycle of $D$ containing all vertices of valency at least 3. Then $O(D:C)$ is non-bipartite. By Theorem 1.6.7 the graph $G$ is non-planar.

2) Let $G$ be a non-planar graph. By Theorem 1.6.7 it contains a subgraph $H$ in which there is a cycle $C$ such that $O(H:C)$ is non-bipartite. Hence $O(H:C)$ contains an induced odd cycle.

By Proposition 1.3.7 the graph $H$ has a cycle $C^*$ such that $O(H:C^*)$ has a triangle and the three $C^*$-bridges are pairwise skew.

By Theorem 1.3.2 the subgraph $C^* \cup B_1 \cup B_2 \cup B_3$ contains a $C^*$-subdivision of $F_i$, $1 \leqq i \leqq 9$. It can easily be proved that each of these graphs contains a subdivision of $K_5$ or $K_{3,3}$. □

## 1.7. Hamiltonian cycles in planar graphs

In this section we present the famous theorem of W.T.Tutte, 1956, on the existence of a Hamiltonian cycle in a 4-connected planar graph and related problems. Tutte's result is a generalization of a theorem of H.Whitney, 1931, that every 4-connected plane triangulation has a Hamiltonian cycle. Tutte did not prove his theorem directly, but proved a more general result. The proof of a more general version of an original assertion is a method often used in mathematics. The general result is an assertion about the properties of bridges of a plane graph with respect to a cycle or path. It implies Tutte's result on the existence of a Hamiltonian cycle in a 4-connected planar graph and other results.

First we formulate Tutte's result on bridges.

**Theorem 1.7.1.** *Let $G$ be a 2-connected plane graph. Let $C_1$ and $C_2$ be facial cycles of $G$ having a common edge $e = (U, V)$ and let $k$ be a different edge of $C_1$. Then $G$ has a cycle $C$ containing $e$ and $k$ such that*

(i) *each $C$-bridge has at most three vertices of attachment,*
(ii) *each $C$-bridge containing an edge of $C_1 \cup C_2$ has at most two vertices of attachment.* □

There are several proofs of this theorem by W. T. Tutte, 1956, 1977 (see also O. Ore, 1967a, and H. Sachs, 1972).

C. Thomassen, 1983b, proved the following sharper version of this theorem (for the proof we refer the reader to C. Thomassen, 1983b), implying that every 4-connected planar graph is Hamiltonian connected.

**Theorem 1.7.2.** *Let $G$ be a 2-connected plane graph with facial cycle $K$. Let $V$ and $k$ be a vertex and an edge, respectively, of $K$ and let $U$ be any vertex of $G$ distinct from $V$. Then $G$ has a $V,U$-path $p$ containing $k$ such that*

(i) *each $p$-bridge has at most three vertices of attachment,*
(ii) *each $p$-bridge containing an edge of $K$ has at most two vertices of attachment.* □

Next we formulate Tutte's main theorem (W. T. Tutte, 1956).

**Theorem 1.7.3.** *Every 4-connected planar graph has a Hamiltonian cycle.* □

We can show even more:

**Theorem 1.7.4.** *Let $G$ be a 4-connected plane graph and let $e$ and $k$ be two edges of a facial cycle of $G$. Then $G$ has a Hamiltonian cycle through $e$ and $k$.*

*Proof.* Let $G$ be a 4-connected plane graph and $C$ a cycle having the properties described in Theorem 1.7.1. We shall show that $C$ is Hamiltonian cycle. The notation of Theorem 1.7.1 is used.

Assume that $C$ does not include all the vertices of $G$. Then $G$ has a regular $C$-bridge $B$. Since $G$ is 4-connected, $B$ has four vertices of attachment, provided that $C$ has length at least 4. This contradicts Theorem 1.7.1.

Let $C$ have length 3. Then $C$ contains no vertex of $C_2 - \{U, V\}$. Hence $G$ has a $C$-bridge $B'$ containing a vertex of $C_2 - \{U, V\}$. Since $G$ is 4-connected, all three vertices of $C$ are vertices of attachment of $B'$, contradicting Theorem 1.7.1. □

With the aid of Theorem 1.7.1 C. Thomassen, 1976, proved a theorem with weaker hypotheses:

**Theorem 1.7.5.** *Let $G$ be a 3-connected planar graph with at most one separating set of three vertices. Then $G$ has a Hamiltonian cycle.* □

Tutte's Theorem 1.7.1 is considerably stronger than Theorem 1.7.5 and has other remarkable corollaries. Here we present two such corollaries.

A connected graph is said to be *cyclically k-edge-connected* if and only if the deletion of any $k - 1$ or fewer edges does not result in a disconnected graph having two components with one cycle each.

**Theorem 1.7.6.** *Every plane cubic cyclically* 4-*edge-connected graph has a cycle* $C$ *such that* $G - V(C)$ *consists of isolated vertices.*

Such a cycle $C$ is called a *dominating cycle.* All $C$-bridges have no or one inner vertex.

*Proof of Theorem 1.7.6.* The cubic graph $G$ is cyclically 4-edge-connected. From this it can easily be deduced that $G$ is 2-connected. By Theorem 1.7.1 the graph $G$ contains a cycle $C$ with the property (i). We shall show that $G - C$ consists of isolated vertices. Assume that $G - C$ has an edge $(X, Y)$. Then $G$ contains a $C$-bridge $B$ with at least two inner vertices $X$ and $Y$. By Theorem 1.7.1 the $C$-bridge $B$ has at most three vertices of attachment. Since $G$ is cubic and $C$ is regular of degree 2, at most three edges $e_1$, $e_2$, $e_3$ join $C$ to the kernel $K_B$ of $B$. Hence $K_B$ has $\left\lceil \frac{1}{2}(3i-3) \right\rceil$ edges, where $i$ denotes the number of its vertices. Since $i \geqq 2$, $\left\lceil \frac{1}{2}(3i-3) \right\rceil \geqq i$, and therefore $K_B$ contains a cycle $C'$. Consequently, the three edges $e_1$, $e_2$, $e_3$ separate the cycles $C$ and $C'$. This contradiction completes the proof. □

D. Nelson observed the following (see M. D. Plummer, 1973):

**Theorem 1.7.7.** *The deletion of any vertex from a* 4-*connected planar graph results in a Hamiltonian graph.*

*Proof.* Let $V$ be any vertex of a 4-connected plane graph $G$. Then $V$ is on a facial cycle. Let $(X, Y)$ be an edge of this facial cycle, not incident with $V$. We apply Theorem 1.7.4 to the graph $G \cup e \cup k$, where $e =_{\text{def}} (V, X)$ and $k =_{\text{def}} (V, Y)$. Hence there is a Hamiltonian cycle $C$ through $e$ and $k$ and $C - \{e, k\} \cup (X, Y)$ is a Hamiltonian cycle of $G - V$. □

As a counterpart of this result C. Thomassen, 1976, has proved the following result, again with the aid of Tutte's theorem.

**Theorem 1.7.8.** *Let* $G$ *be a planar graph with minimum degree at least* 4. *If for each vertex* $V$ *of* $G$ *the vertex-deleted subgraph* $G - V$ *is Hamiltonian, then* $G$ *also has a Hamiltonian cycle.* □

M. D. Plummer, 1973, (see also W. T. Tutte, 1977) conjectured that any 4-connected planar graph is Hamiltonian-connected, i.e., it has a Hamiltonian path connecting any two prescribed vertices. The *main* result of C. Thomassen, 1983b, can be extracted from his Theorem 1.7.2.

**Theorem 1.7.9.** *Every* 4-*connected planar graph is Hamiltonian-connected.*

*Proof.* Let $G$ be a 4-connected graph and $U$, $V$ two of its vertices. Let $C_1$ be a facial cycle of $G$, $U \in V(C_1)$, $V \notin V(C_1)$, and $k$ an edge of $C_1$ which is not incident with $U$. Then by Theorem 1.7.2 the graph $G$ has a path $p$ with the properties described in Theorem 1.7.2. By construction $p$ has at least four vertices, namely $U$, $V$ and the end vertices of $k$.

We shall show that $p$ is a Hamiltonian path. Assume that $p$ does not include all vertices of $G$. Then $G$ has a regular $p$-bridge $B$. Since $G$ is 4-connected, $B$ has four vertices of attachment. This contradicts Theorem 1.7.2. □

The following stronger conjecture of M. D. Plummer, 1973, cannot be proved in this way.

**Conjecture.** *Every 4-connected planar graph is 2-Hamiltonian* (i.e., $G - U - V$ *has a Hamiltonian cycle for every pair of vertices* $U$ *and* $V$).

B. Grünbaum, 1970, and C. St. J. A. Nash-Williams, 1973, conjectured that every 4-connected graph embedded in the torus is Hamiltonian (and B. Grünbaum made an analogous conjecture for other surfaces as well). Theorem 1.7.9. cannot be extended to toroidal graphs, because a 4-connected graph on the torus need not be Hamiltonian-connected. It may even contain an edge which is not contained in a Hamiltonian cycle. This can be illustrated by forming the Cartesian product of two cycles of even length (which is toroidal and bipartite) and then adding an edge such that the resulting graph is toroidal. The resulting graph is non-bipartite.

From Thomassen's Theorem 1.7.2 it can be extracted that any pair of vertices of a planar cubic 3-connected cyclically 4-edge-connected graph is joined by a "dominating" path.

**Theorem 1.7.10.** *Let* $X$ *and* $Y$ *be vertices of a planar cubic 3-connected cyclically 4-edge-connected graph* $G$. *Then* $G$ *has an* $X,Y$*-path* $p$ *such that* $G - p$ *consists of isolated vertices only.* □

This can be proved in the same way as Theorem 1.7.6.

## 1.8. Separation of vertices by a cycle in planar graphs

With the aid of Kuratowski's theorem we shall prove a result of W. T. Tutte, 1960, 1975, on separating cycles, where a cycle $C$ of $G$ is called *separating* if and only if $G - C$ has more components than $G$.

**Theorem 1.8.1.** *Let* $X$ *and* $Y$ *be vertices of a planar graph* $G$. *Then* $G$ *has a plane representation such that* $X$ *and* $Y$ *are adjacent to a common face unless* $G$ *contains a separating cycle* $C$ *such that the* $C$*-bridges of* $G$ *containing* $X$ *and* $Y$ *as inner vertices, respectively, are overlapping.*

*Proof.* If $G$ has no representation as described in the theorem, then $G \cup (X, Y)$ is non-planar and contains a subdivision $H$ of $K_5$ or $K_{3,3}$. Since $G$ is planar, $e =_{\text{def}} (X, Y)$ is contained in $H$, and $H - e$ is planar. It is easily seen that $H - e$ contains a cycle $C$ such that $X$ and $Y$ are inner vertices of distinct overlapping $C$-bridges, say $B_1$ and $B_2$, of $H - e$. If we consider a planar representation of $G$, we see that the images of $B_1$ and $B_2$ are separated by the image of $C$. Hence they are contained in overlapping $C$-bridges of $G$. □

# Chapter 2

# Non-separating cycles in graphs

## 2.0. Introduction

Having introduced the concept of bridge we pose the question: which graphs contain (non-Hamiltonian) cycles with only one regular bridge? Second question: which graphs contain (non-Hamiltonian) paths with only one regular bridge?

A cycle or path with at most one regular bridge is said to be a *non-separating cycle* or *path*, respectively. Hence a cycle $C$ or path $p$ in a connected graph $G$ is a *separating cycle* or *path* if and only if the deletion of $C$ or $p$, respectively, results in a disconnected graph.

An induced non-separating cycle is also called *peripheral.*

### 1° Separating and non-separating cycles in 2-connected graphs

G. Chartrand and H. V. Kronk, 1968, proved the following result:

*Every finite connected graph has a separating path unless the graph is a cycle, a complete graph, or a regular complete bipartite graph.*

However, there are finite 2-connected graphs which have separating paths but no separating cycles. The wheels, $K_{r,r+1}$ and the Petersen graph are examples.

Separating cycles play an important role in the papers of H. Whitney, 1931, and W. T. Tutte, 1960, 1975. In section 1.8 we have presented Tutte's result on vertices of planar graphs which are separated by a cycle such that they are "far apart".

J. A. Bondy and M. D. Plummer, 1975, provide sufficient conditions for graphs to have a separating cycle and characterize the planar graphs with no separating cycles and the cubic graphs with no separating cycles. The complete description of all 2-connected graphs without separating cycles is due to C. Thomassen, 1978. This result will be presented in section 2.1.

C. Thomassen, 1978, states that one can apply his description to give a short alternative proof of the theorem of L. Lovász, 1965 (see also H. Walther and H.-J. Voss, 1974) characterizing all graphs which do not contain two disjoint cycles.

C. Thomassen and B. Toft, 1981, characterized all 2-connected graphs $G$ each induced cycle of which separates $G$. This characterization will be presented in section 2.2.

In their proofs they used the methods of U. Krusenstjerna-Hafstrøm and B. Toft, 1980.

Partial results are due to A. K. Kelmans, 1978b.

Every triangulation $T$ of the plane has the property that each induced cycle $C$ of length at least 4 separates $T$. This property is trivially shared by all chordal graphs because these graphs contain no such cycles at all (a graph is said to be a *chordal graph* if each cycle has at least one diagonal).

N. Robertson called a graph $G$ *strangulated* if each induced cycle of length at least 4 separates $G$.

A graph $G$ is said to be the *clique-sum* of the graphs $G_1$ and $G_2$ if $G_1 \cap G_2$ is complete, $G = G_1 \cup G_2$, and both $G_1$ and $G_2$ have more vertices than $G_1 \cap G_2$.

G. A. Dirac, 1961, characterized all chordal graphs as follows: a graph $G$ is *chordal* if and only if $G$ can be obtained by repeated clique-sums from complete graphs. P. D. Seymour and R. W. Weaver, 1984, generalized this result to strangulated graphs and proved that a 3-connected graph $G$ is *strangulated* if and only if $G$ can be obtained by repeated clique-sums, starting from complete graphs and 4-connected planar triangulations.

A similar characterization of strangulated graphs is due to R. Diestel, 1987.

An interesting class of graphs having non-separating cycles with precisely one regular bridge are the hypohamiltonian graphs (see, e.g., C. Thomassen, 1976).

A non-Hamiltonian graph is called *hypohamiltonian* if and only if each vertex-deleted subgraph has a Hamiltonian cycle. Let $G$ denote the graph in question and $X$ a deleted vertex of $G$. Then for each vertex $X$ there is a cycle $C$ such that $X$ is the kernel of the only regular $C$-bridge of $G$. Obviously, $C$ is a non-separating cycle.

## 2° Separating and non-separating cycles in graphs of higher connectivity

C. Thomassen und B. Toft, 1981, proved that every 3-connected graph has a peripheral cycle. Moreover, if the graph is non-bipartite, then $G$ has even a peripheral cycle of odd length (U. Krusenstjerna-Hafstrøm and B. Toft, 1980).

A. K. Kelmans, 1978b and 1981, (see also W. T. Tutte, 1984) proved that each edge of a 3-connected graph $G$ is in at least two peripheral cycles.

Before formulating a result of W. T. Tutte we introduce the concept of the cycle space of any graph $G$.

Let $E(G) = \{e_1, \ldots, e_m\}$. We represent each subset $S$ of $E(G)$ by a 01-vector $(x_1, \ldots, x_m)$, where $x_i = 1$ if $e_i \notin S$ and $x_i = 0$ if $e_i \notin S$. This way we identify the subsets of $E(G)$ with the elements of an $m$-dimensional vector space $V_G$ over GF(2).

The *cycle space* $W_G$ of $G$ is the subspace generated by all cycles of $G$.

The dimension $z(G)$ of the cycle space is called the *cyclomatic number* of $G$.

For 3-connected graphs W. T. Tutte, 1963, 1984, and C. Thomassen, 1980, proved that the peripheral cycles generate the cycle space.

For graphs of higher connectivity C. Thomassen, 1981a, has proved that every $(k+3)$-connected graph $G$ contains an induced cycle $C$ whose deletion results in a $k$-connected graph $G - C$.

If $G$ has no triangles, then a stronger assertion is true:

*Let $G$ be a $(k+2)$-connected graph without triangles. Then $G$ contains an induced cycle $C$ such that $G - C$ is $k$-connected.*

For $1 \leqq k \leqq 2$ and $k \geqq 3$ the proofs are due to C. Thomassen, 1981a, and Y. Egawa, 1987, respectively.

Similar problems can be considered with the weaker requirement that only the edges of such a cycle should be deleted. For instance, the results of W. Mader, 1974, imply, that every $k$-connected graph of minimum degree at least $k+2$ contains a cycle whose edge-deletion results in a $k$-connected graph. Further results in this direction are due to B. Jackson, 1980b, P. D. Seymour, 1981, H. Fleischner, 1982, H. Fleischner and B. Jackson, 1985, and W. Mader, 1985.

### 3° Peripheral cycles in planar graphs

In 3-connected graphs the well-known characterization of planar graphs by S. MacLane, 1937, is closely related to peripheral cycles.

If $G$ is a 2-connected plane graph, then, obviously, the facial cycles of all finite faces form a basis of the cycle space $W_G$.

S. MacLane, 1937, proved that the converse is also true.

*A graph is planar if and only if $W_G$ has a basis such that each edge belongs to at most two elements of it.*

It is well known that in investigating planar graphs it is sufficient to investigate the 3-connected planar graphs only. But in 3-connected *plane* graphs a cycle $C$ is a facial cycle if and only if $C$ is a peripheral cycle. From this there follows a proof of the theorem of H. Whitney, 1932a, that the embedding of a 3-connected planar graph is uniquely determined up to homeomorphism.

The theorem of S. MacLane, 1937, and the result that the peripheral cycles of a 3-connected graph generate the cycle space imply the theorem of A. K. Kelmans, 1978b and 1981, that a graph $G$ is non-planar if and only if one edge of $G$ is in at least three peripheral cycles.

## 2.1. 2-connected graphs without separating cycles

G. Chartrand and H. V. Kronk, 1968, have determined all connected graphs without separating paths.

In this section we shall present the complete list of all 2-connected graphs without separating cycles, due to C. Thomassen, 1978. The set of these graphs is a proper subset of the set of all 2-connected graphs without separating induced cycles. Such a graph without separating induced cycles having a separating non-induced cycle can be obtained from the graph of an octahedron by joining precisely one pair of non-adjacent vertices by an edge. Similarly, there are connected graphs without separating *induced* paths having a separating (non-induced) path. The complete 5-graph with one edge missing is an example.

First we introduce some notation.

Let $G_1 = (V_1, E_1)$, $G_2 = (V_2, E_2)$ be two finite simple graphs. A 1-1 mapping of $V_1$ onto $V_2$ which preserves adjacency is called an *isomorphism* of $G_1$ onto $G_2$, and $G_1$, $G_2$ are said to be *isomorphic.* This isomorphism relation is an equivalence relation in the family of all simple graphs. The classes with respect to this isomorphism relation are referred to as *abstract graphs.*

In the set of all simple abstract graphs we define two operations. For this purpose $G_1$ and $G_2$ can be chosen to be *disjoint*, i.e., $V_1 \cap V_2 = \emptyset$. Then the *union* of two abstract graphs can be defined with two disjoint representations $G_1$, $G_2$ by

$$G_1 \cup G_2 =_{\text{def}} (V_1 \cup V_2, E_1 \cup E_2).$$

Similarly, the *join* $G_1 + G_2$ is obtained from $G_1 \cup G_2$ by linking each vertex of $G_1$ with each vertex of $G_2$ by an edge.

The *union of h disjoint copies of* $G_1$ is denoted by $hG_1$.

The operations + and $\cup$ are commutative, associative and have other properties which will not be investigated here.

$K_n$ again denotes the *complete graph on n vertices* and $I_n =_{\text{def}} \bar{K}_n$ the *edgeless graph on n vertices.*

$C_n$ is the *cycle with n vertices* and $W_n =_{\text{def}} I_1 + C_n$ the *wheel*, $n \geqq 3$.

A *prism graph* $\mathfrak{P}_6$ consists of the vertices and the edges of a triangular prism, i.e., $\mathfrak{P}_6$ is the only cubic planar graph on six vertices.

We formulate the result of G. Chartrand and H. V. Kronk, 1968, without proof.

**Theorem 2.1.1.** *Any connected graph G has no separating path if and only if G is a cycle or a complete graph or an* $I_r + I_r$, $r \geqq 1$. □

C. Thomassen, 1978, has given a complete list of all finite 2-connected graphs with no separating cycles. Here this list is presented in two theorems without proof. In Theorem 2.1.2 (and 2.1.3) all these graphs are described that have no vertex of valency 2 (or at least one vertex of valency 2).

**Theorem 2.1.2.** *Let G be a 2-connected graph with* $n \geqq 4$ *vertices having minimum degree at least* 3. *Then G has no separating cycles if and only if G is one of the following graphs:*

(i) *G is one of the graphs of Fig. 2.1.1,*

(ii) *G is one of the graphs* $K_n$, $W_{n-1}$, $\mathfrak{P}_6$,

(iii) *n is even and*

$$I_{n/2} + I_{n/2} \subseteqq G \subseteqq (K_2 \cup I_{n/2-2}) + (K_2 \cup I_{n/2-2}),$$

(iv) *n is odd and*

$$(I_1 \cup I_{(n-1)/2}) + I_{(n-1)/2} \subseteqq G \subseteqq (I_1 + I_{(n-1)/2}) + I_{(n-1)/2},$$

*or G is one of the graphs*

$$(K_3 \cup I_{(n-1)/2-2}) + I_{(n-1)/2},$$
$$(2\,K_2 \cup I_{(n-1)/2-3}) + I_{(n-1)/2}.$$ □

Obviously, a cycle has no separating cycle. The same is true for a cycle $C$ with a topological diagonal $t$, where a path $t$ is a *topological diagonal* of $C$ if and only if only the ends of $t$ belong to $C$.

**Theorem 2.1.3.** *Let G be a 2-connected graph with* $n \geqq 4$ *vertices having at least one vertex of valency 2. Then G has no separating cycle if and only if G is one of the following graphs:*

(i) *G is one of the graphs of Fig. 2.1.2.* (In this figure, a wavy line represents a path of any length, and a dotted line indicates an edge that may or may not be present in the graph.)

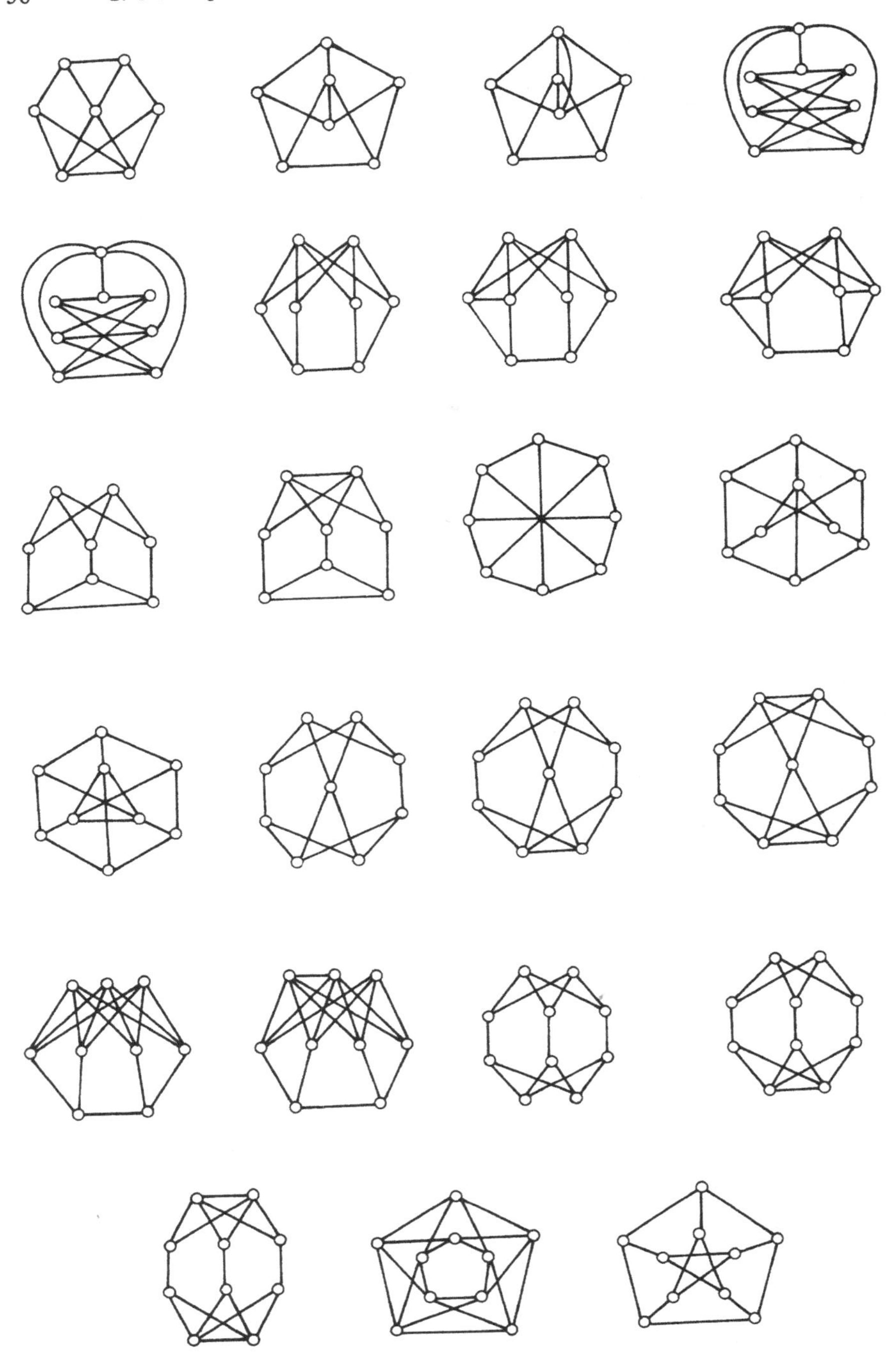

Fig. 2.1.1

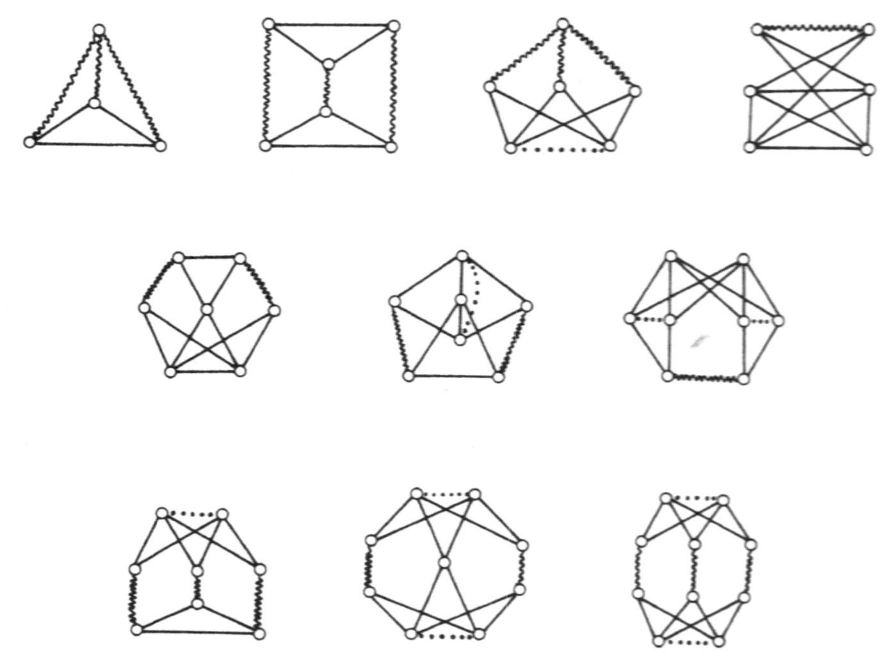

Fig. 2.1.2

(ii) *$G$ is a cycle or a cycle with a topological diagonal or is obtained from a wheel $W_r =_{\text{def}} I_1 + C_r$, $n - 1 > r \geqq 3$, by subdividing $C_r$ by $n - (r + 1)$ vertices of valency* 2. □

## 2.2. 2-connected graphs each induced cycle of which is a separating one. Sufficient conditions for the existence of peripheral cycles

After considering 2-connected graphs without separating cycles, here we shall present a structure theorem for all 2-connected graphs each induced cycle of which is a separating one, due to C. Thomassen and B. Toft, 1981. Sufficient conditions for the existence of a non-separating induced cycle are obtained. A similar question with respect to paths makes no sense because each connected graph contains a non-separating path of length 0.

By a *$k$-rail* $R_k$ in a graph $G$ between two vertices $X$ and $Y$ we shall understand a system of $k$ internally disjoint $X,Y$-paths, all inner vertices of these paths having valency 2 not only in $R_k$, but in the whole of $G$.

In the language of bridges we can formulate the following result: each non-separating cycle $C$ of a graph $G$ is a cycle with at most one regular $C$-bridge; a non-separating induced cycle $C$ of $G$ is a cycle with at most one $C$-bridge $B$. In the latter case $B$ is regular.

Each separating cycle $C$ of a graph $G$ has at least two regular bridges. Besides the 3-rail in which one path is an edge, in all 2-connected graphs each cycle with a diagonal contains a second diagonal or has a regular bridge. Consequently,

*a 2-connected graph has separating induced (separating) cycles only if and only if each cycle C of G has at least two C-bridges (two regular C-bridges).*

First we shall prove two lemmas due to C. Thomassen and B. Toft, 1981. The proof technique of these basic lemmas was first used by U. Krusenstjerna-Hafstrøm and B. Toft, 1980.

**Lemma 2.2.1.** *Let $G$ be a 2-connected graph and let $H$ be a connected subgraph of $G$ such that one component $K$ of $G - H$ contains at least one cycle. Then either $K$ contains an induced cycle $C$ such that $G - C$ is connected or $K$ contains a k-rail $R_k$, $k \geqq 3$, such that $G - R_k$ is connected.*

We remark that $K$ can be regarded as the kernel of an $H$-bridge $B$. Then by Lemma 2.2.1 the kernel of this bridge contains either a non-separating induced cycle of $G$ or a non-separating $k$-rail, $k \geqq 3$.

The hypothesis "$K$ contains a cycle" cannot be replaced by "an $H$-bridge $B$ includes a circuit" as can be seen from the graph of C. Thomassen and B. Toft, 1981, in Fig. 2.2.1: choose $H$ to be the subgraph induced by the lower $Y_1, Y_2$-path and the $k$-rails between $X_1$, $Y_1$ and $X_2$, $Y_2$, respectively.

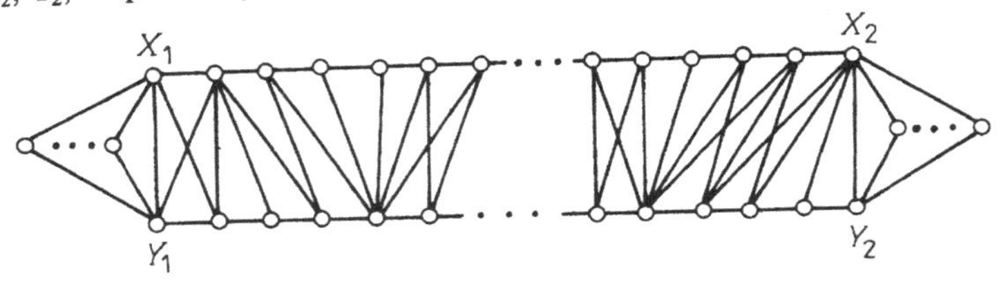


Fig. 2.2.1

*Proof of Lemma 2.2.1.* Let $C$ denote an induced cycle of $K$ with the property that $H$ is contained in the kernel of a $C$-bridge $B_1$ of maximum order.

If $C$ is not separating, the first alternative holds. Hence we shall consider the case where $C$ has a bridge $B_2$ different from $B_1$. Since $G$ is 2-connected, there is a $C,C$-path $p$ in $B_2$. Let its ends be denoted by $X$ and $Y$ and let $a_1$, $a_2$ be the two $X,Y$-arcs of C.

If $a_1 - \{X, Y\}$ contains a vertex of attachment of $B_1$, then $B_1 \cup (a_1 - \{X, Y\})$ is in a bridge of the circuit $a_2 \cup p$, contradicting the maximum property of $C$. This and the 2-connectivity of $G$ imply that $X$, $Y$ are the only vertices of attachment of $B_1$.

If the bridge $B_2$ has a vertex of attachment on $a_1 - \{X, Y\}$, denoted by $A$, then $B_1 \cup [(a_2 \cup a_1[Y, A]) - \{X, A\}]$ is in a bridge of the circuit $a_1[A, X] \cup p'$, where $p'$ is a $C,C$-path of $B_2$ with ends $A$ and $X$. This contradicts the maximum property of $C$. Thus $B_1$ and $B_2$ have the only two vertices of attachment $X$ and $Y$. By the maximum property of $C$ the bridge $B_2$ is a tree, the only two vertices of valency 1 are $X$ and $Y$. Therefore, $B_2$ is an $X,Y$-path.

Correspondingly, all bridges $B_2, B_3, \ldots, B_{k-1}$ of $C$ different from $B_1$ are $X,Y$-paths. Consequently,

$$R_k =_{\text{def}} C \cup B_2 \cup B_3 \cup \ldots \cup B_{k-1}$$

is a $k$-rail between the vertices $X$ and $Y$ such that $G - R_k = B_1 - \{X, Y\}$ is connected. □

The above proof of Lemma 2.2.1 directly implies the following lemma:

**Lemma 2.2.2.** *Let $G$ be a 2-connected graph in which all vertices have valency at least* 3 *(except* *perhaps one vertex of valency 2). Let $H$ be a connected subgraph of $G$ such that $G - H$ contains at least one cycle. If $C$ is an induced cycle in $G - H$ such that $H$ is contained in a bridge $B$ of $C$ of maximum order, then $G - C$ is connected* (i.e., $G - C = B - C$).

Next we can formulate the theorem of C. Thomassen and B. Toft, 1981.

**Theorem 2.2.3.** *Let $G$ be a 2-connected graph in which every induced cycle is separating. Then there exists a pair of vertices $\{X, Y\}$ such that there is a 3-rail between $X$ and $Y$ in $G$. Moreover, either $G$ is equal to a $k$-rail with $k \geqq 4$ or there exists another pair of vertices $\{X', Y'\}$ such that there is also a 3-rail between $X'$ and $Y'$ in $G$.*

A part of Theorem 2.2.3 has been proved independently by A. K. Kelmans, 1978b.

Theorem 2.2.3 is best possible as shown by the type of graph in Fig. 2.2.1. This type of graph is 2-connected, any induced cycle is separating, and it has only two pairs of vertices joined by 3-rails.

In fact, using Lemma 2.2.1, it is possible to prove that any 2-connected graph in which any induced cycle is separating and which has only two pairs of vertices joined by 3-rails can be obtained from the type of graphs of Fig. 2.2.1 by omitting some edges, subdividing others, and replacing some of the non-crossing edges joining the upper $X_1,X_2$-path with the lower $Y_1,Y_2$-path by 2-rails.

Next we shall present the proof of C. Thomassen and B. Toft, 1981.

*Proof.* Lemma 2.2.1 with $H = \emptyset$ implies that there is a 3-rail $R_3$ in $G$ between two vertices $X$ and $Y$. If $G$ is not equal to a $k$-rail between $X$ and $Y$, then either $G - (R_3 - X)$ or $G - (R_3 - Y)$, say $G - (R_3 - X)$, contains a cycle. Then we may apply Lemma 2.2.1 again, this time with $H = R_3 - X$. Hence there is another 3-rail $R'_3$ in $G$ between two vertices $X'$ and $Y'$. Since $R'_3 \subseteqq G - H = G - (R_3 - X)$, it is clear that $\{X, Y\} \neq \{X', Y'\}$. □

If the first alternative of Lemma 2.2.1 is made weaker by removing the word "induced", then the second alternative may be made stronger by adding the condition that each of the paths of the $k$-rail has length at least 2. It follows that if $G$ is a 2-connected graph in which every cycle is separating, then either $G$ is a $k$-rail or there exist two different pairs of vertices $\{X, Y\}$ and $\{X', Y'\}$ such that there is a 3-rail between $X$ and $Y$ and another 3-rail between $X'$ and $Y'$, and each of the six paths of the two 3-rails has length at least 2.

Theorem 2.2.3 implies the following theorem of C. Thomassen and B. Toft, 1981.

**Theorem 2.2.4.** *Any 2-connected graph $G$ such that all vertices (except perhaps two, or except perhaps three if $|V(G)| \geqq 6$) have valency at least 3 contains a non-separating induced cycle $C$, i.e. a peripheral cycle $C$.* □

Next we present results of U. Krusenstjerna-Hafstrøm and B. Toft, 1980, on separating induced *odd* cycles. We start with the proof of a lemma on odd cycles, which corresponds to Lemma 2.2.1 on arbitrary cycles.

**Lemma 2.2.5.** *Let $G$ be a 2-connected graph and let $H$ be a connected subgraph of $G$ such that one component $K$ of $G - H$ contains at least one odd cycle. Then either $K$ contains an induced odd cycle $C$ such that $G - C$ is connected or $K$ contains a cut set $\{X, Y\}$ of $G$ such that all bridges of $\{X, Y\}$ not containing $H$ are bipartite.*

With an odd cycle $C$ each graph $G$ also contains an induced odd cycle completely contained in $G\langle C\rangle$. The same is true with regard to arbitrary cycles. For even cycles this assertion is not valid, as chordal graphs show, where a chordal graph is defined as a graph in which each cycle of length at least 4 has at least one chord (F. Gavril, 1974). Since the graphs described above include no induced even cycle, a lemma on even cycles similar to Lemma 2.2.6 on odd cycles cannot be proved.

*Proof of Lemma 2.2.5.* Let $C$ denote an induced odd cycle of $K$ with the property that $H$ is contained in the kernel of a $C$-bridge $B_1$ of maximum order. An obvious consequence is that $G - K \subseteqq B_1$.

If $C$ is not separating, the first alternative holds. Hence we shall consider the case, where $C$ has a bridge $B_2$ different from $B_1$.

(1) *$B_2$ is bipartite.*

*Proof of (1).* Suppose that $B_2$ is non-bipartite. Then $B_2$ contains an induced odd cycle $C^*$ through at most one vertex $A$ of attachment of $B_2$, i.e., $C \cap C^* = \{A\}$ or $C \cap C^* = \emptyset$. In the latter case let $A$ denote an arbitrary vertex of attachment of $B_2$. Since $B_1$ is connected, it is completely contained in a bridge $B_1^*$ of $C^*$. By the 2-connectedness of $G$ the bridge $B_1^*$ has a vertex of attachment in $C^* - A$ which is not in $B_1$. Therefore, $B_1 \subseteqq B_1^*$, $B_1 \neq B_1^*$ and $H \subseteqq B_1 - C \subseteqq B_1^* - C^*$. This contradicts the maximum property of $B_1$! □

Let the vertices $A_1, \ldots, A_n$ of attachment of $B_2$ be in this cyclic order on $C$. Since $G$ is 2-connected, we have $n \geqq 2$. These vertices subdivide $C$ into $n$ arcs $a_1, a_2, \ldots, a_n$ with end vertices $A_1$ and $A_2$, $A_2$ and $A_3, \ldots, A_n$ and $A_1$, respectively, and all interior vertices of these arcs are not in $B_2$.

Since $C$ is an odd cycle, either there exists an arc $a_i$ of odd length joining two vertices of the same bipartition class or there exists an arc $a_i$ of even length joining two vertices of different bipartition classes (otherwise a 2-colouring of $B_2$ could be extended to $B_2 \cup C$; the result would be an odd cycle suitably coloured by two colours). We say that $a_i$ has property $\mathfrak{p}$.

Without loss of generality, we put $i = 1$, i.e., $a_1$ has property $\mathfrak{p}$. Then either $A_1, A_2$ belong to the same bipartition class of $B_2$ and $a_1$ is odd or $A_1, A_2$ belong to different bipartition classes of $B_2$ and $a_1$ is even. Hence a $C,C$-path $p$ of $B_2$ with ends $A_1, A_2$ having no diagonals and $a_1$ constitute an induced odd cycle $C^*$.

By the maximum property of $B_1$

(2) *all vertices of attachment of $B_1$ are on $a_1$*

(otherwise $C^*$ would have a bridge containing $H$, which has more vertices than $B_1$). Let the vertices $Y_1, \ldots, Y_m$ of attachment of $B_1$ be such that $A_1, Y_1, \ldots, Y_m, A_2$ are on $a_1$ in this order. Since $G$ is 2-connected, we have $m \geqq 2$.

By (2) only $a_1$ has property $\mathfrak{p}$. Therefore $B_2 \cup a_2 \cup \ldots \cup a_n$ and also $B_2 \cup C[Y_m, Y_1]$ are bipartite, where $C[Y_m, Y_1]$ denotes the $Y_m,Y_1$-arc containing all $A_1, \ldots, A_n$.

By (1) and (2) each bridge $B^*$ of $C$ different from $B_1$ is bipartite, does not overlap $B_1$, and $B^* \cup C[Y_m, Y_1]$ is also bipartite.

Consequently, $\{Y_1, Y_m\}$ is a cut set each bridge of which not containing $H$ is bipartite. □

The following lemma gives a more precise description of the second alternative of Lemma 2.2.5.

A block of $G$ is either a $K_2$ or a maximal 2-connected subgraph of $G$.

Let $m \geqq 2$ be an integer. An induced subgraph $G'$ of $G$ is called an *m*-ring $\mathfrak{R}_m$ of bipartite blocks through $m$ pairwise distinct vertices $Y_1, \ldots, Y_m$ in $G$ if and only if 1) $G'$ is a sequence $Y_1\mathfrak{B}_1Y_2\mathfrak{B}_2Y_3 \ldots Y_m\mathfrak{B}_mY_1$, where $\mathfrak{B}_1, \mathfrak{B}_2, \ldots, \mathfrak{B}_m$ are bipartite blocks of $G$ with the following properties: $\mathfrak{B}_i \cap \mathfrak{B}_{i+1} = \{Y_{i+1}\}$ for all $1 \leqq i \leqq m-1$, $\mathfrak{B}_m \cap \mathfrak{B}_1 = \{Y_1\}$, $\mathfrak{B}_i \cap \mathfrak{B}_j = \emptyset$ for all $j \in \{i-1, i, i+1\}$; 2) no vertex of $\mathfrak{B}_i - \{Y_i, Y_{i+1}\}$ is joined to a vertex of $G - \mathfrak{B}_i$ by an edge of $G$.

$Y_1, \ldots, Y_m$ is referred to as the main vertices. $\mathfrak{R}_m$ is called *odd* if and only if all cycles through $Y_1, \ldots, Y_m$ have an odd length.

The above proof of Lemma 2.2.5 directly implies the following two assertions.

**Lemma 2.2.6.** *Let $G$ be a 2-connected graph and let $H$ be a connected subgraph of $G$ such that one component $K$ of $G - H$ contains at least one odd cycle. Then either $K$ contains an induced odd cycle $C$ such that $G - C$ is connected or $K$ contains an odd m-ring $\mathfrak{R}_m$ of bipartite blocks such that $G - \mathfrak{R}_m$ is connected.* □

**Theorem 2.2.7.** *Let $G$ be a 2-connected non-bipartite graph in which every induced odd cycle is separating. Then $G$ contains an odd m-ring $\mathfrak{R}_m$ of bipartite blocks such that $G - \mathfrak{R}_m$ is connected, i.e., $G$ has only one $\mathfrak{R}_m$-bridge. Moreover, if this $\mathfrak{R}_m$-bridge is non-bipartite, then $G$ contains a second odd m'-ring $\mathfrak{R}_{m'}$ of bipartite blocks such that $G - \mathfrak{R}_{m'}$ is connected and $\mathfrak{R}_m$ and $\mathfrak{R}_{m'}$ have at most one main vertex in common.*

A graph $G$ is called 4-*critical* if $G$ has chromatic number 4 and each proper subgraph of $G$ has a chromatic number at most 3.

With the aid of Lemma 2.2.5 we can prove two theorems due to U. Krusenstjerna-Hafstrøm and B. Toft, 1980.

**Theorem 2.2.8.** *Every 4-critical graph $G$ contains an induced odd cycle $C$ such that $G - C$ is connected.*

*Proof.* Choose $H = \emptyset$. Since $G$ has a chromatic number at least 3, it has an odd cycle. By Lemma 2.2.5 either $G$ contains an induced odd cycle $C$ such that $G - C$ is connected or $G$ contais a cut set $\{X, Y\}$ of $G$ such that apart from perhaps one bridge all bridges of $\{X, Y\}$ are bipartite. Let $B$ be a bipartite bridge of $\{X, Y\}$. Since $G$ is 4-critical, the graph $G - (B - \{X, Y\})$ can be suitably coloured by the colours 1, 2, 3.

If $X$ and $Y$ have the same colour, say 1, then colouring $B - \{X, Y\}$ by 2 and 3 we obtain a suitable 3-colouring of a 4-critical graph. Contradiction!

If $X$ and $Y$ have different colours, say 1 and 2, respectively, then colouring $B - Y$ by 1 and 3 we again obtain a suitable 3-colouring of a 4-critical graph. Contradiction!

Thus, for $G$ the second alternative of Lemma 2.2.5 is not valid. Consequently, $G$ contains an induced odd cycle $C$ such that $G - C$ is connected. □

**Theorem 2.2.9.** *Every non-bipartite 3-connected graph $G$ contains an induced odd cycle $C$ such that $G - C$ is connected.*

*Proof.* Choose $H$ to be the empty graph. Since $G$ has no cut set of size at most 2, Lemma 2.2.5 implies the validity of Theorem 2.2.9. □

In Theorem 2.2.9 it is not sufficient to assume only that $G$ is 2-connected. This can easily be seen from an odd *m*-ring $\mathfrak{R}_m$ of bipartite blocks $K_{r,r}$, $r \geqq 3$, $m \geqq 2$. This shows that even if $G$ is 2-connected with large minimum degree the conclusion of Theorem 2.2.9 need not hold.

## 2.3. Peripheral cycles in 3-connected graphs

By the results of section 2.2 we know that every 3-connected graph contains a peripheral cycle.

Here we present the following results on 3-connected graphs $G$:

1) each edge of $G$ is in at least two peripheral cycles,

2) the peripheral cycles generate the cycle space of $G$,

3) the graph $G$ is non-planar if and only if one edge of $G$ is in at least three peripheral cycles.

Theorem 1.4.5 implies the following result, which will be applied in this section.

**Lemma 2.3.1.** *Let $G$ be a 3-connected graph with a cycle $C$ having at least two $C$-bridges; let $B$ be one of them. Then $G$ has a second $C$-bridge $B'$ overlapping $B$.* □

Next we shall prove the following result:

**Theorem 2.3.2.** *Let $G$ be a 3-connected graph with a given edge $e$. Let $C$ be any cycle through $e$ and $K$ a component of $G - C$. Then there is a peripheral cycle through $e$ avoiding $K$.*

*Proof.* In the set of all cycles of $G - K$ through $e$, let $C'$ be chosen so that the component $K'$ of $G - C'$ containing $K$ has the largest possible number of vertices. If $K' = G - C'$, then the theorem is valid with the peripheral cycle $C'$ through $e$.

Now let $K' \neq G - C'$. Then $G$ has at least two $C'$-bridges, one of them, say $B'$, has the kernel $K'$. A consequence of Lemma 2.3.1 is that there is a $C'$-bridge $B''$ overlapping $B'$. Then $B''$ has a path $p$ with end vertices $P$ and $Q$ in $C'$ containing no other vertex of $C'$ with the following property: the cycle $C'$ is split into two arcs $a_1$, $a_2$ by $P$ and $Q$ so that $e$ is an edge of $a_1$ and an inner vertex $V$ of $a_2$ is a vertex of attachment of $B'$. Hence $C'' =_{\text{def}} a_1 \cup p$ is a cycle of $G - K' \subseteqq G - K$ through $e$ such that $K' \cup \{V\}$ is contained in one component of $G - C''$. This contradicts the maximality of $C'$ and $K'$. Thus the proof of Theorem 2.3.2 is complete. □

Next we prove a theorem of A. K. Kelmans, 1978b and 1981 (see also W. T. Tutte, 1984).

**Theorem 2.3.3.** *Let $G$ be a 3-connected graph and $e$ an edge of $G$. Then $G$ has two peripheral cycles through $e$, and the intersection of them is only the edge $e$ with its two end vertices.*

*Proof.* Let $e = (X, Y)$ be an edge of a 3-connected graph $G$. The edge $e$ is contained in a cycle. By Theorem 2.3.2 the edge $e$ is contained in a peripheral cycle $C$ of $G$. The graph $G$ has only one $C$-bridge $B$. Hence $X$ and $Y$ are vertices of attachment of $B$, and $B$ contains a $C,C$-path $p$ with ends $X$ and $Y$. Then the cycle $C' =_{\text{def}} p \cup e$ has a bridge $B$ containing $C - \{e\}$. By Theorem 2.3.2 the graph $G$ has a peripheral cycle $C^*$ through $e$ containing no vertex of $C - \{X, Y\}$. Thus the intersection of $C$ and $C^*$ is only $e$ and $X$, $Y$. □

The peripheral cycles of any 3-connected graph $G$ generate its cycle space. This has been proved by W. T. Tutte, 1963, 1984, and C. Thomassen, 1980.

**Theorem 2.3.4.** *Let $G$ be any 3-connected graph. Then the peripheral cycles generate the cycle space of $G$.*

*Proof.* We have to find $z(G)$ linearly independent peripheral cycles of $G$, whre $z(G)$ is the dimension of the cycle space. By Theorem 2.3.3 there are two linearly independent peripheral cycles of $G$. Suppose we have found a set $M$ of $m$ linearly independent cycles, where $2 \leqq m < z(G)$. We shall construct an $(m+1)$-th peripheral cycle, linearly independent of them all. Since $m < z(G)$, there is a cycle of $G$ linearly independent of $M$. In the set of all cycles of $G$ linearly independent of $M$ let $C$ be chosen so that one component $K$ of $G - C$ has the largest possible number of vertices.

If $C$ is peripheral, then $m + 1$ linearly independent cycles are obtained.

Now let $G$ have two $C$-bridges, one of them, say $B$, has the kernel $K$. By Lemma 2.3.1 there is a $C$-bridge $B'$ overlapping $B$. We consider two cases:

Case 1. Suppose first that *$B$ and $B'$ are skew*. Then $B'$ has two vertices $X$, $Y$ of attachment such that in both arcs $C[X, Y]$ and $C[Y, X]$ there are inner vertices $V_1$ and $V_2$, respectively, belonging to $B$. Let $p'$ denote a $C$,$C$-path of $B'$ with ends $X$ and $Y$. We consider the cycles

$$C_1 =_{\text{def}} p' \cup C[X, Y] \quad \text{and} \quad C_2 =_{\text{def}} p' \cup C[Y, X].$$

Their mod 2 sum is $C$. Hence $C_1$ or $C_2$ is linearly independent of $M$, say $C_1$. The graph $G - C_1$ has a component containing $K \cup \{V_1\}$, contradicting the maximality property of $C$ and $K$.

Case 2. In the remaining case *$B$ and $B'$ have precisely three common vertices of attachment $X$, $Y$, $Z$* lying on $C$ in this cyclic order. Since $G$ is 3-connected, an inner vertex $P$ of $B'$ is joined in $B'$ by three internally disjoint paths $p_X$, $p_Y$, $p_Z$ to $X$, $Y$, $Z$, respectively. We consider the cycles

$$C_X =_{\text{def}} C[Y, Z] \cup p_Z \cup p_Y,$$
$$C_Y =_{\text{def}} C[Z, X] \cup p_X \cup p_Z,$$
$$C_Z =_{\text{def}} C[X, Y] \cup p_Y \cup p_X.$$

Their mod 2 sum is $C$. Hence one of the three cycles, say $C_X$, is linearly independent of $M$. The graph $G - C_X$ has a component containing $K \cup \{X\}$, contradicting the maximality property of $C$ and $K$.

Consequently, in both cases we arrive at a contradiction! Hence $C$ is peripheral and $M \cup \{C\}$ is a set of $m + 1$ linearly independent cycles.

We can repeat the procedure until we have $z(G)$ linearly independent peripheral cycles of $G$. □

## 2.4. Peripheral cycles in 3-connected planar graphs

Peripheral cycles play an important role in planar graphs. That can be seen first by the following theorem due to W. T. Tutte, 1963.

**Theorem 2.4.1.** *Let $G$ be a 3-connected plane graph. Then $C$ is a facial cycle if and only if $C$ is peripheral.*

*Proof.* 1) If $C$ is peripheral, then $G$ has precisely one bridge completely drawn either inside or outside $C$. Hence $C$ is a facial cycle.

2) Let $C$ be a facial cycle of $G$. If $G$ has two $C$-bridges, then by Lemma 2.3.1 the graph $G$ has two overlapping $C$-bridges $B$ and $B'$. Since $C$ is the facial cycle of some face $F$, then both $C$-bridges $B$ and $B'$ lie outside $C$. This contradiction proves that $C$ is peripheral. □

For brevity we make the following definition: two 2-connected plane graphs $G_1$ and $G_2$ are said to be *homeomorphic* if and only if there are 1-1 mappings of the vertex set, the edge set, and the face set of $G_2$, respectively, such that all incidences are preserved.

By induction on the number of faces the well-known result of H. Whitney, 1932a, can be derived.

**Theorem 2.4.2.** *Let $G$ be a 3-connected planar graph. Then all plane representations of $G$ in the plane are homeomorphic.* □

Thus each 3-connected planar graph has precisely one plane representation up to homeomorphism.

For (2-connected) plane graphs it can easily be shown that their facial cycles generate the cycle space. S. MacLane, 1937, has proved that the converse is also true.

**Theorem 2.4.3.** *A graph $G$ is planar if and only if the cycle space of $G$ has a basis such that each edge belongs to at most two basic cycles.* □

By Theorem 2.3.3 any edge of a 3-connected graph is contained in at least two peripheral cycles. A. K. Kelmans, 1978b and 1981 (see also W. T. Tutte, 1984) proved that if each edge is in precisely two peripheral cycles, then $G$ is planar. This result can be proved with the help of MacLane's Theorem 2.4.3. It also follows immediately from Tutte's theorem about planar meshes in a 3-connected graph (W. T. Tutte, 1963, Theorem 2.6). We formulate Kelmans' result as follows:

**Theorem 2.4.4.** *A 3-connected graph $G$ is non-planar if and only if one edge is in at least three peripheral cycles.*

Instead of Theorem 2.4.4 we shall prove the following equivalent assertion.

**Theorem 2.4.5.** *Let $G$ be a 3-connected graph.*

(i) *If one edge of $G$ is in at least three peripheral cycles, then $G$ is non-planar.*

(ii) *If each edge of $G$ is in precisely two peripheral cycles, then $G$ is planar.*

*Proof.* (i) Let $e$ be an edge which is in three peripheral cycles. Assume $G$ to be a planar graph. Let $\hat{G}$ be a plane representation of $G$. Then $e$ is in precisely two facial cycles of $\hat{G}$ and by Theorem 2.4.1 in precisely two peripheral cycles. This contradiction proves assertion (i).

(ii) Let each edge be in precisely two peripherical cycles. By Theorem 2.3.4 the peripheral cycles generate the cycle space. Therefore the cycle space has a basis consisting of peripheral cycles only. Each edge of $G$ is in at most two cycles of this basis. By MacLane's Theorem 2.4.3 the graph $G$ is planar. □

A 3-cut $S$ of $G$ is said to be *non-essential* if $S$ has precisley two $S$-bridges, one of them is a $K_1 + \underline{I}_3$. Otherwise a 3-cut is an *essential* one. A. K. Kelmans, 1986, proved the following supplement of Theorem 2.4.5.:

Let $G$ be a 3-connected graph without essential 3-cuts or triangle. Then

(i) if $G$ is planar, then *each* edge belongs to exactly two peripheral cycles,

(ii) if $G$ is non-planar, then *each* edge belongs to at least three peripheral cycles.

# Chapter 3

## Bridges of longest cycles. Intersection pattern of longest cycles and paths

### 3.0. Introduction

#### 1° Bridges of longest cycles

In investigating bridges of longest cycles we may ask: how long can a longest cycle of such a bridge be? Is the length of each cycle of such a bridge smaller than the circumference of the whole graph? The answer is "Yes!". This was first proved by K. Hauschild, H. Herre and W. Rautenberg, 1972. In section 3.6 we shall present an improved version of this result (H.-J. Voss, 1982b).

The result of K. Hauschild, H. Herre and W. Rautenberg, 1972, is helpful to proving theorems by induction on the circumference. So in Chapter 4 properties of the class $\mathfrak{B}_l^2$ of all 2-connected graphs having circumference at most $l$ will be proved by induction on $l$.

K. Hauschild, H. Herre and W. Rautenberg, 1972, have used their theorem for proving metamathematical theorems, for instance they have proved the recursive decidability of the theory of all graphs with "universally bounded circumference" (i.e. of all graphs with circumference at most $l$, where $l$ is a fixed integer).

Similar results are obtained for bridges of longest paths in connected graphs. K. Hauschild was the first to prove that if $B$ is a bridge of a longest path $p$, then each path of $B$ is smaller than $p$.

Only partial results could be obtained in answering the following question: consider the bridges of a longest *odd* cycle $L_{\mathrm{odd}}$ in a 2-connected non-bipartite graph (or the bridges of a longest *even* cycle $L_{\mathrm{even}}$ in a 3-connected non-bipartite graph). Is each *odd* cycle $C'$ (*even* cycle $C''$) of a bridge of $L_{\mathrm{odd}}$ (of $L_{\mathrm{even}}$) smaller than $L_{\mathrm{odd}}$ ($L_{\mathrm{even}}$)?

The answer is "Yes!" if $L_{\mathrm{odd}}$ and $C'$ (or $L_{\mathrm{even}}$ and $C''$) have at most four common vertices. If the set of common vertices has cardinality greater than 4, then the question is an open problem! In the theory of matroids "connected matroid $M$", "longest circuit $L$ of $M$" and "$L$-bridge $B$" are well-defined concepts (see W. T. Tutte, 1971). Again we ask: are all circuits of $B$ smaller than $L$?

J. Kincses, 1988, conjectured that this is true for all connected real linear matroids.

By the result of K. Hauschild, H. Herre and W. Rautenberg, 1972, this conjecture is true for connected graphic matroids. (Note: the underlying graph of a connected matroid is 2-connected).

J. Kincses, 1988 and 1987, has proved the conjecture in the affirmative for all cographic matroids and for all real linear matroids containing only circuits of length at most 5.

In the proof of an improved version of the theorem of K. Hauschild, H. Herre and

W. Rautenberg and of related problems we shall introduce the concept "length of a bridge". The length of bridges of longest cycles, longest odd cycles, shortest cycles, shortest odd cycles will be investigated. The results obtained on this "length" of one, two and three overlapping bridges of a longest cycle, respectively, are used in this chapter and in the following chapters, especially in Chapter 11.

### 2° Intersection pattern of longest cycles and paths

The problems described above are closely connected with the investigation of the intersection pattern of two longest cycles or of two longest paths or of a longest cycle and a longest path etc.

It is well known that two longest cycles of a 2-connected graph have at least two common vertices. This is also true for two longest odd cycles in 2-connected graphs and for two even cycles in 3-connected graphs.

M. Grötschel, 1984, investigated the intersection pattern of two longest cycles with precisely $k$ common vertices for $2 \leqq k \leqq 5$. Moreover, M. Grötschel presented a problem of Scott Smith: have any two longest cycles of a $k$-connected graph at least $k$ common vertices?

There is a partial answer by T. Zamfirescu (private communication), namely that any two longest cycles of an $s$-connected graph, $s \geqq k^2 - 4k + 4$, $k \geqq 4$, have $k$ common vertices.

It is also a well-known fact that two longest paths of a connected graph have a common vertex.

T. Gallai, 1968, asked: is it true that in any connected graph there exists a vertex such that every longest path of the graph contains this vertex? And H. Sachs, 1968, asked: is it true that in any 3-connected graph there exist two vertices such that every longest cycle of the graph contains at least one of them? Both conjectures have been disproved by H. Walther, 1969, 1970. Further contributions to this problem are due to B. Grünbaum, 1974, W. Schmitz, 1975, C. Thomassen, 1974b, 1976, and T. Zamfirescu, 1972, 1974, 1975, 1976.

The following similar problem was investigated by G. A. Dirac, 1952a, H.-J. Voss, 1973, J. A. Bondy and S. C. Locke, 1980, 1981, and S. C. Locke, 1982; how long can a longest path be in a $k$-connected graph, $k \geqq 2$, with given circumference?

## 3.1. The length of bridges of subgraphs

In Chapter 1 we have discussed the overlap graph of a cycle $C$ in detail. From the structure of the overlap graph properties of $G$ have been derived, in particular we have found characteristic subgraphs formed by the cycle $C$ and certain parts of the bridges considered.

Here we shall study bridges of longest cycles in 2-connected graphs. These bridges are bounded in a certain sense. This can be expressed by defining an appropriate notion "the length of a bridge".

**Definition 3.1.1.** Let $B$ be a bridge of a subgraph $H$ of a graph $G$. The bridge $B$ contains a tree whose end vertices are precisely the vertices of attachment of $B$. The number of edges of a longest such tree is called the *length* $\lambda(B)$ of $B$. The number of edges of a smallest such tree is denoted by $\underline{\lambda}(B)$.

**Remark.** If $B$ has at most one vertex of attachment, then $\underline{\lambda}(B) = 0$.

We shall prove some properties of the length of bridges in graphs. For instance, we shall show that $\lambda(B) \leqq \lfloor l/2 \rfloor$ for each bridge $B$ of a longest cycle $L$ of length $l$ in a 2-connected graph.

Similar questions will be investigated for bridges of longest paths, longest odd and even cycles, smallest cycles, smallest odd and even cycles.

## 3.2. Some lemmas

Let $L$ be a longest cycle of a graph $G$. A path with ends on $L$ which is internally disjoint from $L$ is called an *$L,L$-path*. In this section we shall prove some simple lemmas relating to $L,L$-paths of longest cycles $L$.

**Lemma 3.2.1.** *Let $p$ be an $L,L$-path of a longest cycle $L$ with ends $P$ and $Q$* (see Fig. 3.2.1 a). *Then*

$$\lambda(p) \leqq \lambda L[P, Q] \quad \textit{and} \quad \lambda(p) \leqq \lambda L[Q, P]. \ \square$$

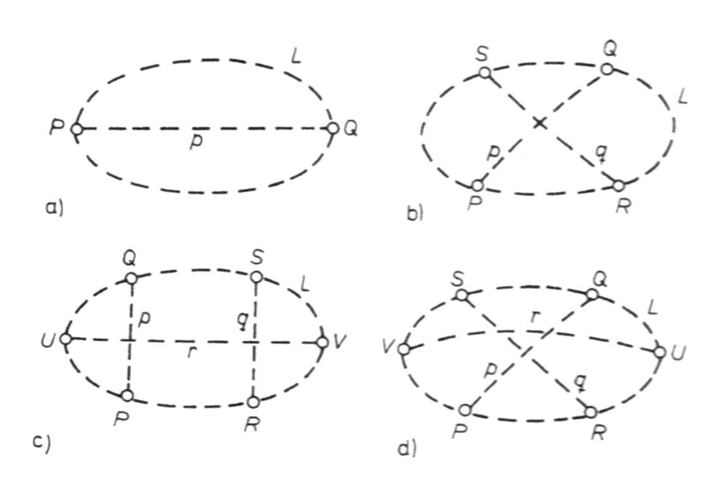


Fig. 3.2.1

**Lemma 3.2.2.** *Let $p$ and $q$ be two $L, L$-paths of a longest cycle $L$ with ends $P$, $Q$ and $R$, $S$, respectively, where, $P$, $R$, $Q$, $S$ are on $L$ in this cyclic order* (see Fig. 3.2.1 b). *Then*

$$\lambda(p) + \lambda(q) \leqq \lambda L[P, R] + \lambda L[Q, S]$$

*and*

$$\lambda(p) + \lambda(q) \leqq \lambda L[R, Q] + \lambda L[S, P].$$

*Proof.* $p \cup L[Q, S] \cup q \cup L[P, R]$ is a cycle of length at most $l$. Therefore,

$$\lambda(p) + \lambda L[Q, S] + \lambda(q) + \lambda L[P, R]$$
$$\leqq l = \lambda L[Q, S] + \lambda L[S, P] + \lambda L[P, R] + \lambda L[R, Q].$$

The second inequality is proved in the same way. □

**Lemma 3.2.3.** *Let p, q, r be three L,L-paths of a longest cycle L with ends P, Q and R, S and U, V, respectively, where P, R, V, S, Q, U are on L in this cyclic order* (Fig. 3.2.1c). *Then*

$$\lambda(p) + \lambda(q) + \lambda(r) \leqq \lambda L[V, S] + \lambda L[Q, U] + \lambda L[P, R]$$

*and*

$$\lambda(p) + \lambda(q) + \lambda(r) \leqq \lambda L[S, Q] + \lambda L[U, P] + \lambda L[R, V].$$

*Proof.* $p \cup L[S, Q] \cup q \cup L[R, V] \cup r \cup L[U, P]$ is a cycle of length at most $l$. Therefore,

$$\lambda(p) + \lambda L[S, Q] + \lambda(q) + \lambda L[R, V] + \lambda(r) + \lambda L[U, P]$$
$$\leqq l = \lambda L[S, Q] + \lambda L[Q, U] + \lambda L[U, P] + \lambda L[P, R] + \lambda L[R, V] + \lambda L[V, S].$$

The second inequality is proved in the same way. □

Similarly the following lemma can be proved:

**Lemma 3.2.4.** *Let p, q, r be three L,L-paths of a longest cycle L with ends P, Q and R, S and U, V, respectively, where P, R, U, Q, S, V are on L in this cyclic order* (see Fig. 3.2.1d). *Then*

$$\lambda(p) + \lambda(q) + \lambda(r) \leqq \lambda L[P, R] + \lambda L[V, Q] + \lambda L[S, U]$$

*and*

$$\lambda(p) + \lambda(q) + \lambda(r) \leqq \lambda L[R, V] + \lambda L[Q, S] + \lambda L[U, P]. \ \square$$

## 3.3. The length of bridges of longest cycles

Bridges with at most one vertex of attachment have length 0. Therefore, we concentrate our attention on bridges with at least two vertices of attachment.

If a graph $G$ is 2-connected, then each bridge with respect to a subgraph $H$ of $G$ has at least two vertices of attachment. Therefore, from now on only 2-connected graphs will be considered.

In this section we shall prove theorems on the lengths of bridges of longest cycles depending on their overlap properties.

**Theorem 3.3.1.** *Let G be a 2-connected graph with a longest cycle L; let l be the length of L. Then each bridge B of L has length* $\lambda(B) \leqq \lfloor l/2 \rfloor$.

First we state a simple lemma:

**Lemma 3.3.2.** *Let T be a tree with end vertices* $E_1, \ldots, E_s$. *Then the system* $\{T[E_i, E_{i+1}] \mid 1 \leqq i \leqq s \text{ and } E_{s+1} = E_1\}$ *of paths of T covers each edge of T at least twice.* □

Extended versions of Lemma 3.3.2 have been proved in H.-J. Voss, 1982b.

*Proof of Theorem 3.3.1.* Let $A_1, \ldots, A_k$ denote the vertices of attachment of $B$ which are on

$L$ in this cyclic order. Let $T$ be a longest tree of $B$ with end vertices $A_1, \ldots, A_k$. By Lemma 3.2.1, $\lambda T[A_i, A_{i+1}] \leqq \lambda L[A_i, A_{i+1}]$ for all $1 \leqq i \leqq k$ with $A_{k+1} =_{\text{def}} A_1$. By Lemma 3.3.2 the system $\{T[A_i, A_{i+1}] \mid 1 \leqq i \leqq k\}$ of paths of $T$ covers each edge of $T$ at least twice. Consequently,

$$2\lambda(B) = 2\lambda(T) \leqq \sum_{i=1}^{k} \lambda T[A_i, A_{i+1}] \leqq l. \ \square$$

The application of Theorem 3.3.1 yields the following result.

**Corollary 3.3.3.** *Let $G$ be a 2-connected graph with the longest cycle $L$; let $l$ be the length of $L$. Then each bridge $B$ of $L$ has at most $l/2$ blocks and the sum of the number $r$ of the vertices of attachment and of the number $c$ of the cut vertices is $r + c \leqq l/2 + 1$. For each block $G_i$ of $B$ we have $r_i + c_i \leqq l/2$.*

*Proof.* Let $T$ be a tree of $B$ whose end vertices are the vertices of attachment of $B$. Obviously, $T$ contains all vertices of attachment and all cut vertices of $B$ and at least one edge of each block. Hence Theorem 3.3.1 implies the validity of Corollary 3.3.3. $\square$

**Theorem 3.3.4.** *Let $G$ be a 2-connected graph with a longest cycle $L$; let $l$ be the length of $L$. Let $B_1$, $B_2$ be two overlapping vertex-disjoint bridges of $L$. Then*

$$\lambda(B_1) + \lambda(B_2) \leqq \lfloor l/2 \rfloor.$$

Before proving Theorem 3.3.4 we state a simple lemma:

**Lemma 3.3.5.** *Let $T$ be a tree with and vertices $E_1, \ldots, E_s$. Let $\begin{pmatrix} 1 & 2 & \ldots & s \\ i_1 & i_2 & \ldots & i_s \end{pmatrix}$ be a permutation of the integers $1, 2, \ldots, s$. Then the system*

$$\{T[E_i, E_{i+1}] \mid 1 \leqq i \leqq s-1\} \cup \{T[E_{i_j}, E_{i_{j+1}}] \mid 1 \leqq j \leqq s-1\} \cup \{T[E_s, E_{i_1}], T[E_{i_s}, E_1]\}$$

*of paths of $T$ covers each edge of $T$ at least four times.*

*Proof of Theorem 3.3.4.* A sketch of the proof will be given. The details are left to the reader.

Let

$$P_1^1, \ldots, P_{p_1}^1, Q_1^1, \ldots, Q_{q_1}^1, \ldots, P_1^i, \ldots, P_{p_i}^i, Q_1^i, \ldots, Q_{q_i}^i, P_1^{i+1}, \ldots, P_{q_{i+1}}^{i+1},$$
$$Q_1^{i+1}, \ldots, Q_{q_{i+1}}^{i+1}, \ldots, P_1^t, \ldots, P_{p_t}^t, Q_1^t, \ldots, Q_{q_t}^t$$

be vertices of $G$ which are on $L$ in this order, where the $P$'s and the $Q$'s are the vertices of attachment of $B_1$ and $B_2$, respectively (see Fig. 3.3.1).

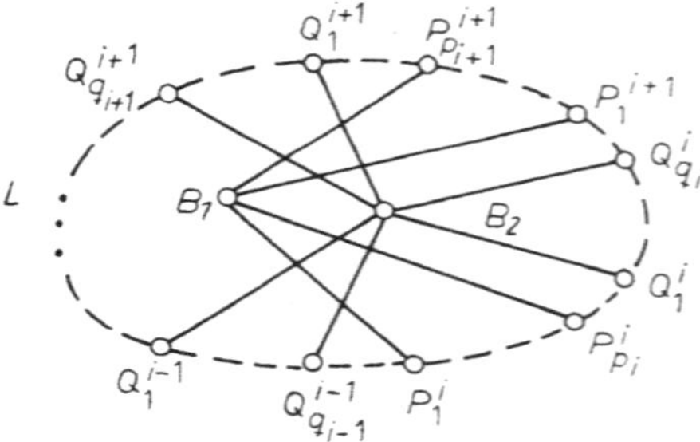


Fig. 3.3.1

Let $T_1$ and $T_2$ be longest trees of $B_1$ and $B_2$ such that the $P$'s are the end vertices of $T_1$ and the $Q$'s are the end vertices of $T_2$, respectively. By Lemma 3.2.1

(1) $$\lambda T_1[P^i_\varrho, P^i_{\varrho+1}] \leqq \lambda L[P^i_\varrho, P^i_{\varrho+1}] \quad \text{for} \quad 1 \leqq \varrho \leqq p_i - 1,$$
$$\lambda T_2[Q^i_\mu, Q^i_{\mu+1}] \leqq \lambda L[Q^i_\mu, Q^i_{\mu+1}] \quad \text{for} \quad 1 \leqq \mu \leqq q_i - 1.$$

Lemma 3.2.2 implies the validity of (2) and (3) (see Fig. 3.3.1).

(2) $$\lambda T_1[P^i_{p_i}, P^{i+1}_{p_{i+1}}] + \lambda T_2[Q^i_1, Q^{i+1}_1] \leqq \lambda L[P^i_{p_i}, Q^i_1] + \lambda L[P^{i+1}_{p_{i+1}}, Q^{i+1}_1]$$
for all $1 \leqq i \leqq t$ $(t+1 = 1 \bmod t)$.

(3) $$\lambda T_1[P^i_1, P^{i+1}_1] + \lambda T_2[Q^{i-1}_{q_{i-1}}, Q^i_{q_i}] \leqq \lambda L[Q^{i-1}_{q_{i-1}}, P^i_1] + \lambda L[Q^i_{q_i}, P^{i+1}_1]$$
for all $1 \leqq i \leqq t$ $(t+1 = 1 \bmod t)$.

(4) If $t$ is an odd number, then we consider the paths of $T_1$ joining two neighbours of the following sequence:
$$P^1_1, \ldots, P^1_{p_1}, P^2_{p_2}, \ldots, P^2_1, P^3_1, \ldots, P^3_{p_3}, \ldots, P^t_1, \ldots, P^t_{p_t}, P^1_{p_1}, \ldots, P^1_1, P^2_1, \ldots, P^2_{p_2},$$
$$P^2_{p_3}, \ldots, P^3_1, \ldots, P^t_{p_t}, \ldots, P^t_1, P^1_1.$$

(5) If $t$ is an even number, then we consider the paths of $T_1$ joining two neighbours of the following sequences:
$$P^1_1, \ldots, P^1_{p_1}, P^2_{p_2}, \ldots, P^2_1, \ldots, P^t_{p_t}, \ldots, P^t_1, P^1_1$$
and
$$P^1_{p_1}, \ldots, P^1_1, P^2_1, \ldots, P^2_{p_2}, \ldots, P^t_1, \ldots, P^t_{p_t}, P^1_{p_1}.$$

In both cases ((4) and (5)) each end vertex of $T_1$ appears twice. By Lemma 3.3.5 (if $t$ is odd) and by Lemma 3.3.2 (if $t$ is even) these paths cover each edge of $T_1$ at least four times.

The corresponding result for $B_2$ and $T_2$ can be obtained in the same way.

Adding the lengths of these paths of $T_1$ and $T_2$ and applying (1), (2) and (3) we obtain that
$$4\lambda(B_1) + 4\lambda(B_2) = 4\lambda(T_1) + 4\lambda(T_2) \leqq 2l. \ \square$$

**Theorem 3.3.6.** *Let $G$ be a 2-connected graph with longest cycle $L$, and let $l$ be the length of $L$. Let $B_1$, $B_2$, $B_3$ be three bridges of $L$ forming the induced path $[B_1, B_3, B_2]$ in the overlap graph $O(G:L)$. Let $B_i$, $B_j$ be vertex-disjoint for all $1 \leqq i < j \leqq 3$. Then*
$$\lambda(B_1) + \lambda(B_2) + \lambda(B_3) \leqq \lfloor l/2 \rfloor.$$

*Proof.* Only a sketch of the proof will be given. The details are left to the reader.

Since $[B_1, B_3, B_2]$ is an induced path of the overlap graph $O(G:L)$, we have: the vertices of attachment of $B_1$, $B_2$ and $B_3$ can be denoted by
$$R^0_1, \ldots, R^0_{r_0}, P^1_1, \ldots, P^1_{p_1}, R^1_1, \ldots, R^1_{r_1}, \ldots, P^t_1, \ldots, P^t_{p_t},$$
$$S^0_1, \ldots, S^0_{s_0}, Q^1_1, \ldots, Q^1_{q_1}, S^1_1, \ldots, S^1_{s_1}, \ldots, S^{t'-1}_1, \ldots, S^{t'-1}_{s_{t'-1}}, Q^{t'}_1, \ldots, Q^{t'}_{q_{t'}},$$
so that they are on $L$ in this cyclic order and the $P$'s and the $Q$'s are the vertices of attachment of $B_1$ and $B_2$, respectively, and the $R$'s and the $S$'s are the vertices of attachment of $B_3$ (see Fig. 3.3.2).

Let $T_i$ be a longest tree of $B_i$ such that the vertices of attachment of $B_i$ are the end ver-

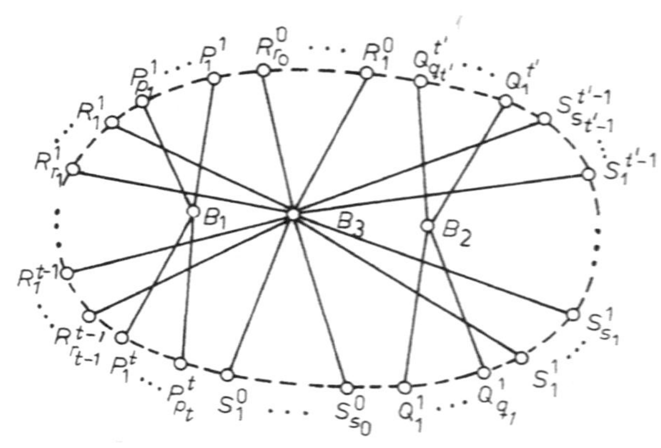


Fig. 3.3.2

tices of $T_i$ $(1 \leqq i \leqq 3)$. We consider three cases depending on whether $r_0 = 0$ or $r_0 > 0$ and $s_0 = 0$ or $s_0 > 0$.

Case 1. Assume that $r_0 > 0$ and $s_0 > 0$. Let $T_3^1$ and $T_3^2$ be subtrees of $T_3$ such that the end vertices of $T_3$ on $L[R_1^0, S_1^0]$ and on $L[S_1^0, R_1^0]$ are the end vertices of $T_3^1$ and $T_3^2$, respectively. Obviously, $T_3 = T_3^1 \cup T_3^2$, where $T_3[R_1^0, S_1^0]$ is contained in both $T_3^1$ and $T_3^2$, more precisely: $T_3[R_1^0, S_1^0] \subseteqq T_3^1 \cap T_3^2$.

Considering $T_1$ and $T_3^1$, we obtain a covering $\Sigma_1$ of $T_1$ and $T_3^1$ by paths of type (4) or (5) of the proof of Theorem 3.3.4.

Correspondingly, a path covering $\Sigma_2$ of $T_2$ and $T_3^2$ is obtained. As in the proof of Theorem 3.3.4 we conclude that $T_1$ and $T_2$ are covered four times by $\Sigma_1$ and $\Sigma_2$, respectively.

The subtree $T_3^i$ is also covered four times, where $T_3^i[R_1^0, S_1^0] = T_3[R_1^0, S_1^0]$ appears twice in the covering $\Sigma_i$ of $T_i \cup T_3^i$. We delete the path $T_3[R_1^0, S_1^0]$ from $\Sigma_i$ twice. The new path covering of $T_i \cup T_3^i$ is denoted by $\Sigma^i$.

If we apply the arguments of the proof of Theorem 3.3.4 to $\Sigma_i$, we arrive at the inequality $\lambda(\Sigma_i) \leqq 2l$ for $i = 1, 2$, where $\lambda(\Sigma_i)$ is the sum of the lengths of all paths of $\Sigma_i$. If we apply these arguments to $\Sigma^i$, then the following two inequalities cannot be used any longer:

(A) $\lambda T_3^1[S_1^0, R_1^0] \leqq \lambda L[S_1^0, R_1^0]$,

(B) $\lambda T_3^2[R_1^0, S_1^0] \leqq \lambda L[R_1^0, S_1^0]$.

With these arguments we can only derive the following assertions (C) and (D):

(C) $\lambda(\Sigma^1) \leqq 2\lambda(L) - 2\lambda L[S_1^0, R_1^0] = 2\lambda L[R_1^0, S_1^0]$,

(D) $\lambda(\Sigma^2) \leqq 2\lambda(L) - 2\lambda L[R_1^0, S_1^0] = 2\lambda L[S_1^0, R_1^0]$.

Let $\Sigma$ be the collection of all paths of $\Sigma^1$ and $\Sigma^2$. The assertions (C) and (D) imply that

(E) $\lambda(\Sigma) = \lambda(\Sigma^1) + \lambda(\Sigma^2) \leqq 2l$.

$T_i$ and $T_3^i$ are covered four times by their path system $\Sigma_i$. Hence $T_3 = T_3^1 \cup T_3^2$ is covered four times by $\Sigma_1$ and $\Sigma_2$, where each edge of $T_3[R_1^0, S_1^0]$ is covered eight times. Hence the paths of $\Sigma$ cover each edge of $T_3$ (and also of $T_1$ and $T_2$) at least four times. With (E) we can conclude that

$$4(\lambda(T_1) + \lambda(T_2) + \lambda(T_3)) \leqq \lambda(\Sigma) \leqq 2l.$$

Case 2. Assume that $r_0 = 0$ and $s_0 > 0$. We subdivide an edge of $L[Q^{t'}_{q_{t'}}, P^1_1]$ by a new vertex $R^0_1$ and join it by an edge to a vertex $X$ of valency at least 3 of $T_3[R^1_{r_1}, R^{t-1}_{r_{t-1}}, S^1_1, S^{t'-1}_1]$ (this is the subtree of $T_3$ having precisely four end vertices, namely, $R^1_{r_1}, R^{t-1}_{r_{t-1}}, S^1_1, S^{t'-1}_1$). The new tree is denoted by $\overline{T}_3$ (see Fig. 3.3.3).

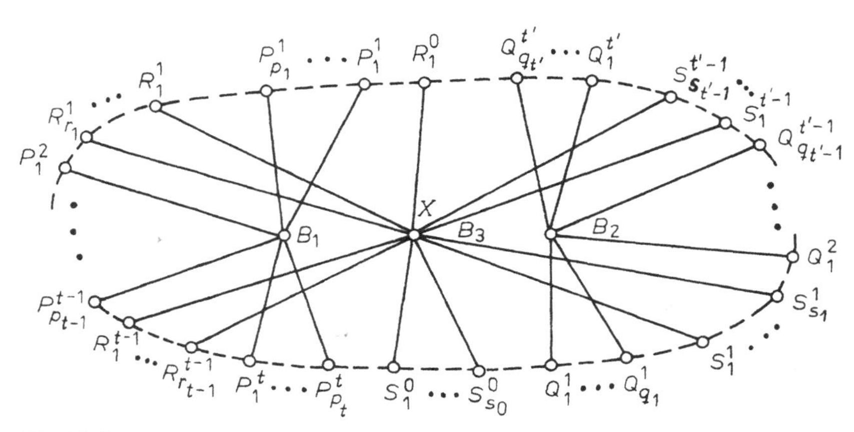


Fig. 3.3.3

We consider the path covering $\Sigma$ of $T_1$, $T_2$ and $\overline{T}_3$ found in Case 1.

Some paths of $\Sigma$ contain $R^0_1$. For these paths the inequalities (1), (2), (3) of the proof of Theorem 3.3.4 are no longer true. The non-valid inequalities are listed in

$$\text{(F)} \qquad \begin{aligned} &\lambda\overline{T}_3[R^1_{r_1}, R^0_1] + \lambda T_1[P^1_1, P^2_1] \leqq \lambda L[R^0_1, P^1_1] + \lambda L[R^1_{r_1}, P^2_1],\\ &\lambda\overline{T}_3[R^{t-1}_{r_{t-1}}, R^0_1] + \lambda T_1[P^t_1, P^1_1] \leqq \lambda L[R^0_1, P^1_1] + \lambda L[R^{t-1}_{r_{t-1}}, P^t_1],\\ &\lambda\overline{T}_3[S^1_1, R^0_1] + \lambda T_2[Q^1_{q_1}, Q^{t'}_{q_{t'}}] \leqq \lambda L[Q^{t'}_{q_{t'}}, R^0_1] + \lambda L[Q^1_{q_1}, S^1_1],\\ &\lambda\overline{T}_3[S^{t'-1}_1, R^0_1] + \lambda T_2[Q^{t'-1}_{q_{t'-1}}, Q^{t'}_{q_{t'}}] \leqq \lambda L[Q^{t'}_{q_{t'}}, R^0_1] + \lambda L[Q^{t'-1}_{q_{t'-1}}, S^{t'-1}_1].\end{aligned}$$

*Let* $\Sigma'$ denote the path system of $T_1 \cup T_2 \cup \overline{T}_3$ obtained from $\Sigma$ by deleting the set $M$ of all paths of $\Sigma$ used in (F). The application of the arguments of the proof of Theorem 3.3.4 to $\Sigma'$ gives $\lambda(\Sigma') \leqq 2l - \varrho$, where $\varrho$ is the sum of the right-hand sides of the inequalities of (F). We shall show that $\Sigma'$ can be supplemented by adding some paths so that this new path system $\Sigma''$ covers $T_1 \cup T_2 \cup T_3$ four times and the total length of these added paths is not greater than $\varrho$. Then

$$4(\lambda(T_1) + \lambda(T_2) + \lambda(T_3)) \leqq \lambda(\Sigma'') \leqq 2l.$$

By Lemma 3.2.3 we have (G) and (H):

$$\text{(G)} \qquad \begin{aligned} &\lambda T_3[R^1_{r_1}, S^{t'-1}_1] + \lambda T_1[P^1_1, P^2_1] + \lambda T_2[Q^{t'-1}_{q_{t'-1}}, Q^{t'}_{q_{t'}}]\\ &\quad \leqq \lambda L[Q^{t'-1}_{q_{t'-1}}, S^{t'-1}_1] + \lambda L[Q^{t'}_{q_{t'}}, P^1_1] + \lambda L[R^1_{r_1}, P^2_1];\\ &\lambda T_3[S^1_1, R^{t-1}_{r_{t-1}}] + \lambda T_1[P^1_1, P^t_1] + \lambda T_2[Q^1_{q_1}, Q^{t'}_{q_{t'}}]\\ &\quad \leqq \lambda L[Q^1_{q_1}, S^1_1] + \lambda L[Q^{t'}_{q_{t'}}, P^1_1] + \lambda L[R^{t-1}_{r_{t-1}}, P^t_1].\end{aligned}$$

$$\text{(H)} \qquad \begin{aligned} &\lambda T_3[R^1_{r_1}, S^1_1] + \lambda T_1[P^1_1, P^2_1] + \lambda T_2[Q^1_{q_1}, Q^{t'}_{q_{t'}}]\\ &\quad \leqq \lambda L[Q^1_{q_1}, S^1_1] + \lambda L[Q^{t'}_{q_{t'}}, P^1_1] + \lambda L[R^1_{r_1}, P^2_1];\end{aligned}$$

$$\lambda T_3[R^{t-1}_{r_{t-1}}, S^{t'-1}_1] + \lambda T_1[P^1_1, P^t_1] + \lambda T_2[Q^{t'-1}_{q_{t'-1}}, Q^{t'}_{q_{t'}}]$$
$$\leqq \lambda L[Q^{t'-1}_{q_{t'-1}}, S^{t'-1}_1] + \lambda L[Q^{t'}_{q_{t'}}, P^1_1] + \lambda L[R^{t-1}_{r_{t-1}}, P^t_1].$$

Let $\dot\cup$ denote the disjoint union of two sets, i.e. we write $M_1 \dot\cup M_2$ for $M_1 \cup M_2$ if and only if $M_1 \cap M_2 = \emptyset$. It can easily be shown that at least one of the following two assertions (I), (J) is true:

(I) $\quad T_3[R^1_{r_1}, S^{t'-1}_1] = \bar T_3[R^1_{r_1}, X] \,\dot\cup\, \bar T_3[X, S^{t'-1}_1];$

$\quad T_3[S^1_1, R^{t-1}_{r_{t-1}}] = \bar T_3[S^1_1, X] \,\dot\cup\, \bar T_3[X, R^{t-1}_{r_{t-1}}].$

(J) $\quad T_3[R^1_{r_1}, S^1_1] = \bar T_3[R^1_{r_1}, X] \,\dot\cup\, \bar T_3[X, S^1_1];$

$\quad T_3[R^{t-1}_{r_{t-1}}, S^{t'-1}_1] = \bar T_3[R^{t-1}_{r_{t-1}}, X] \,\dot\cup\, \bar T_3[X, S^{t'-1}_1].$

Assume that (I) is valid. Let $\bar M$ and $\Sigma''$ be obtained from $M$ and $\Sigma$, respectively, by replacing the two paths $\bar T_3[R^1_{r_1}, R^0_1]$, $\bar T_3[R^0_1, S^{t'-1}_1]$ by the path $T_3[R^1_{r_1}, S^{t'-1}_1]$ and the two paths $\bar T_3[S^1_1, R^0_1]$, $\bar T_3[R^0_1, R^{t-1}_{r_{t-1}}]$ by the path $T_3[S^1_1, R^{t-1}_{r_{t-1}}]$. By (I) the path system $\Sigma''$ covers each edge of $T_1 \cup T_2 \cup T_3$ four times. $\Sigma'$ is obtained from $\Sigma$ by deleting all paths of $M$. The path system $\Sigma''$ is obtained from $\Sigma'$ by adding all paths of $\bar M$. Above we have proved that $\lambda(\Sigma') \leqq 2l - \varrho$. Since $\bar M$ is the set of all paths occuring on the left-hand sides of the inequalities of (G), the length $\lambda(\bar M)$ is not greater than $\varrho$ (note that the sum of the right-hand sides of the inequalities of (F) is equal to the sum of the right-hand sides of the inequalities in (G)). Consequently,

$$\lambda(\Sigma'') = \lambda(\Sigma') + \lambda(\bar M) \leqq 2l - \varrho + \varrho = 2l.$$

Since $\Sigma''$ covers each edge of $T_1 \cup T_2 \cup T_3$ at least four times, we obtain

(K) $\quad 4(\lambda(T_1) + \lambda(T_2) + \lambda(T_3)) \leqq \lambda(\Sigma'') \leqq 2l.$

If (J) is valid, then with the help of (H) this inequality is obtained in the same way. Consequently, in all cases we arrive at (K).

Case 3. Assume that $r_0 = 0$, $s_0 = 0$. This case can be handled in the same way as Case 2. □

If $B_1$, $B_2$, $B_3$ form a triangle in $O(G:L)$, then Theorem 3.3.6 is no longer true. One possible counterexample is the Petersen graph of Fig. 3.3.4 a, where

$$\lambda(B_1) + \lambda(B_2) + \lambda(B_3) = 5 > \lfloor l/2 \rfloor = 4.$$

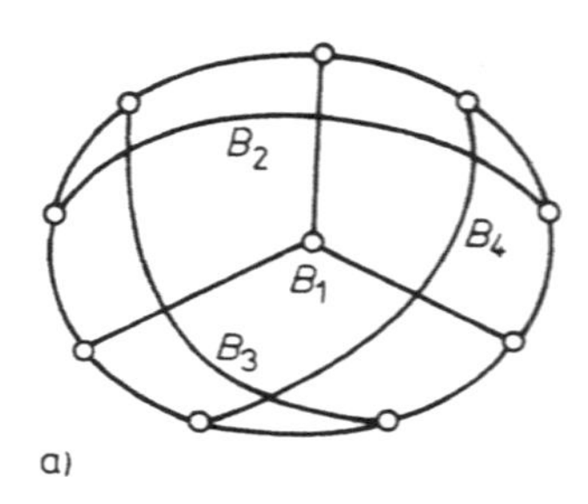


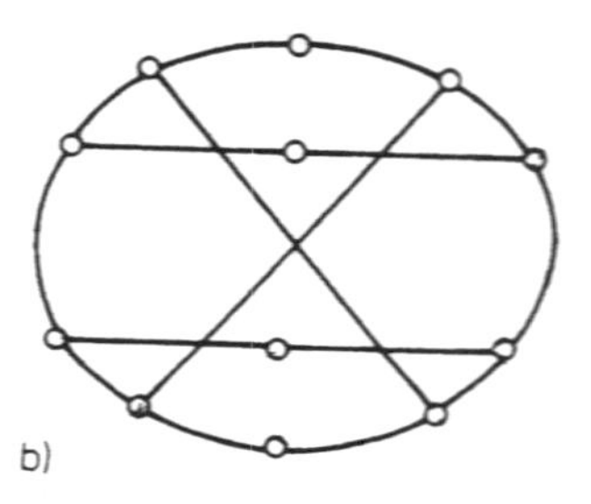


Fig. 3.3.4

We say that a configuration is multiplied by a positive integer $\alpha$ if each edge is replaced by a path of length $\alpha$. Multiplying the configuration of Fig. 3.3.4 a by 1, 2, 3, ... we obtain an infinite sequence of counterexamples.

Theorem 3.3.6 remains true if $B_1$, $B_2$, $B_3$ form a triangle in $O(G:L)$ and all $B_1$, $B_2$, $B_3$ have precisely two vertices of attachment. Lemmas 3.2.3 and 3.2.4 imply the following result:

**Theorem 3.3.7.** *Let G be a 2-connected graph with a longest cycle L, and let l denote the length of L. Let $B_1$, $B_2$, $B_3$ be three pairwise vertex-disjoint bridges of L, each of them having precisely two vertices of attachment. Let $B_1$, $B_2$, $B_3$ induce a connected subgraph of $O(G:L)$. Then*

$$\lambda(B_1) + \lambda(B_2) + \lambda(B_3) \leqq \lfloor l/2 \rfloor.$$

The assertion of Theorem 3.3.7 does not hold for four bridges with precisely two vertices of attachment. This can be seen by the example graph of Fig. 3.3.4b. The graph of Fig. 3.3.4b has circumference $l = 10$ and the sum of the lengths of the four bridges of the longest cycle $L$ is 6, where $\lfloor l/2 \rfloor = 5$.

I make the following conjecture:

*Let G be a 2-connected graph with a longest cycle L; let l be the length of L. Let $B_1, B_2, \ldots, B_n$ be bridges of L forming an induced tree in the overlap graph $O(G:L)$. Let $B_i$, $B_j$ be pairwise vertex-disjoint. Then*

$$\sum_{i=1}^{n} \lambda(B_i) \leqq \lfloor l/2 \rfloor.$$

This inequality, if true, is best possible even if $B_i$ has many vertices of attachment.

In sections 3.1.-3.3 the length $\lambda(B)$ of bridges $B$ has been considered. We conclude this section with a second measure $\tilde{\lambda}(B)$ of $B$ defined as follows: Let $A_1, A_2, \ldots, A_s$, $s \geqq 2$, denote the vertices of attachment of $B$, and let $\mu_{i,j}$ be the length of the longest $H,H$-paths of $B$ connecting $A_i$ and $A_j$. Then we define

$$\tilde{\lambda}(B) = \frac{1}{2} \min\,(\mu_{i_1 i_2} + \mu_{i_2 i_3} + \ldots + \mu_{i_s i_1}),$$

where the minimum is taken over all permutations $i_1, i_2, \ldots, i_s$ of $1, 2, \ldots, s$.

Obviously, $\lambda(B) \leqq \tilde{\lambda}(B)$, and $\lambda(B) = \tilde{\lambda}(B)$ if $B$ is a tree. The Theorems 3.1.1, 3.1.4, 3.1.6-7 remain to be true if $\lambda(B_i)$ is replaced by $\tilde{\lambda}(B_i)$. The result so obtained can be summarized as follows:

**Theorem 3.3.8.** *Let L be a longest cycle of a 2-connected graph G, and let l be its length. Let $B_1, \ldots, B_s$ be s pairwise non-intersecting bridges of L in G, $1 \leqq s \leqq 3$, such that the subgraph of the overlap graph $O(G:L)$ induced by $B_1, \ldots, B_s$ is a tree. Then*

$$\sum_{i=1}^{s} \tilde{\lambda}(B_i) \leqq \lfloor l/2 \rfloor.$$

Theorems 3.3.1, 3.3.4, 3.3.6-8 are also true for bridges of a longest cycle some of its vertices of attachment are deleted. Such bridges are sometimes called *reduced* bridges. Their lengths can be obviously smaller than the lengths of the original bridges.

## 3.4. The length of bridges of longest odd and even cycles

To begin our investigations of the length of the bridges of an odd cycle we present a lemma.

**Lemma 3.4.1.** *Let $L_{odd}$ be a longest odd cycle of a graph G and let $l_{odd}$ denote its length. Then each $L_{odd}$,$L_{odd}$-path of G has length at most $l_{odd} - 1$*; i.e. *each bridge B of $L_{odd}$ with precisely two attaching vertices has length $\lambda(B) \leqq l_{odd} - 1$.* □

Next we formulate the main result of this section.

**Theorem 3.4.2.** *Let $L_{odd}$ be a longest* **odd** *cycle of a graph G, and let $l_{odd}$ denote its length. Let B be an $L_{odd}$-bridge, and let $\alpha$ denote the number of vertices of attachment of B. Then*

$$\lambda(B) \leqq l_{odd} \qquad \text{if} \quad \alpha \geqq 3$$

*and*

$$\lambda(B) \leqq l_{odd} - 1 \quad \text{if} \quad \alpha = 2.$$

Equality is attained by the following infinite families of graphs. For $\alpha \geqq 3$, $\alpha$ odd, these examples can be obtained from $K_{\alpha+1}$ by replacing each edge by a path of length $2k+1$, $k \in N$. In a word, we consider the set $\{(2k+1)\, K_{\alpha+1} \mid k \in N\}$. Each longest odd cycle $L_{odd}$ contains $\alpha$ vertices of $K_{\alpha+1}$ and has length $\alpha(2k+1)$. The bridge of $L_{odd}$ containing the only vertex of degree $\alpha \geqq 3$ of $K_{\alpha+1} - L_{odd}$ has length $\alpha(2k+1) = l_{odd}$. For $\alpha = 2$ these examples can be formed from a cycle $C$ of length $4k+4$, $k \in N$, by joining a pair of vertices of $C$ of distance $2k+2$ by an edge.

*Proof of Theorem 3.4.2.* The assertion of Theorem 3.4.2 is obvious for $\alpha = 2$. Hence we assume that $\alpha \geqq 3$. Let $A_1, \ldots, A_\alpha$ denote the vertices of attachment of $B$ on $L_{odd}$ in this cyclic order.

Let $T$ be a tree of $B$ with the end vertices $A_1, \ldots, A_\alpha$ and let

$$C_i =_{def} A_i L_{odd} A_{i+1} T A_i$$

for all $1 \leqq i \leqq \alpha$, where $A_{\alpha+1} = A_1$.

Since $L_{odd} = \underset{i}{\triangle}\, C_i$ has odd length, at least one $C_i$ is an odd cycle. Without loss of generality, let $C_1$ be an odd cycle.

If $C_1$ is the only odd cycle among the $C_j$'s, then $L_{odd} \triangle C_i$ is an odd cycle for all $i \geqq 2$. Hence $\lambda\,(L_{odd} \triangle C_i) \leqq l_{odd}$.

These $\alpha - 1$ cycles cover $T$ at least once and $L_{odd}$ at least $\alpha - 2$ times. Therefore,

$$\lambda(T) + (\alpha - 2)\, \lambda(L_{odd}) \leqq (\alpha - 1)\, l_{odd}.$$

Consequently,

$$\lambda(T) \leqq l_{odd}.$$

Next let at least three of the cycles $C_j$ have an odd length. Let $C_{i_1}, C_{i_2}, \ldots, C_{i_{2k+1}}$, $i_1 < i_2 < \ldots < i_{2k+1}$, $k \geqq 1$, be these cycles. We define the following set of odd cycles:

$$C'_j =_{def} L_{odd} \triangle C_j = A_j T A_{j+1} L_{odd} A_j \quad \text{if } C_j \text{ is an even cycle, i.e. } j \notin \{i_1, \ldots, i_{2k+1}\},$$

$$C'_{i_s} =_{def} L_{odd} \triangle \underset{i_s \leqq i \leqq i_{s+1}}{\triangle} C_i = A_{i_s} T A_{i_{s+1}+1} L_{odd} A_{i_s} \text{ for all } 1 \leqq s \leqq 2k+1.$$

The paths of $T$ contained in the $C'_j$, $1 \leqq j \leqq \alpha$, form a closed walk on $T$ including all end vertices of $T$, namely,

$$\begin{aligned} A_1TA_2T\ldots\ & TA_{i_1}TA_{i_1+1}TA_{i_1+2}T \\ & \ldots TA_{i_3}TA_{i_4+1}TA_{i_4+2}T \\ & \ldots TA_{i_{2k+1}}TA_{i_1+1}TA_{i_1+2}T \\ & \ldots TA_{i_2}TA_{i_3+1}TA_{i_3+2}T \\ & \ldots TA_{i_{2k}}TA_{i_{2k+1}+1}TA_{i_{2k+1}+2}T\ldots TA_1. \end{aligned}$$

It is obvious that in such a closed walk each edge of $T$ occurs at least twice. Therefore, these $\alpha$ odd cycles $C'_j$ cover $T$ at least twice and $L_{\text{odd}}$ at least $\alpha - 2$ times. Hence

$$2\lambda(T) + (\alpha - 2)\,\lambda\,(L_{\text{odd}}) \leqq \sum_j \lambda(C'_j) \leqq \alpha l_{\text{odd}}$$

and, consequently,

$$\lambda(T) \leqq l_{\text{odd}}. \ \square$$

It may be that Theorem 3.4.2 can be generalized to two or three bridges, like Theorems 3.3.4 and 3.3.6. I have not dealt with this question.

If in the proof of Theorem 3.4.2 the cycle $L_{\text{odd}}$ is replaced by a longest even cycle $L_{\text{even}}$, then the number of odd cycles among the set $\{C_i \mid 1 \leqq i \leqq \alpha\}$ is even. The arguments of the proof do not work if the number of these odd cycles is precisely 2. In fact, in this case an infinite class of possible counterexamples can be constructed. Such graphs are obtained from $K_{\alpha,\alpha+1}$ by joining two vertices of the bipartition class of $\alpha + 1$ vertices by a path of length $\beta$, $\beta \in N - \{0\}$, $\beta$ odd, no inner vertex of which belongs to $K_{\alpha,\alpha+1}$. The longest even cycles $L_{\text{even}}$ of these graphs have length $2\alpha$; each of them has $L_{\text{even}}$-bridges of length $\alpha + \beta$ (for further examples see the class $\mathfrak{D}(r)$ in Chapter 7, page 139).

It would also be of interest to investigate the lengths of bridges of longest cycles or of longest odd and even cycles in 3-connected graphs.

## 3.5. The length of bridges of smallest cycles

In section 3.3 we have considered bridges of longest cycles. Those bridges cannot be too big. In fact, in a 2-connected graph the lengths of bridges of a longest cycle $L$ are restricted by the upper bound $\lfloor \lambda(L)/2 \rfloor$, which depends *only on* $\lambda(L)$. If the bridges of a smallest cycle $L$ of $G$ are studied, then it is obvious that the lengths of these bridges are restricted by a lower bound depending only on $\lambda(L)$. Besides $\lambda(B)$ we also use $\underline{\lambda}(B)$ (see Definition 3.1.1). We make the following definition: if $G$ is any graph and $\beta \geqq 1$ an integer, then $\beta G$ denotes the graph obtained from $G$ by replacing each edge of $G$ by a path of length $\beta$.

First we deal with the case when $B$ has at least two vertices of attachment.

**Proposition 3.5.1.** *Let $L$ be a smallest cycle of a graph $G$ and $l$ its length. Then each $L,L$-path of $G$ has length at least* $\lceil l/2 \rceil$, i.e. *each bridge $B$ of $L$ with at least two vertices of attachment has length*

$$\lambda(B) \geqq \underline{\lambda}(B) \geqq \lceil l/2 \rceil.$$

For even $l$ equality is attained by the graph $l/2\ \Theta$, where $\Theta$ denotes the theta-graph (that is the graph consisting of two vertices which are joined by three parallel edges).

In general we can prove the following result:

**Theorem 3.5.2.** *Let $L$ be a smallest cycle of a graph $G$ and $l$ its length. Let $B$ be an $L$-bridge and $\alpha \geqq 2$ its number of vertices of attachment. Then*

$$\lambda(B) \geqq \underline{\lambda}(B) \geqq \left\lfloor \frac{\alpha-1}{2} \right\rfloor \left\lceil \frac{l}{2} \right\rceil + \varepsilon,$$

*where $\varepsilon = -1$ if both $\alpha$ and $l$ are odd integers and $\varepsilon = 0$ otherwise.*

For $\alpha = 2$ and $\alpha = 3$ equality is attained at $\beta\Theta$, $\beta \geqq 2$, and $\gamma K_4$, $\gamma \geqq 1$, respectively. Thus in both cases equality is attained for infinitely many values of $l$.

In the case $\alpha = 4$ a slightly improved result can be proved.

**Corollary 3.5.3.** $\lambda(B) \geqq \left(1 + \frac{1}{4}\right) l$ *if* $\alpha = 4$. □

Equality is attained by $\beta K_{3,3}$, $\beta \geqq 1$. Thus equality holds for infinitely many $l$.

*Proof of Theorem 3.5.2.* By induction on $\alpha$. In the cases $\alpha = 2$ and $\alpha = 3$ the assertion of Theorem 3.5.2 can easily be derived. Hence we may assume that $\alpha \geqq 4$. Let $T$ be a tree of $B$, all $\alpha$ end vertices of which are vertices of attachment of $B$. In $T$ an $L,L$-path $p$ with the following property can easily be determined: if $X$ and $Y$ denote the end vertices of $p$, then no $L - \{X, Y\}, L - \{X, Y\}$-path of $T$ contains an edge of $p$. Theorem 3.5.1 implies that $\lambda(p) \geqq \lceil l/2 \rceil$. We delete all edges of $p$ and all vertices of $T$ that are not on any $L - \{X, Y\}$, $L - \{X, Y\}$-path of $T$. The new tree will be denoted by $T'$. Then $\lambda(T) \geqq \lambda(T') + \lambda(p) \geqq \lambda(T') + \lceil l/2 \rceil$. The tree $T$ will be recursively reduced by this method until the resulting tree $T''$ has only two or three end vertices if $\alpha$ is even or odd, respectively. But in these cases the assertion of Theorem 3.5.2 is true and $\lambda(T'') \geqq \lceil l/2 \rceil$ or $\lambda(T'') \geqq l$, respectively. Now the proof can easily be completed. □

For $\alpha \geqq 4$ the inequality of Theorem 3.5.2 seems not to be best possible. For $\alpha = 4$ this is obvious by Corollary 3.5.3.

Next we shall present an infinite family of 2-connected planar graphs $G$ having the following property: $G$ has a smallest cycle $L$ with an $L$-bridge $B$ of length $\lambda(B) = \frac{\alpha-1}{2} l$, where $\alpha$ is the number of attaching vertices of $B$. These graphs are obtained from the $(2k+1)$-wheels, $k \geqq 1$, by subdividing each spoke by $k-1$ vertices. Let $H_k$ denote the resulting graph (for fixed $k$). Then for $k \geqq 1$ the set $\{\beta H_k \mid \beta \in N, \beta \geqq 1\}$ is an infinite family of planar graphs with smallest cycle length $l = \beta(2k+1)$ and a bridge $B$ of length $\lambda(B) = \beta(2k+1)\ k = \frac{\alpha-1}{2} l$. The length of these bridges is the bound for the length of bridges of a smallest cycle in *planar* graphs.

**Theorem 3.5.4.** *Let $L$ be a smallest cycle of a planar graph $G$ and $l$ its length. Let $B$ be an $L$-bridge and $\alpha \geqq 2$ its number of vertices of attachment. Then*

$$\lambda(B) \geqq \underline{\lambda}(B) \geqq \frac{\alpha-1}{2} l.$$

*Proof.* Let $A_1, \dots, A_\alpha$ denote the vertices of attachment of $B$ on $L$ in this cyclic order. Let $T$ be a tree of $B$ with end vertices $A_1, \dots, A_\alpha$ and let

$$C_i =_{\text{def}} A_i L A_{i+1} T A_i$$

for all $1 \leqq i \leqq \alpha$, where $A_{\alpha+1} = A_1$. Then $C_1, \dots, C_\alpha$ cover $L$ precisely once. Since $L \cup T$ is planar, the cycles $C_1, \dots, C_\alpha$ cover $T$ precisely twice. Hence

$$2\lambda(T) + \lambda(L) = \sum_i \lambda(C_i) \geqq \alpha l.$$

Consequently, $\lambda(T) \geqq \dfrac{\alpha - 1}{2} l$. □

Since the graphs $\beta H_k$ are planar, the bound in Theorem 3.5.4 is attained by an infinite family of graphs, i.e. for infinitely many values of both $\alpha$ and $l$. If $L \cup T$ is not planar, then possibly $\lambda(T) < \dfrac{\alpha - 1}{2} l$. By Corollary 3.5.3 this is the case for $\alpha = 4$. Thus the determination of the greatest lower bound of $\lambda(B)$ in the non-planar case is an open problem.

In a similar way the lengths of bridges of smallest odd and even cycles can be investigated. But here this problem is not dealt with.

## 3.6. The circumference of bridges of longest cycles

K. Hauschild, H. Herre and W. Rautenberg, 1972, have shown that in a 2-connected graph the circumference of the bridge of a longest cycle is smaller than the circumference of the whole graph. In 1982 I proved a somewhat sharper version of their theorem (H.-J. Voss, 1982b).

**Theorem 3.6.1.** *Let $G$ be a 2-connected graph and $L$ a longest cycle of $G$. Then each cycle of an $L$-bridge in $G$ has length at most $\lambda(L) - 2$.*

Here I present an improved result (H.-J. Voss, 1982b):

**Theorem 3.6.2.** *Let $G$ be a 2-connected graph with circumference $l$ and $L$ a longest cycle of $G$. Let $B$ denote an $L$-bridge in $G$ with $r$ vertices of attachment. Then each longest cycle $L'$ of $B$ has length $\lambda(L') \leqq l - 2(r-1)$.*

Since $G$ is 2-connected, each $L$-bridge has at least two vertices of attachment, i.e. $r \geqq 2$. Then, by Theorem 3.6.2, each cycle of an $L$-bridge has length at most $\lambda(L) - 2$. This verifies the assertion of Theorem 3.6.1.

Each $L$-bridge for which equality is attained in Theorem 3.6.2 is called an *extremal $L$-bridge* of Theorem 3.6.2. Next we shall present some extremal $L$-bridges of Theorem 3.6.2.

If $l$ is even and if $2 \leqq r \leqq l/2 - 1$, then there are $L$-bridges for which equality is attained in Theorem 3.6.2. This is shown by the graphs of Fig. 3.6.1, where $\lambda(L') = 2\alpha_1 \geqq 2$, $l = 2(\alpha_1 + 1) + 2(r-2)$ and $\lambda(L') = l - 2(r-1)$. The bridges of $L$ in Fig. 3.6.1a contain vertices of attachment not belonging to $L'$.

If $l$ is even and if $2 \leqq r \leqq l/4$, there are always $L$-bridges, all vertices of attachment of which belong to $L'$ and for which in Theorem 3.6.2 equality is attained. This is shown by

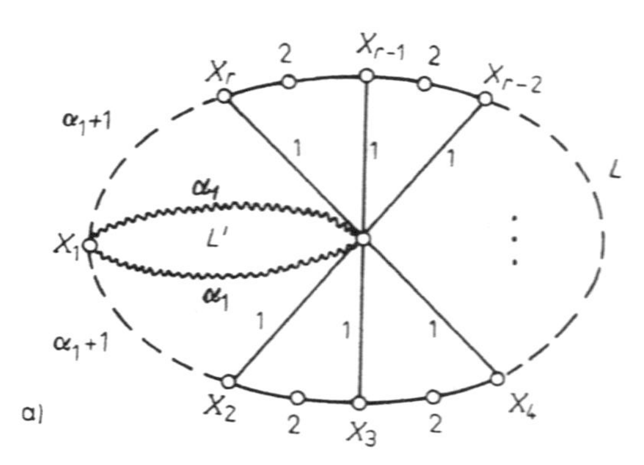


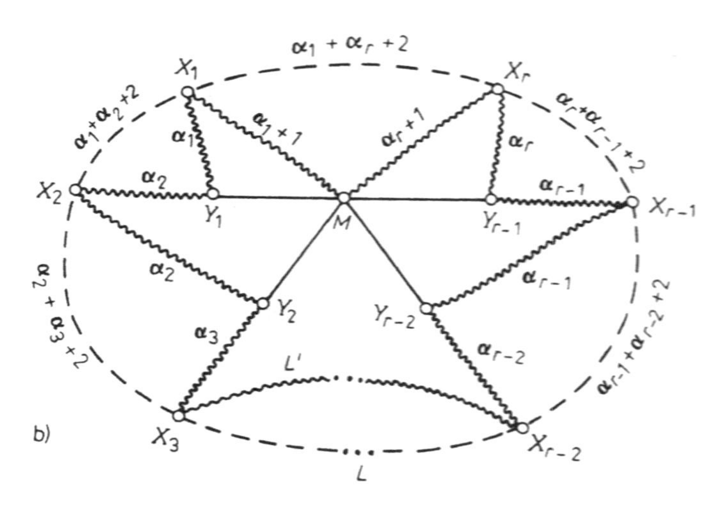


Fig. 3.6.1

the graphs of Fig. 3.6.1b, where $\alpha_i \geqq 1$ for $1 \leqq i \leqq r$ and

$$\lambda(L') = \left(\sum_{i=1}^{r} 2\alpha_i\right) + 2, \quad l = \left(\sum_{i=1}^{r} 2\alpha_i\right) + 2r,$$

$$\lambda(L') = l - 2(r-1).$$

A characterization of all extremal bridges of Theorem 3.6.2 is given in H.-J. Voss, 1982b.

If $G$ is 3-connected, then Theorem 3.6.2 implies (note that $r \geqq 3$) that $l' \leqq l - 4$. But this result is far from being best possible because the following fact is not used: the overlap graph $O(G:L)$ of a cycle $L$ in a 3-connected graph $G$ is connected. Such bridges were investigated by J. A. Bondy and M. Simonovits, 1980, in the proof of their main theorem.

*Proof of Theorem 3.6.2.* Let $G, L, l, B, r, L'$ be defined as in Theorem 3.6.2. Let $K$ denote the kernel of $B$ and let $A_1, \ldots, A_s, A_{s+1}, \ldots, A_r$ be the vertices of attachment of $B$ such that $A_1, \ldots, A_s$ are on $L'$ in this cyclic order and $A_{s+1}, \ldots, A_r$ are not on $L'$. We consider two cases.

Case 1. Assume that $s \geqq 1$. Then

$$F =_{\text{def}} (L' - \{A_1, \ldots, A_s\}) \cup \{A_{s+1}, \ldots, A_r\}$$

is a forest. By recursively joining components of $F$ by paths of $B - \{A_1, \ldots, A_s\}$ we shall construct a tree $T$ of $B - \{A_1, \ldots, A_s\}$ which contains $F$ and is such that all its end vertices belong to $F$. The construction is described in two steps.

Step 1. Obviously, $L' - \{A_1, \ldots, A_s\} \subseteqq K$ consists of $s$ components. Let $\pi_1$ denote one of them. Since $K$ is connected, there is a path $p_1$ joining $\pi_1$ to another component of $L' - \{A_1, \ldots, A_s\}$, say $\pi_2$. Also there is a path $p_2$ joining $\pi_1 \cup p_1 \cup \pi_2$ to another component of $L' - \{A_1, \ldots, A_s\}$, say $\pi_3$, etc. Thus we obtain a tree

$$T^* =_{\text{def}} \pi_1 \cup p_1 \cup \pi_2 \cup p_2 \cup \ldots \cup p_{s-1} \cup \pi_s$$

of $K$ having all end vertices in $\pi_1 \cup \pi_2 \cup \ldots \cup \pi_s$. We remark that

(1) *$\pi_i$ contains a vertex $X$ of valency $v(T^* \cup L' : X) \geqq 3$.*

Step 2. Since $B$ is a bridge, there is a path $q_1$ of $B$ joining $A_{s+1}$ *to* $\pi_1 \cup p_1 \cup \ldots \cup \pi_s$ such that $q_1 \cap L = \{A_{s+1}\}$. There is also a path $q_2$ of $B$ joining $A_{s+2}$ to $\pi_1 \cup p_1 \cup \ldots \cup \pi_s \cup q_1$ such that $q_2 \cap L = \{A_{s+2}\}$, etc. Thus we obtain a tree

$$T =_{\text{def}} \pi_1 \cup p_1 \cup \ldots \cup p_{s-1} \cup \pi_s \cup q_1 \cup q_2 \cup \ldots \cup q_{r-s}$$

which contains $F$ and is such that all its end vertices belong to $F$, where $A_{s+1}, \ldots, A_r$ are end vertices of $T$.

We consider $T \cup L'$. By (1) the arc $L'[A_i, A_{i+1}]$ (indices modulo $s$) contains a vertex $Y_i$ of valency $v(T \cup L' : Y_i) \geqq 3$. Let $L'[A_i, P_i]$ and $L'[Q_i, A_{i+1}]$ be maximal partial paths whose inner vertices have valency 2 in $T \cup L'$. Then by (1) we have:

(2) $\lambda L'[A_i, P_i] + \lambda L'[Q_i, A_{i+1}] \leqq \lambda L'[A_i, A_{i+1}]$ *for all* $1 \leqq i \leqq s$ (indices modulo $s$).

By (2) the subgraphs

$$T_1 =_{\text{def}} (T \cup L') - \bigcup_{i=1}^{s} (L'[A_i, P_i] - \{A_i, P_i\})$$

and

$$T_2 =_{\text{def}} (T \cup L') - \bigcup_{i=1}^{s} (L'[Q_i, A_{i+1}] - \{Q_i, A_{i+1}\})$$

are trees of $B$ with end vertices $A_1, \ldots, A_r$. Therefore,

(3) $\lambda(T_1) \leqq \lambda(B)$ *and* $\lambda(T_2) \leqq \lambda(B)$. □

By construction,

(4) $L' \subseteqq T_1 \cup T_2$ *and* $p_1 \cup p_2 \cup \ldots \cup p_{s-1} \cup q_1 \cup \ldots \cup q_{r-s} \subseteqq T_i$ *for* $i = 1, 2$. □

By Theorem 3.3.1 and assertions (3) and (4) it follows that

$$\lambda(L') + 2 \sum_{i=1}^{s-1} \lambda(p_i) + 2 \sum_{j=1}^{r-s} \lambda(q_j) \leqq \lambda(T_1) + \lambda(T_2) \leqq l.$$

Since $\lambda(p_i) \geqq 1$ and $\lambda(q_j) \geqq 1$, we obtain

$$l' + 2(r-1) \leqq l.$$

Case 2. Assume that $s = 0$. Since $G$ is 2-connected, there are two disjoint paths $q_1$ and $v$ joining $L'$ to two vertices of $L$, say $A_1$ and $A_j$, respectively. Let the end vertices of $q_1$ and $v$ on $L$ be denoted by $X_1$ and $X_2$, respectively. Since $B$ is a bridge, there is a path $q_2$ of $B$ joining $A_2$ to $L' \cup q_1 \cup v$ such that $q_2 \cap L = \{A_2\}$. There is also a path $q_3$ of $B$ joining $A_3$ to $L' \cup q_1 \cup q_2 \cup v$ such that $q_3 \cap L = \{A_3\}$, etc. Thus we obtain a subgraph

$$H =_{\text{def}} L' \cup q_1 \cup q_2 \cup \ldots \cup q_r,$$

where $q_j = v$.

Let $a$ denote a smallest arc of $L'$ with inner vertices of valency 2 and ends of valency at least 3 in $H$. Let $T$ denote the tree obtained from $H$ by deleting all inner vertices and all edges of $a$. The tree $T$ of $B$ has ends $A_1, \ldots, A_r$. Therefore,

$$\lambda(T) \leqq \lambda(B).$$

Since $\lambda(a) \leqq \lfloor l'/2 \rfloor$, at least $\lceil l'/2 \rceil$ edges of $L'$ are in $T$. The tree $T$ contains paths $q_i$ of lengths $\lambda(q_i) \geqq 1$ for all $1 \leqq i \leqq r$. Consequently,

$$\lceil l'/2 \rceil + r \leqq \lfloor l/2 \rfloor \quad \text{and} \quad l' + 2r \leqq l. \ \square$$

As usual, a block of a graph $G$ is either a 2-connected subgraph of $G$ which is maximal with respect to this property or a $K_2$ not belonging to a cycle of length at least 3. In the first case the block has longest cycles and, therefore, a well-defined circumference. For technical reasons a block that is a $K_2$ will be considered as a cycle of length 2, i.e. it has "circumference 2".

Using the same arguments as in the proof of Theorem 3.6.2, I proved (H.-J. Voss, 1982b) the following generalization of Theorem 3.6.2:

**Theorem 3.6.3.** *Let $L$ denote a longest cycle in a 2-connected graph, and let $\mathfrak{B}_1, \mathfrak{B}_2, \ldots$ denote the blocks of an $L$-bridge $B$ in $G$. Then*

$$\sum c(\mathfrak{B}_i) \leqq \lambda(L) - 2 \sum_{r_i \geqq 2} (r_i - 1),$$

*where $r_i$ denotes the number of vertices of attachment of $B$ belonging to $\mathfrak{B}_i$.* $\square$

With the help of Theorem 3.3.4 and 3.3.6 the results of Theorem 3.6.2 and 3.6.3 can be generalized to two or three bridges:

**Theorem 3.6.4.** *Let $L$ denote a longest cycle in a 2-connected graph and let $\mathfrak{B}_1, \mathfrak{B}_2, \ldots$ denote the blocks of at most three pairwise vertex-disjoint $L$-bridges, which induce a path in the overlap graph $O(G:L)$. Then*

$$\sum c(\mathfrak{B}_i) \leqq \lambda(L) - 2 \sum_{r_i \geqq 2} (r_i - 1),$$

*where $r_i$ denotes the number of vertices of $\mathfrak{B}_i$ belonging to $L$.* $\square$

## 3.7. The length of longest paths in bridges of longest cycles

In the investigation of bridges $B$ of longest cycles $L$ in 2-connected graphs we ask for the properties of such bridges. One of these questions we ask is: how long can the longest cycles and the longest paths of $B$ be? In particular, we ask for longest $L,L$-paths, longest $L$-paths, and longest paths in general. The question relating to $L,L$-paths is the simplest one because, obviously, each $L,L$-path $p$ has length not greater than the lengths of the two arcs of $L$ between the end vertices of $p$. Hence each $L,L$-path has length at most $\lfloor c(G)/2 \rfloor$. This implies that each vertex $P$ of $G$ is joined to $L$ by an $L$-path of length at most $\lfloor c(G)/2 \rfloor/2$.

The question of a precise bound for the lengths of all cycles of $B$ is a more complicated problem. We solved it in section 3.6 by showing that $c(B) \leqq c(G) - 2$ (Theorem 3.6.1). The problem with respect to $L$-paths will be solved in this section. We shall prove that each $L$-path of $B$ has length at most $\binom{\lfloor c(G)/2 \rfloor}{2}$, and this bound is sharp.

As corollaries we derive bounds for the lengths of all paths in $B$ and in the whole graph. First we shall prove the following result:

**Lemma 3.7.1.** *Let $G$ be a 2-connected graph and $L$ one of its longest cycles. Let $p$ denote an $L$-path with end vertex $P \in L$ and second end vertex $P_1$. Let $Q_0$ denote the predecessor of $P_1$ on $p$. Then $G$ contains $s$ ($s \geqq 1$) pairwise internally disjoint paths $p_1, \ldots, p_s$ having precisely their end vertices in $p \cup L$ with the following properties:*

*$p_1$ joins $P_1$ to a vertex $Q_1 \in PpQ_0 - Q_0$;*

*for all $2 \leqq i \leqq s-1$ the path $p_i$ joins a vertex $P_i \in Q_{i-1}pQ_{i-2} - Q_{i-1}$ to a vertex $Q_i \in PpQ_{i-1} - Q_{i-1}$;*

*$p_s$ joins a vertex $P_s \in Q_{s-1}pQ_{s-2} - Q_{s-1}$ to a vertex $Q_s \in L - p$* (see Fig. 3.7.1a, b).

*Proof.* Since $G$ is 2-connected, there is a path joining $Q_0pP_1 - Q_0$, i.e. the vertex $P_1$, to a vertex of $(L \cup PpQ_0) - Q_0$ and having no inner vertex in $L \cup p$. If one of these paths joins $P_1$ to a vertex of $L - p$, then the lemma is true with $s = 1$.

Now let each such path join $P_1$ to a vertex of $PpQ_0 - Q_0$. Let $p_1$ be one of these paths with ends $P_1$, $Q_1$ such that $Q_1pP_1$ has maximum length. Since $G$ is 2-connected, there is a path joining $(Q_1pP_1 \cup p_1) - Q_1$ to a vertex of $(L \cup PpQ_1) - Q_1$ and having no inner vertex in $L \cup p \cup p_1$. By the maximum property of $p_1$ the end vertex of each such path lying in $(Q_1pP_1 \cup p_1) - Q_1$ is on $Q_1pQ_0 - Q_1$. If one of these paths joins $Q_1pQ_0 - Q_1$ to a vertex of $L - p$, then the lemma is true with $s = 2$.

Now let each such path join $Q_1pQ_0 - Q_1$ to a vertex of $PpQ_1 - Q_1$. Let $p_2$ be one of these paths with ends $P_2$, $Q_2$ such that $Q_2pP_1$ has maximum length and $P_2 \in Q_1pQ_0 - Q_1$. This construction can be continued until a path $p_s$ is obtained joining $Q_{s-1}pQ_{s-2} - Q_{s-1}$ to a vertex of $L - p$ and having no inner vertex in $L \cup p \cup p_1 \cup \ldots \cup p_{s-1}$. Such a path $p_s$ will be encountered, because $G$ is a finite graph. □

**Theorem 3.7.2.** *Let $G$ be a 2-connected graph and $L$ one of its longest cycles. Then each $L$-path of $G$ has length at most $\binom{\lfloor l/2 \rfloor}{2}$, where $l = \lambda(L)$.*

*Proof.* Let $p$ be an $L$-path of $G$ and $B$ the $L$-bridge containing $p$. Then by Lemma 3.7.1 the $L$-path $p$ is contained in a configuration as depicted in Fig. 3.7.1a, b. A simple calculation shows that the maximum of the length of $p$ in all such configurations is achieved if

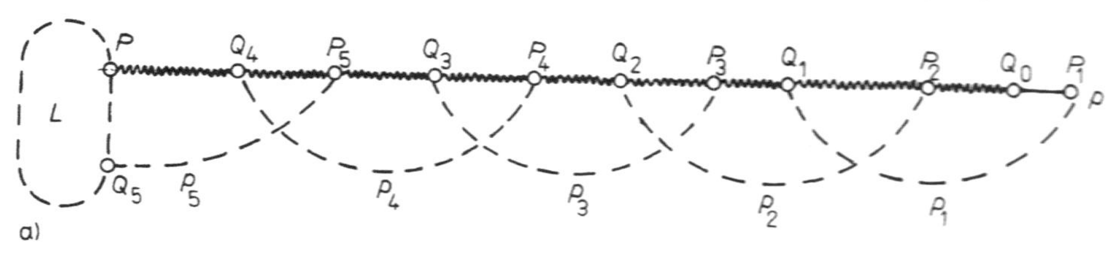


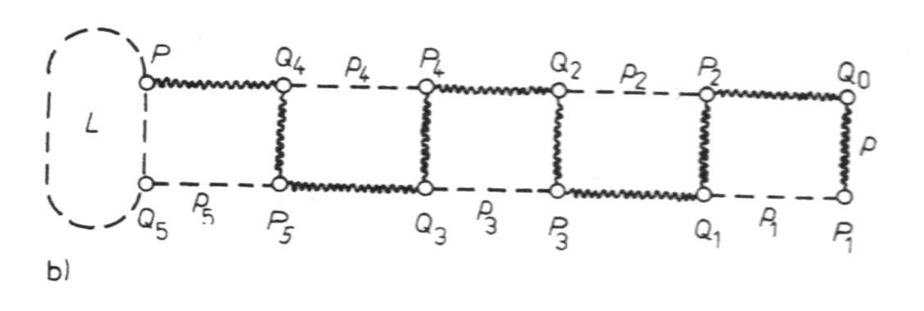


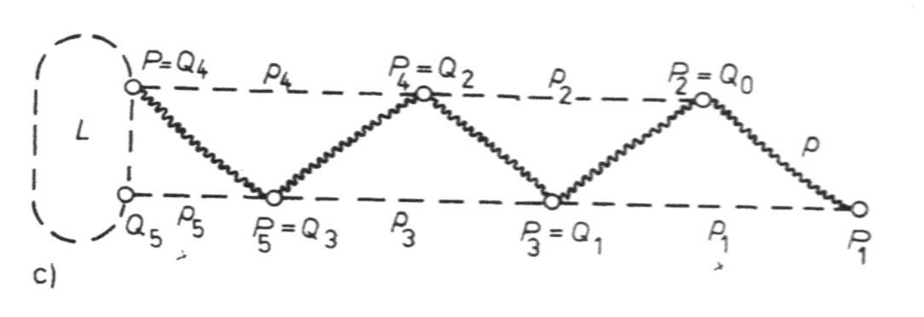


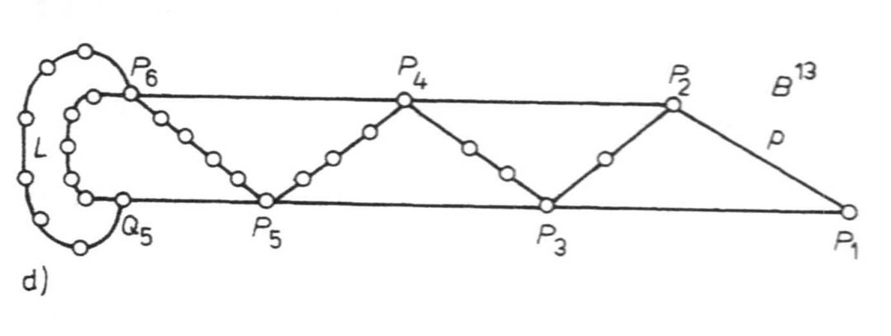


Fig. 3.7.1

$Q_{s-1} = P$ and $Q_i = P_{i+2}$ for all $0 \leqq i \leqq s-2$ (see Fig. 3.7.1c). For convenience let

$$P_{s+1} =_{\text{def}} P.$$

Then $v = \ldots P_7 p_5 P_5 p_3 P_3 p_1 P_1 P_2 p_2 P_4 p_4 P_6 \ldots$ is an $L,L$-path. Hence $\lambda(v) \leqq \lfloor l/2 \rfloor$. Therefore $s \leqq \lfloor l/2 \rfloor - 1$.

The path $P_i p P_{i+1}$ and the two partial paths of $v$ joining $P_i$, $P_{i+1}$ to $P$, $Q_s$, respectively, form an $L,L$-path of length at most $\lfloor l/2 \rfloor$. These partial paths of $v$ contain altogether at least $s - i + 1$ edges. Hence

(1) $\qquad \lambda(P_i p P_{i+1}) + s - i + 1 \leqq \lfloor l/2 \rfloor$ *for all* $1 \leqq i \leqq s$, *where* $s \leqq \lfloor l/2 \rfloor - 1$.

Since $p = \bigcup_{i=1}^{s} \lambda(P_i p P_{i+1})$, the length of $p$ is

$$(2) \qquad \lambda(p) \leqq \sum_{i=1}^{\lfloor l/2 \rfloor - 1} i = \binom{\lfloor l/2 \rfloor}{2}. \ \square$$

The bridge $B$ contains an $L$-path $p$ of length (precisely) $\binom{\lfloor l/2 \rfloor}{2}$ if equality holds in (1), i.e. $s = \lfloor l/2 \rfloor - 1$, $\lambda(P_i p P_{i+1}) + s - i + 1 = \lfloor l/2 \rfloor$, i.e. $\lambda(P_i p P_{i+1}) = i$, and $\lambda(p_i) = 1$ for all $1 \leqq i \leqq s$ (see Fig. 3.7.1d). In this case the only vertices of attachment of $B$ are $P_{s+1}$ and $Q_s$ with $\lambda L[P_{s+1}, Q_s]$, $\lambda L[Q_s, P_{s+1}] \geqq \lfloor l/2 \rfloor$. If $B$ coincides with $p \cup p_1 \cup \ldots \cup p_{\lfloor l/2 \rfloor - 1}$, then $B$ is a bridge containing a path $p$ of length $\lambda(p) = \binom{\lfloor l/2 \rfloor}{2}$. From now on such a bridge will be denoted by $B^l$. Thus in Theorem 3.7.2 equality is attained by the bridge $B^l$ (see Fig. 3.7.1d).

Lemma 3.7.1 is also valid in the case when $L$ is replaced by a single vertex $P$ and $p$ is replaced by any $P$-path of $G$. Applying the arguments of the proof of Theorem 3.7.2 to the $P$-path $p$, we arrive at the following conclusions: $P = Q_s = Q_{s-1}$ and $P_{i+2} = Q_i$ for all $0 \leqq i \leqq s - 2$, $s = l$, and consequently $\lambda(p) \leqq \lfloor l^2/4 \rfloor$.

**Corollary 3.7.3.** *Let $G$ be a 2-connected graph and $L$ one of its longest cycles. Then each path of $G$ has length at most $\lfloor l^2/4 \rfloor$, where $l = \lambda(L)$.* $\square$

This corollary is an improved version of a result of G. A. Dirac, 1952a, by H.-J. Voss, 1973.

In Corollary 3.7.3 equality is attained by a graph $G^l$. Whe shall construct this graph $G^l$.

Let $H_1, H_2, H_3, H_4$ be four pairwise disjoint graphs, where $H_1 \cong H_2 \cong B^l$ and $H_3, H_4$ are paths of length $\lfloor l/2 \rfloor$. Let $P^i, Q^i$, $i = 1, 2$, denote the vertices of attachment of $H_i$ so that $P^i$ is the end vertex of a path of $H_i$ of length $\binom{\lfloor l/2 \rfloor}{2}$. Let $P^j, Q^j, j = 3, 4$, denote the end vertices of the path $H_j$.

If $l$ is even, then we consider $H_1 \cup H_2 \cup H_3$. The graph $G^l$ is obtained from this graph by identifying $P^1, Q^2, P^3$ and $Q^1, P^2, Q^3$.

If $l$ is odd, then we consider $H_1 \cup H_2 \cup H_3 \cup H_4 \cup \{(P^3, P^4)\}$. The graph $G^l$ is obtained from this graph by identifying $Q^1, Q^2, Q^3, Q^4$, identifying $P^1$ with $P^3$ and identifying $P^2$ with $P^4$.

The graph $G^l$ contains a path of length $\lfloor l^2/4 \rfloor$. Consequently, in Corollary 3.7.3 equality is attained by $G^l$.

If the graph $G$ of Theorem 3.7.2 is even 3-connected, then a better upper bound for the $L$-paths of $G$ have been proved by C.-Q. Zhang, 1989a.

**Theorem 3.7.4.** *Let $G$ be a 3-connected non-Hamiltonian graph and $v(G:X) + v(G:Y) \geqq m$ for any pair of non-adjacent vertices $X$ and $Y$. Let $L$ denote a longest cycle of $G$ of length $l$. Then the length of a longest $L$-path is at most $l - m + 1$.*

The graphs $4K_s + K_1$, $s \geqq 1$, show that this bound is sharp.

## 3.8. The length of maximum paths in bridges of longest paths

There is a close connection between the problems about longest paths and longest cycles discussed here (see G. Kopylow, 1977).

Let $G + X$ denote the graph obtained from a graph $G$ by adding a new vertex $X$ to $G$ and joining $X$ to all vertices of $G$ by edges. $p \oplus X$ denotes the cycle of $G + X$ which consists of a path $p$, the vertex $X$, and the two edges joining $X$ to the two end vertices of $p$.

Next we consider a connected graph $G$ with a longest path $w$ and a bridge $B$ of $w$ with $\varrho$ vertices of attachment. Let $w'$ denote a longest path of $B$.

We transform this path problem into a cycle problem by forming the graph $G + X$. The graph $G + X$ is 2-connected; $w \oplus X$ is a longest cycle of $G + X$ of length $\lambda(w) + 2$; $B + X$ is a bridge of the cycle $w \oplus X$ with $\varrho + 1$ vertices of attachment containing the longest cycle $w' \oplus X$ of $B + X$ having length $\lambda(w') + 2$.

Applying Theorem 3.6.2 to $G + X$ we obtain the following result (which, according to B. Dahn, 1973, is due to K. Hauschild).

**Theorem 3.8.1.** *Let $w$ denote a longest path of length $\lambda$ of a connected graph $G$ and let $B$ denote a bridge of $w$ with $\varrho$ vertices of attachment. Then each longest path $w'$ of the bridge $B$ has a length $\lambda(w') \leqq \lambda - 2\varrho$.*

Each $w$-bridge for which equality is attained in Theorem 3.8.1 is called an extremal $w$-bridge of Theorem 3.8.1. The set of all extremal $w$-bridges of Theorem 3.8.1 can be obtained in the following way: we start with a cycle $L$ of length $\lambda(w) + 2$. Then let $K$ denote the class of all possible extremal $L$-bridges $B$ of Theorem 3.6.2 which have a vertex $X$ of attachment adjacent to all vertices of $B$. Deleting $X$ in each $L$-bridge of $K$, we obtain the set of all $w$-bridges which are extremal with respect to Theorem 3.8.1.

The graphs of Fig. 3.6.1a and 3.6.1b show that such $L$-bridges exist. In Fig. 3.6.1a the vertex $X_1$ can be adjacent to all vertices of $L$ and $B$, and in Fig. 3.6.1b the vertex $X_i$ can be adjacent to all vertices of $L$ and $B$ if and only if $\alpha_i \geqq \alpha_j$ for all $1 \leqq j \leqq r$. In Theorem 3.8.1 equality holds for $G - X_1$ and $G - X_i$ in the first and the second cases, respectively. Similar results may be obtained by replacing the concept of "a longest cycle" by "a longest system of two disjoint paths", etc. The transformation of the latter problem to the problem of longest cycles can be carried out by considering $G + X + Y$ $(X, Y \notin G)$.

## 3.9. The "odd circumference" in bridges of longest odd cycles

Above we considered the circumference of bridges of longest cycles. This was done without considering the parity of these longest cycles. Now we shall include the parity of the cycles under consideration, in particular, we deal with longest odd cycles.

I make the following conjecture (see H.-J. Voss, 1976b):

**Conjecture 3.9.1.** *Let $G$ be a non-bipartite 2-connected graph. If $B$ denotes a non-bipartite bridge of a longest odd cycle $L_{\text{odd}}$ of length $l_{\text{odd}}$, then each longest odd cycle $L'_{\text{odd}}$ in $B$ has length $l'_{\text{odd}} < l_{\text{odd}}$.* □

I can prove this conjecture under the additional assumption that $L'_{\text{odd}}$ contains at most four vertices of $L_{\text{odd}}$. In H.-J. Voss, 1982b, I proved the conjecture only for the case when $L'_{\text{odd}}$ and $L_{\text{odd}}$ have at most three common vertices.

**Theorem 3.9.2.** *Let $G$, $B$, $L_{odd}$ meet the hypotheses of the conjecture. If there exists a longest odd cycle $L'_{odd}$ in $B$ containing at most three vertices of $L_{odd}$, then*

$$l'_{odd} < l_{odd}. \square$$

The graph of Fig. 3.9.1 shows that such an assertion concerning even cycles is not true (see also the class $\mathfrak{D}(r)$ of graphs in section 7.6). Thus such a problem with respect to longest even cycles can be formulated for 3-connected graphs only.

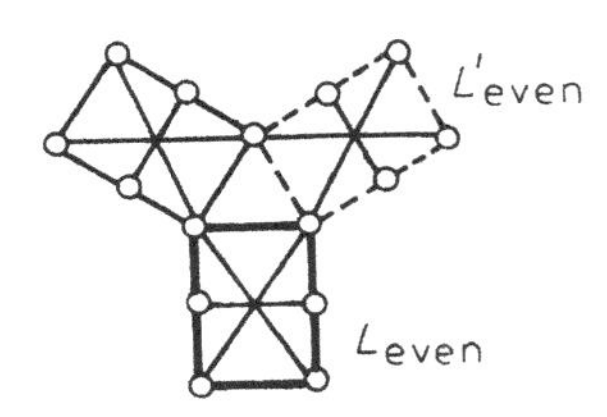


Fig. 3.9.1

I make the following conjecture (see H.-J. Voss, 1976b):

**Conjecture 3.9.3.** *Let $G$ be a 3-connected graph. If $B$ denotes a bridge of a longest even cycle $L_{even}$ of length $l_{even}$, then each longest even cycle $L'_{even}$ in $B$ has length $l'_{even} < l_{even}$.* $\square$

I can only prove this conjecture under the additional assumption that $L'_{even}$ contains at most four vertices of $L_{even}$. Here I prove this conjecture for the case when $L'_{even}$ and $L_{even}$ have at most three common vertices.

**Theorem 3.9.4.** *Let $G$, $B$, $L_{even}$ meet the hypotheses of the conjecture. If there exists a longest even cycle $L'_{even}$ in $B$ containing at most three vertices of $L_{even}$, then $l'_{even} < l_{even}$.*

*Proof.* Let $G$, $B$ and $L_{even}$ meet the hypotheses of the conjecture. Let $L'_{even}$ be a longest even cycle of $B$ containing at most three vertices of $L_{even}$. Since $L'_{even}$ is completely contained in a bridge of $L_{even}$, the two cycles have no common edge. Since $G$ is 3-connected, there are three disjoint paths $w_1$, $w_2$, $w_3$ connecting $L_{even}$ and $L'_{even}$ (see Fig. 3.9.2a). If $L'_{even}$ and $L_{even}$ have $i$ common vertices, $i \geqq 1$, then let $\lambda(w_1) = \ldots = \lambda(w_i) = 0$. With the notation of Fig. 3.9.2a the symmetric difference $C_1 \triangle C_2 \triangle C_3 = L_{even} \cup L'_{even}$ is even. Hence at least one of the cycles $C_i$, $1 \leqq i \leqq 3$, is even, say $C_j$. Then $C_j \triangle L_{even}$ and $C_j \triangle L'_{even}$ are two even cycles of length at most $l_{even}$ covering $L'_{even}$, $L_{even}$ once and $w_j$, $w_{j+1}$ twice. Consequently,

$$l_{even} + l'_{even} + 2\lambda(w_j) + 2\lambda(w_{j+1}) \leqq \lambda(L_{even}) + \lambda(L'_{even}) + 2\lambda(w_j) + 2\lambda(w_{j+1}) \leqq 2\, l_{even}.$$

We arrive at a contradiction if

$$(*) \qquad \lambda(w_j) > 0 \quad \text{or} \quad \lambda(w_{j+1}) > 0.$$

If $i \leqq 1$, the condition (*) is satisfied for each $j$. Hence $\lambda(L'_{even}) < \lambda(L_{even})$ and, therefore, $l'_{even} < l_{even}$ if $L'_{even}$ and $L_{even}$ have at most one common vertex. If $i = 2$, the condition (*) is satisfied for $j = 2$ and $j = 3$ (see Fig. 3.9.2b). Hence $\lambda(L'_{even}) < \lambda(L_{even})$ if $L'_{even}$ and $L_{even}$ have precisely two common vertices and $C_2$ or $C_3$ is an even cycle.

Let $M$ denote the set of common vertices of $L_{even}$ and $L'_{even}$. We consider two cases:

(i) $|M| = 2$ and $C_2$, $C_3$ are odd cycles.

(ii) $|M| = 3$.

First we shall prove the following result:

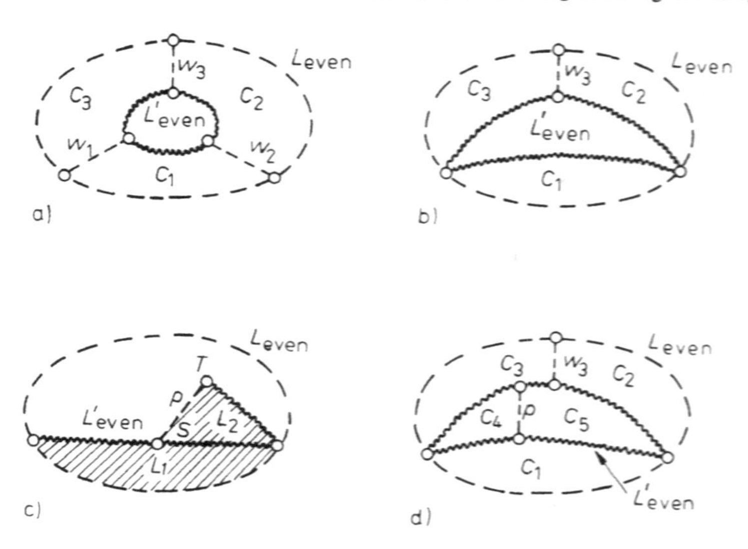


Fig. 3.9.2

**Lemma 3.9.5.** *Let $P, Q \in M$ with the property: $L_{even}$ and $L'_{even}$ can be oriented so that* 1) *both $L_{even}[P, Q]$ and $L'_{even}[P, Q]$ contain no vertex of $M - \{P, Q\}$,* 2) *an inner vertex $S$ of $L'_{even}[P, Q]$ is joined to an inner vertex $T$ of $L'_{even}[Q, P]$ by a path $p$ so that $p$ meets no vertex of $(L_{even} \cup L'_{even}) - \{S, T\}$.*

*Let the cycles $L'_{even}[P, Q] \cup L_{even}[P, Q]$ and $L'[S, T] \cup p$ be denoted by $L_1$ and $L_2$, respectively* (see Fig. 3.9.2c). *If the cycle $L_1 \triangle L_2$ has even length, then $\lambda(L'_{even}) < l_{even}$.*

*Proof of Lemma 3.9.5.* The even cycles $L_1 \triangle L_2 \triangle L_{even}$ and $L_1 \triangle L_2 \triangle L'_{even}$ cover $L'_{even}$ and $L_{even}$ once and $p$ twice, where $\lambda(p) \geqq 1$. Hence

$$\lambda(L'_{even}) + \lambda(L_{even}) + 2\lambda(p) \leqq 2l_{even}.$$

Thus the lemma is proved. □

Case (i). Let $|M| = 2$ and let $C_2$, $C_3$ be odd cycles (see Fig. 3.9.2d). Since $C_1 \triangle C_2 \triangle C_3 = L'_{even} \cup L_{even}$ is even, $C_1$ is an even cycle. Since $L'_{even}$ is in a bridge $B$ of $L_{even}$, there is a path $p$ in $B - L_{even}$ joining an inner vertex of $L'[P, Q]$ to an inner vertex $V$ of $L'[Q, P] \cup w_3$, where $L'$ is short for $L'_{even}$.

If $V \in L'$, then $p$ is an $L',L'$-path; if $V \in w_3$, then $p \cup w_3$ contains an $L',L'$-path. This $L',L'$-path splits $L'_{even}$ into two cycles $C_4$ and $C_5$ (see Fig. 3.9.2d). The lemma 3.9.5 implies that $C_4$ and $C_5$ are odd cycles. By the hypotheses $C_2$ and $C_3$ are also odd cycles. The three even cycles $C_1 \triangle C_4 \triangle C_3$, $C_1 \triangle C_5 \triangle C_2$ and $C_2 \triangle C_3$ cover $L'_{even}$ once and $L_{even} \cup p$ twice. Hence

$$\lambda(L'_{even}) + 2\lambda(L_{even}) + 2\lambda(p) \leqq 3l_{even}. \ \square$$

Case (ii). Let $L'_{even}$ contain precisely three vertices $X_1$, $X_2$, $X_3$ of $L_{even}$. Since the kernel $K$ of $B$ is a connected subgraph of $G$, there are two paths $p$, $p'$ connecting the three (non-empty) components of $L'_{even} - \{X_1, X_2\ X_3\}$ to form one component. Then $L_{even}$, $L'_{even}$ and $p$, $p'$ form one of the following four configurations in $G$. One is represented by the graph $H$ of Fig. 3.9.3.

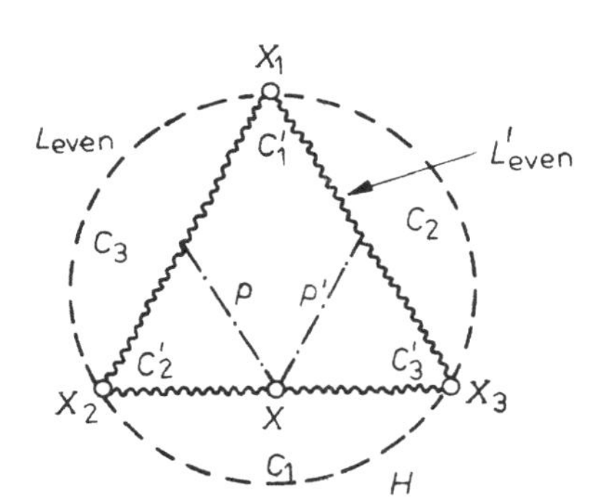


Fig. 3.9.3

The three different possible configurations are obtained from $H$ in the following way: we split the vertex $X$ into two vertices so that each of the two new vertices is incident to precisely two of the old edges, and then we join the two new vertices by a path $w$.

Next we prove Theorem 3.9.4 for the case described in Fig. 3.9.3; there the symbol $C_i$ (and $C_i'$) denotes the cycle bounding the face where the sign $C_i$ (or $C_i'$) is put in. By Lemma 3.9.5 only one case remains to be considered, namely that $C_2'$, $C_3'$ are odd cycles and $C_1$, $C_2$, $C_3$, $C_1'$ are even cycles. But then $C_1 \triangle L_{\text{even}}$, $C_2 \triangle C_1' \triangle L_{\text{even}}$, $C_3 \triangle C_1' \triangle L_{\text{even}}$ are three even cycles covering $L'_{\text{even}}$ once and $L_{\text{even}} \cup p \cup p'$ twice. Consequently,

$$l'_{\text{even}} + 2\, l_{\text{even}} + 4 \leqq 3\, l_{\text{even}}.$$

Thus in the case described in Fig. 3.9.3 our proof is complete. New we deal with the three cases different from that of Fig. 3.9.3. These three cases were obtained by splitting up the vertex $X$ in it and introducing a new path $w$. If the edges of $C_i'$ form a cycle in the resulting graph, then this cycle is again denoted by $C_i'$. If the edges of $C_i'$ form only an open path, then this path and $w$ form a cycle again denoted by $C_i'$. The same may be done for $C_i$. Now the proof is exactly the same as in the case of Fig. 3.9.3: we only need to check that the cycles used above in the proof are again cycles and these cycles cover $L_{\text{even}}$, $L'_{\text{even}}$ and $p$, $p'$ so that our inequalities remain true.

## 3.10. The vertices of a longest cycle $L$ which are neighbours of an $L$-bridge

In sections 3.1–3.3 we have discussed properties of bridges $B$ of a longest cycle $L$. In this section we shall deal with vertices on $L$ outside $B$, namely, with the vertices of $L$ which are neighbours of the vertices of attachment of $B$ on $L$. Let $A_1, \ldots, A_\alpha$ be the vertices of attachment of a regular bridge $B$ of a longest cycle $L$. Let $A_i^+$ and $A_i^-$ be the successor and the predecessor of $A_i$ on $L$, respectively.

**Proposition 3.10.1.** *$\{A_1^+, \ldots, A_\alpha^+\}$ and $\{A_1^-, \ldots, A_\alpha^-\}$ are independent sets of vertices.*

This proposition implies the results of V. Chvátal and P. Erdős, 1972, D. Amar, I. Fournier, A. Germa and R. Häggkvist, 1984, and P. Erdős.

Before proving Proposition 3.10.1 we shall show a simple lemma.

**Lemma 3.10.2.** *For all $i$, $1 \leqq i \leqq \alpha$, the vertex $A_i^+$ is not a vertex of attachment of $B$,* i.e. $\lambda(L[A_i, A_{i+1}]) \geqq 2$.

*Proof.* Suppose that $A_i^+$ is a vertex of attachment of $B$. Then there is an $L,L$-path $p$ of $B$ of length at least 2 with ends $A_i$ and $A_i^+$. Then $p \cup L[A_i^+, A_i]$ is a cycle of length greater than $\lambda(L)$. Contradiction! □

*Proof of Proposition 3.10.1.* Suppose that $(A_i^+, A_j^+)$ is an edge of $G$. By Lemma 3.10.2 the edge $(A_i^+, A_j^+)$ is a diagonal of $L$. Let $p$ denote an $L,L$-path of $B$ with ends $A_i$ and $A_j$. Then $p \cup L[A_i^+, A_j] \cup [A_i^+, A_j^+] \cup L[A_j^+, A_i]$ is a cycle of length greater than $\lambda(L)$, a contradiction. Hence $\{A_1^+, \ldots, A_\alpha^+\}$ is an independent set of vertices. Similarly, it can be shown that $\{A_1^-, \ldots, A_\alpha^-\}$ is an independent set of vertices. □

**Corollary 3.10.3.** *Let $G$ be a graph not containing $k+1$ independent vertices. Let $L$ be a longest cycle of $G$. Then every regular $L$-bridge has fewer than $k$ vertices of attachment.*

This corollary was proved by C. St. J. A. Nash-Williams, 1971a.

*Proof.* Suppose that $G$ contains a regular $L$-bridge $B$ with vertices of attachment $A_1, \ldots, A_\alpha$ in this cyclic order, where $\alpha \geqq k$. Then by Proposition 3.10.1 the vertices $A_1^+, \ldots, A_k^+$ are pairwise independent. Let $X$ be an inner vertex of $B$. By Lemma 3.10.2 the vertex $X$ is adjacent to none of the vertices $A_1^+, \ldots, A_k^+$. Hence $A_1^+, \ldots, A_k^+, X$ is an independent set of $k+1$ vertices. This contradiction proves the validity of Corollary 3.10.3. □

If a graph $G$ is non-Hamiltonian, then the requirement that $G$ does not contain $k+1$ independent vertices is a strong restriction on the number of bridges of a longest cycle $L$. In fact, if $B_1, \ldots, B_s$ are the $L$-bridges in $G$, then $s \leqq k$, because each set of vertices $X_i$, $X_i \in B_i - L$, is independent. By Corollary 3.10.3 each $L$-bridge has at most $k-1$ vertices of attachment.

Next we present three theorems on graphs not containing $k+2$ independent vertices. V. Chvátal and P. Erdős, 1972, proved the following result:

**Theorem 3.10.4.** *Let $G$ be a $k$-connected graph, $k \geqq 0$, not containing $k+1$ independent vertices. Then $G$ has a Hamiltonian cycle.*

*Proof.* Suppose that $G$ is not Hamiltonian. Then each longest cycle $L$ does not include all vertices of $G$. Hence $G$ contains a regular $L$-bridge $B$. Since $G$ is $k$-connected, $B$ has $k$ vertices of attachment. This contradicts Corollary 3.10.3. □

Let $n(G)$ denote the number of vertices, $\alpha(G)$ the independence number, and $\varkappa(G)$ the connectivity of $G$. Then the Chvátal-Erdős condition

$$\alpha(G) \leqq \varkappa(G)$$

implies the existence of a Hamiltonian cycle in $G$.

In an interesting paper J. A. Bondy, 1971, proved that any simple graph $G$ which satisfies Ore's condition

$$v(G:P) + v(G:Q) \geqq n(G) \quad \text{for all pairs of non-adjacent vertices } P, Q \text{ of } G$$

also satisfies the Chvátal-Erdős condition. Thus the Chvátal-Erdős theorem is seen to be a generalization of Ore's theorem, O. Ore, 1960 (see section 7.0, 1°). D. Amar, I. Fournier, A. Germa and R. Häggkvist, 1984, extended the result of V. Chvátal and P. Erdős to the case where $G$ can contain $k+1$ independent vertices.

**Theorem 3.10.5.** *Let $G$ be a $k$-connected graph, $k \geqq 0$, not containing $k+2$ independent ver-*

*tices. Let L be a longest cycle of G. Then G − L is complete,* i.e. *L has at most one regular bridge; the kernel of this unique bridge is a complete subgraph of G.*

*Proof.* First we prove that

(1) *the kernel of each bridge is a complete graph.*

*Proof of (1).* Suppose that the kernel of an $L$-bridge contains two distinct vertices $X$, $Y$ such that $(X, Y)$ is not an edge of $G$. Since $G$ is $k$-connected, $G$ has at least $k$ vertices of attachment $A_1, \ldots, A_k$. Then by Proposition 3.10.1 and Lemma 3.10.2 the $k+2$ vertices $A_1^+, \ldots, A_k^+, X, Y$ are independent. This contradiction proves (1). □

Next we show that

(2) *G does not contain two distinct regular L-bridges.*

*Proof of (2).* Suppose that in $G$ there are two distinct regular $L$-bridges $B$ and $B'$. Let $A_1, \ldots, A_k$ again denote $k$ vertices of attachment of $B$ in this cyclic order, and let $X$ be an inner vertex of $B$ and $Y$ an inner vertex of $B'$. If $(Y, A_i^+) \notin E(G)$ for all $i$, $1 \leqq i \leqq k$, then by Proposition 3.10.1 and Lemma 3.10.2 the $k+2$ vertices $A_1^+, \ldots, A_k^+, X, Y$ are independent. Without loss of generality, suppose that $(Y, A_1^+) \in E(G)$. For all $j \geqq 2$ the edge $(Y, A_j^+) \notin E(G)$, because otherwise $p \cup L[A_1^+, A_j] \cup [A_1^+, Y, A_j^+] \cup L[A_j^+, A_1]$ is a cycle of length greater than $\lambda(L)$, where $p$ is an $L,L$-path of $B$ with ends $A_1^+, A_j^+$. Hence $A_2^+, \ldots, A_k^+, X, Y$ are $k+1$ independent vertices of $G$.

The vertex $A_1^{++}$ is the successor of the vertex $A_1^+$, which is a vertex of attachment of $B'$. Then Lemma 3.10.2 implies that $(A_1^{++}, Y) \notin E(G)$.

If $(A_1^{++}, A_j^+) \in E(G)$ or $(A_1^{++}, X) \in E(G)$, then let $C_1 =_{\text{def}} p \cup L[A_1^{++}, A_j] \cup [A_1^{++}, A_j^+] \cup L[A_j^+, A_1]$ or $C_2 =_{\text{def}} q \cup L[A_1^{++}, A_1]$, respectively, where $p$ and $q$ are $L,L$-paths of $B$ with ends $A_1$, $A_j$ and $A_1, A_1^{++}$, respectively. Then $C_i$, $i = 1, 2$, is either a cycle of length greater than $\lambda(L)$ or a longest cycle of length $\lambda(L)$ with a $C_i$-bridge $\bar{B}$ including both $B'$ and $[A_1^+, Y]$.

In the first case we arrive at a contradiction!

In the second case $B'$ has $k$ vertices of attachment because $G$ is $k$-connected. Consequently, $\bar{B}$ has at least $k+1$ vertices of attachment. This contradicts Theorem 3.10.4! □

Two further results are Theorem 3.10.5 by J. A. Bondy and P. Erdős, and Theorem 3.10.6 by P. Erdős (Some problems in graph theory; pp. 187-190). These two theorems are also proved by using Proposition 3.10.1.

**Theorem 3.10.5.** *If $k$, $n$, $l$ are positive integers such that $n \geqq (k-1)(l-1)+1$ and $n \geqq k^2 - 2$, then every graph with $n$ vertices contains a cycle of length $l$ or $k$ independent vertices.* □

**Theorem 3.10.6.** *Let $n > 4k^4$ and let G be a Hamiltonian graph of order n which does not contain a set of k pairwise independent vertices. Then G is pancyclic,* i.e. *G contains a cycle of length m for all m, $3 \leqq m \leqq n$.* □

If $S \subseteqq V(G)$, then $N(S)$ denotes the set of all neighbours of $S$ in $G$.

We conclude this section with the so-called "hopping lemma" of D. R. Woodall, 1973, which is undoubtedly one of the deeper results on cycles in graphs. This lemma is also devoted to vertices of a cycle $C$ which are not in a bridge of $C$.

**Theorem 3.10.8** (Hopping lemma). *Let $C = (V_1, V_2, \ldots, V_s, V_1)$ be a cycle of length s, with as few regular C-bridges as possible, in a graph G. Suppose that G has no cycle of length $s+1$, and that $G - C$ contains an isolated vertex V. Set $\mathfrak{Y}_0 = \emptyset$ and, for $j \geqq 1$, define*

$$\mathfrak{X}_j =_{\text{def}} N(\mathfrak{Y}_{j-1} \cup \{V\}),$$
$$\mathfrak{Y}_j =_{\text{def}} \{V_i \in V(C) \mid V_{i-1}, V_{i+1} \in \mathfrak{X}_j\}.$$

*Then, for all $j \geqq 1$, the set $\mathfrak{X}_j$ is a subset of $V(C)$ but does not include consecutive vertices of C.* □

By Theorem 3.10.8 we have $\mathfrak{X}_j \subseteqq \mathfrak{X}_{j+1}$ and $\mathfrak{Y}_j \subseteqq \mathfrak{Y}_{j+1}$ for all $j \geqq 1$ and no vertex of $\mathfrak{X}_j$ is on a $C$-bridge.

## 3.11. On intersections of two longest cycles

Let $L$ and $L'$ denote two distinct longest cycles of a 2-connected graph. By Theorem 3.6.1 of H. Herre, K. Hauschild and W. Rautenberg, 1972, the cycle $L'$ is not a subgraph of an $L$-bridge, i.e. $L'$ crosses at least two $L$-bridges or it contains an arc of $L$. In both cases $L$ and $L'$ have at least two common vertices.

M. Grötschel, 1984, and M. Grötschel and G. L. Nemhauser, 1984, first studied the intersection pattern of longest cycles. Such results are of interest, for instance, in designing recursive algorithms (e.g. in combinatorial optimization) in which in every step a longest cycle is shrunk or deleted and one wants to know a bound on the length of the longest cycle in the resulting graph.

The case where two longest cycles meet in two vertices was first analyzed by M. Grötschel and G. L. Nemhauser, 1984 (to obtain a polynomial algorithm for the max-cut problem in graphs without long odd cycles). M. Grötschel and G. L. Nemhauser, 1984, also applied their results to vertex transitive graphs.

The above stated implication of the theorem of H. Herre, K. Hauschild and W. Rautenberg is formulated as follows:

**Theorem 3.11.1.** *Every two longest cycles of a 2-connected graph meet in at least two common vertices.* □

Let $L_{\text{odd}}$ and $L'_{\text{odd}}$ denote two distinct longest odd cycles of a (non-bipartite) 2-connected graph. By Theorem 3.9.2, $L'_{\text{odd}}$ either contains at least four vertices of $L_{\text{odd}}$ or it is not a subgraph of a bridge of $L_{\text{odd}}$, i.e. $L'_{\text{odd}}$ either crosses at least two bridges of $L_{\text{odd}}$ or it contains an arc of $L_{\text{odd}}$. In all cases $L_{\text{odd}}$ and $L'_{\text{odd}}$ have at least two common vertices.

**Theorem 3.11.2.** *Every two longest odd cycles of a 2-connected graph have at least two common vertices.* □

A similar assertion on longest *even* cycles in 2-connected graphs is not true. An infinite family of counterexamples is $\{\beta H_{2k} \mid \beta \in N - \{0\}, \beta \text{ odd}\}$, $k \geqq 2$, where $H_{2k}$ is the following graph: $H_{2k}$ consists of two vertex-disjoint cycles $C_1$, $C_2$ of length $2k$ and two additional edges joining two vertices of $C_1$ of odd distance to two vertices of $C_2$ of even distance.

If the graphs under consideration are 3-connected, then Theorem 3.9.4 implies the following:

*Every two longest even cycles of a 3-connected graph have at least two common vertices.* □

We can even prove the following result:

**Theorem 3.11.3.** *Every two longest even cycles of a 3-connected graph have at least three common vertices.* □

Let $\Gamma_s$, $s \geqq 1$, denote the following graph: $\Gamma_s = 6K_{2s} + I_3$ is obtained from six pairwise disjoint $K_{2s}$ and three additional vertices which are joined by edges to all vertices of the six $K_{2s}$. This 3-connected graph $\Gamma_s$ has two longest even cycles of length $6s + 2$ meeting in precisely three vertices.

*Proof.* Let $L_{\text{even}}$, $L'_{\text{even}}$ be two longest even cycles having at most two common vertices. We consider two cases.

Case (i). Let $L_{\text{even}}$ and $L'_{\text{even}}$ have at most one common vertex. Then $L'_{\text{even}}$ is completely contained in a bridge of $L_{\text{even}}$. By Theorem 3.9.4 the cycle $L'_{\text{even}}$ has length less than $\lambda(L_{\text{even}})$. This contradiction proves the theorem in Case (i).

Case (ii). Let $L_{\text{even}}$ and $L'_{\text{even}}$ have precisely two common vertices, say $P$ and $Q$. If $(P, Q)$ is a common edge of $L_{\text{even}}$ and $L'_{\text{even}}$, i.e. if $[P, Q] = L_{\text{even}}[P, Q] = L'_{\text{even}}[P, Q]$, then $L_{\text{even}}[Q, P] \cup L'_{\text{even}}[Q, P]$ is an even cycle of length $2(\lambda(L_{\text{even}}) - 1) > \lambda(L_{\text{even}})$. This contradicts the definition of $L_{\text{even}}$ as a longest even cycle.

Suppose that $(P, Q) \notin E(L_{\text{even}})$. Then $L_{\text{even}}$ and $L'_{\text{even}}$ are edge-disjoint cycles.

Case 1. Let $L_{\text{even}}[P, Q]$ and $L'_{\text{even}}[P, Q]$ have the same parity. Then it is obvious that $\lambda(L_{\text{even}}[P, Q]) = \lambda(L'_{\text{even}}[P, Q]) = \lambda(L_{\text{even}}[Q, P]) = \lambda(L'_{\text{even}}[Q, P])$. Hence any two of these arcs form a longest even cycle $L$ of $G$, the two remaining arcs forming a second longest even cycle $\bar{L}$ of $G$. By Theorem 3.9.4, $\bar{L}$ is not in a bridge of $L$. Consequently, $G - \{P, Q\}$ consists of at least four components. This contradicts the 3-connectivity of $G$.

Case 2. Let $L_{\text{even}}[P, Q]$ and $L'_{\text{even}}[P, Q]$ have different parity. Then $L_{\text{even}}[P, Q]$ and $L_{\text{even}}[Q, P]$ ($L'_{\text{even}}[P, Q]$ and $L'_{\text{even}}[Q, P]$) have equal parity. Since $G$ is 3-connected, there is a path $p$ of $G - \{P, Q\}$ joining two arcs of different parity (see Fig. 3.11.1). These two arcs form an odd cycle which is split by $p$ into an even cycle $C_1$ and an odd cycle $C_2$ (see Fig. 3.11.1). The cycles $L'_{\text{even}} \triangle C_1$ and $L_{\text{even}} \triangle C_1$ are two even cycles covering $L_{\text{even}}$ and $L'_{\text{even}}$ once and $p$ twice. Consequently, with

$$\lambda(L'_{\text{even}}) + \lambda(L_{\text{even}}) + 2 \leqq \lambda(C_1) + \lambda(C_2) \leqq 2\lambda(L_{\text{even}})$$

we arrive at a contradiction!

This completes the proof of Theorem 3.11.3. □

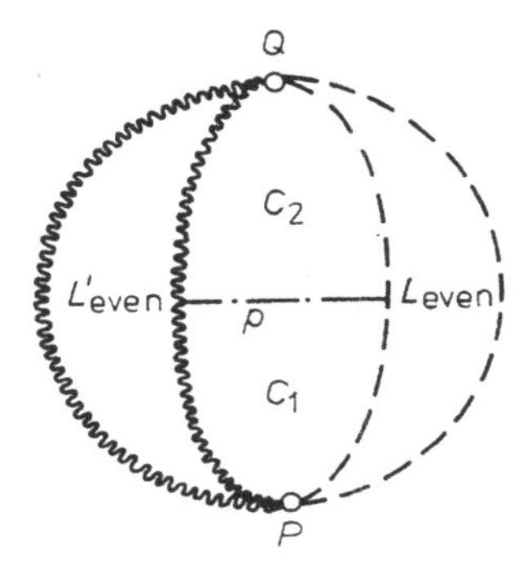


Fig. 3.11.1

Next we investigate the intersection pattern of two longest cycles which have *precisely k* common vertices, $2 \leqq k \leqq 5$. In the case $k = 2$, M. Grötschel and G. L. Nemhauser, 1984, proved the following result:

**Theorem 3.11.4.** *Let G be a 2-connected graph, and let L and L′ be two longest cycles of G intersecting in precisely two vertices P and Q of G. Then G contains four distinct* $\{P,Q\}$*-bridges, each containing precisely one of the arcs* $L[P, Q]$, $L[Q, P]$, $L'[P, Q]$ *and* $L'[Q, P]$; *all arcs have the same length and the circumference of G is even.* $\{P, Q\}$ *is a separating set of G.*

*Moreover, every longest cycle L″ contains P and Q, and the arcs* $L''[P, Q]$ *and* $L''[Q, P]$ *are in different* $\{P,Q\}$*-bridges and have equal lengths.*

*Proof.* If $L$ and $L'$ have a common edge $e = (P, Q)$, then $L \cup L' - e$ is a cycle of length $2l - 2 > l$, a contradiction! Let $L$ and $L'$ be edge-disjoint cycles. The four arcs $L[P, Q]$, $L[Q, P]$, $L'[P, Q]$ and $L'[Q, P]$ obviously have the same length. Suppose that two of the four arcs, say $a_1$, $a_2$, are joined by a path $p$ of $G - \{P, Q\}$ meeting none of the two remaining arcs, say $a_3$, $a_4$. Then $a_1 \cup a_2 \cup p$ is in a bridge of the longest cycle $a_3 \cup a_4$. This contradicts Theorem 3.6.1 of H. Herre, K. Hauschild and W. Rautenberg, 1972, that no bridge of a longest cycle of $G$ completely contains a different longest cycle of $G$. Hence no two of these arcs are joined by a path of $G - \{P, Q\}$, i.e. every two of them are in distinct regular $\{P,Q\}$-bridges of $G$. The regularity of the bridges implies that $\{P, Q\}$ is a separating set of the graph $G$.

A further consequence is that each $\{P,Q\}$-bridge is a bridge of $L$ or $L'$. Hence by Theorem 3.6.1 of H. Herre, K. Hauschild and W. Rautenberg, 1972, no longest cycle of $G$ is completely contained in a $\{P,Q\}$-bridge. Consequently, each longest cycle $L''$ of $G$ contains $P$ and $Q$, and both arcs $L''[P, Q]$ and $L''[Q, P]$ are in different $\{P,Q\}$-bridges. □

An immediate consequence of Theorem 3.11.4 is the following (compare with Theorem 3.11.1):

**Corollary 3.11.5.** *Let G be a 2-connected graph with an odd circumference. Then every two longest cycles have at least three common vertices.* □

Next we investigate the intersection pattern of two longest cycles having precisely three common vertices. We present a slightly improved theorem of M. Grötschel, 1984.

**Theorem 3.11.6.** *Let G be a 2-connected graph, and let L and L′ be two longest cycles of G intersecting in precisely three vertices P, Q, R. Let these three vertices be on L and L′ in this cyclic order. Then*

(i) $\lambda(L[P, Q]) = \lambda(L'[P, Q])$, $\lambda(L[Q, R]) = \lambda(L'[Q, R])$, *and* $\lambda(L[R, P]) = \lambda(L'[R, P])$.

(ii) *Any two arcs of* $\{L[P, Q], L[Q, R], L[R, P], L'[P, Q], L'[Q, R], L'[R, P]\}$ *coincide or they are contained in distinct* $\{P,Q,R\}$*-bridges.*

(iii) $\{P, Q, R\}$ *is a separating set.*

(iv) *Moreover, if none of the cycles* $L[P, Q] \cup L'[P, Q]$, $L[Q, R] \cup L'[Q, R]$ *and* $L[R, P] \cup L'[R, P]$ *is a longest one, then every longest cycle L″ contains P, Q, R. Let these vertices be on L″ in this cyclic order. Then* $L''[P, Q]$, $L''[Q, R]$, $L''[R, P]$ *are in three pairwise distinct* $\{P,Q,R\}$*-bridges.*

*Proof.* (i) is obviously true.

(ii) Assume that two of the six arcs, say $a_1$, $a_2$, have length at least 2 and are joined by a path $p$ of $G - \{P, Q, R\}$ meeting none of the four remaining arcs. We consider two cases.

Case 1. The arcs $a_1$, $a_2$ have the same end vertices. Without loss of generality, let these two arcs be $L[P, Q]$ and $L'[P, Q]$. Let $S$ and $T$ denote the end vertices of $p$ on $L[P, Q]$ and $L'[P, Q]$, respectively. Then $L[P, S] \cup p \cup L'[T, Q]$ or $L'[P, T] \cup p \cup L[S, R]$ is a $P,Q$-path $q$ of length greater than $\lambda(L[P, Q])$ meeting no inner vertex of $L[Q, P]$ and $L'[Q, P]$. Thus the $P,Q$-path $q$ and $L[Q, P]$ form a cycle of length greater than $\lambda(L)$. Contradiction!

Case 2. The arcs $a_1$, $a_2$ have precisely one common end vertex. Without loss of generality, let these arcs be $L'[P, Q]$ and $L'[Q, R]$. Let $S$ and $T$ denote the end vertex of $p$ on $L'[P, Q]$ and $L'[Q, R]$, respectively. Then the cycles

$$C_1 = L'[P, S] \cup p \cup L'[Q, T] \cup L[Q, P]$$

and

$$C_2 = L[P, Q] \cup L'[S, Q] \cup p \cup L'[T, P]$$

cover $L$ and $L'$ once and $p$ twice. Hence $\lambda(C_1) + \lambda(C_2) \geqq 2\lambda(L) + 2$, contradicting $\lambda(C_1) + \lambda(C_2) \leqq 2\lambda(L)$! Thus in both cases the assumption fails, and no two of the arcs in question are joined by a path of $G - \{P, Q, R\}$, i.e. any two different arcs are in (pairwise) distinct $\{P, Q, R\}$-bridges of $G$.

(iii) The only 2-connected graph with circumference 3 is the triangle. Since $G$ has two distinct (longest) cycles, $G$ is no triangle and its circumference is at least 4. Hence at least one of the arcs $L[P, Q]$, $L[Q, R]$, $L[R, P]$ has length at least 2, say $L[P, Q]$. Since $\lambda(L[P, Q]) = \lambda(L'[P, Q]) \geqq 2$, the two (distinct!) $\{P,Q,R\}$-bridges containing these arcs are regular. Consequently, (ii) implies that $\{P, Q, R\}$ is a separating set of $G$.

(iv) A consequence of (ii) is that each regular $\{P,Q,R\}$-bridge is a bridge of $L$ or $L'$. By Theorem 3.6.1 of H. Herre, K. Hauschild and W. Rautenberg, 1972, no longest cycle of $G$ is completely contained in a $\{P,Q,R\}$-bridge. Hence an arbitrarily chosen longest cycle $L''$ of $G$ contains at least two of the vertices $P$, $Q$, $R$. We consider two cases.

Case 1. Let $L''$ contain precisely two of the vertices $P$, $Q$, $R$, say $P$ and $Q$. Then $L''[P, Q]$ is in precisely one $\{P,Q,R\}$-bridge $B$. By (ii) the $\{P,Q,R\}$-bridge $B$ is a bridge of $L$ or $L'$, say $L$. Therefore $\lambda(L''[P, Q]) \leqq \lambda(L[P, Q])$ (note also that $\lambda(L[P, Q]) = \lambda(L'[P, Q])$ by assertion (i)). Similarly, $\lambda(L''[Q, P]) \leqq \lambda(L'[P, Q])$. Summing the two inequalities, we obtain

$$l = \lambda(L'') \leqq \lambda(L[P, Q]) \cup \lambda(L'[P, Q]).$$

Hence $L[P, Q] \cup L'[P, Q]$ is a longest cycle of $G$. This contradicts the hypotheses of (iv)!

Case 2. Let $L''$ contain $P$, $Q$, $R$. Let the orientation of $L''$ be chosen so that $P$, $Q$, $R$ are on $L''$ in this cyclic order. By the same arguments as in Case 1 we can show that

$$\lambda(L''[P, Q]) \leqq \lambda(L[P, Q]), \quad \lambda(L''[Q, R]) \leqq \lambda(L[Q, R]),$$

$$\lambda(L''[R, P]) \leqq \lambda(L[R, P]).$$

Summing these inequalities, we obtain

$$l = \lambda(L'') \leqq \lambda(L) = l.$$

Hence in all these three inequalities equality is attained.

If $L''[P, Q]$, $L''[Q, R]$ and $L''[R, P]$ are in three distinct $\{P,Q,R\}$-bridges, then the proof of (iv) is complete. Now let these arcs of $L''$ be in precisely two $\{P,Q,R\}$-bridges. Then two of the arcs, say $\dot{L}''[P, Q]$ and $L''[Q, R]$, are in the same bridge $B$. By (ii), without loss of generality $B$ is also a bridge of $L$. Therefore there is a path $p$ of $B - L$ joining an inner vertex of $L''[P, Q]$ to an inner vertex of $L''[Q, R]$. Using the same arguments as in the proof of Case 2 of (ii), we arrive at a contradiction!

Thus the proof of Theorem 3.11.6 is complete. □

Let $\Gamma_1 = I_2 + (2K_1 \cup K_2)$ (see Fig. 3.11.2a). M. Grötschel, 1984, also investigated the intersection pattern of two longest cycles having precisely four or five common vertices.

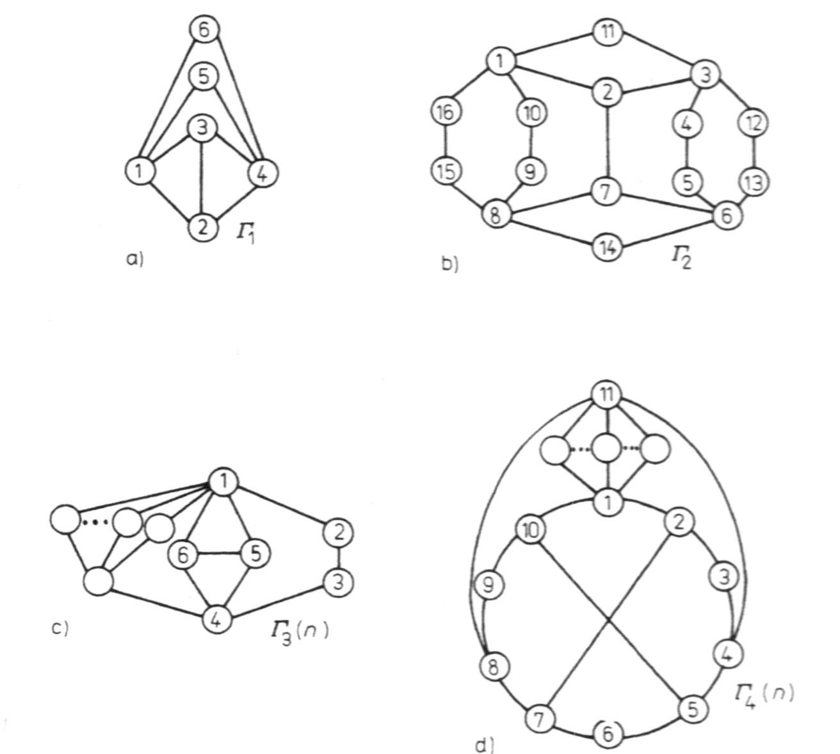


Fig. 3.11.2

**Theorem 3.11.7.** *Suppose that $k = 4$ or 5, and let $G$ be a 2-connected graph with at least $k + 1$ vertices, $G \neq \Gamma_1$. Suppose that $L$ and $L'$ are two distinct longest cycles meeting in a set $W$ of exactly $k$ vertices. Then $W$ is a separating set of $G$.*

For $k = 4$ Theorem 3.11.6 (ii) is not true, as is shown by the graph of Fig. 3.11.2b. The circumference of $\Gamma_2$ is 10. The cycle $L$ through the vertices 1, ..., 10 and the cycle $L'$ through 1, 11, 3, 12, 13, 6, 14, 8, 15, 16 meet in $W = \{1, 3, 6, 8\}$.

The graph $G$, however, has only seven $W$-bridges. One can of course ask what is the largest number $k$ with the property that if a graph $G$ contains two longest cycles meeting in a set $W$ of precisely $k$ vertices, then $W$ is a separating set of $G$.

In fact, $k = 4$ is best possible by Theorem 3.11.7. According to Theorem 3.11.7, for $k = 5$ the graph $\Gamma_1$ is the only counterexample: $\Gamma_1$ contains the cycles $L$ through 1, 2, 3, 4, 5 and $L'$ through 1, 3, 2, 4, 5 meeting in five vertices; their removal results in a connected graph.

For $k = 6$ there is an infinite set of counterexamples $\Gamma_3(n)$ with $n \geqq 8$ vertices shown in Fig. 3.11.2c. The cycles $L$ through 1, 2, 3, 4, 5, 6 and $L'$ through 1, 2, 3, 4, 6, 5 meet in six vertices; their removal results in a connected graph.

It seems, however, sensible to add to Theorem 3.11.7 the condition that the circumference of $G$ should be at least $k + 1$ (for $k \leqq 5$ this slightly weakens the theorem but this is the only interesting case). Then the graphs $\Gamma_1$ and $\Gamma_3(n)$ of Fig. 3.11.2a, c are no longer counterexamples.

Indeed M. Grötschel, 1984, conjectures that this version of Theorem 3.11.7 is also true for $k = 6$ or 7.

For $k = 8$ the Petersen graph $P$ is a counterexample. If $R$ and $S$ are two adjacent vertices of $P$, then $P - R$ (respectively $P - S$) contains a Hamiltonian cycle $L$ (respectively $L'$). The intersection of $L$ and $L'$ is the eight vertices of $P$ different from $R$ and $S$. These eight vertices do not form a separating set by construction. It is not known whether there are infinitely many graphs in which two longest cycles meet in a set of eight vertices which is not a separating set of $G$.

For $k = 9$ we present such a class. Consider Fig. 3.11.2d. The graph $\Gamma_4(n)$ shown therein has $n \geqq 12$ vertices and circumference 10. The cycles $L$ through the vertices 1, 2, ..., 10 and $L'$ through 2, 3, 4, 11, 8, 9, 10, 5, 6, 7 are longest cycles of length 10 and meet in the set $W = \{2, 3, ..., 10\}$. The graph $G - W$ consists of one component.

It seems to be clear that connectivity plays a role in the problem we are considering. M. Grötschel, 1984, initiates the investigation of the following two functions:

*For every integer $k \geqq 2$ let $f(k)$ denote the largest integer with the following property: if $G$ is a $k$-connected (non-Hamiltonian) graph and if two longest cycles of $G$ meet in at most $f(k)$ vertices, then these vertices form a separating set.*

*Let $f'(k)$ be defined as $f(k)$, making the additional assumption that the $k$-connected graph $G$ has circumference larger than $f'(k)$.*

Clearly, we have $f'(k) \geqq f(k)$. The results discussed above show that $f(2) = 4$, $f'(2) \geqq 5$. The Petersen graph shows that $f'(3) \leqq 7$. M. Grötschel conjectures that $f'(2) = f'(3) = 7$. It seems to be a very hard problem to determine the functions $f$ and $f'$. M. Grötschel, 1984, makes the following conjecture:

**Conjecture 3.11.8.** $f(k) \geqq k$. □

In other words, if $G$ is $k$-connected and two longest cycles meet in precisely $k$ vertices, then these vertices form a separating set of $G$.

Conjecture 3.11.8, if true, would imply the following result:

**Conjecture 3.11.9.** *In a $k$-connected graph two longest cycles meet in at least $k$ vertices.* □

This conjecture is presented by M. Grötschel, 1984, who states that, by a communication of R. Häggkvist and J. A. Bondy, Conjecture 3.11.9 was first conjectured in 1979 by Scott Smith (then a high-school student), and that the truth of the conjecture has probably been verified up to $k = 10$ (up to $k = 6$ it follows from Grötschel's Theorems 3.11.6 and 3.11.7). The general case is still unsettled.

S. Burr and T. Zamfirescu (private communication) proved the following:

**Theorem 3.11.10.** *Any two longest cycles of a $(k^2 - 4k + 4)$-connected graph have at least k common vertices, $k \geqq 4$. Moreover, if $4 < \alpha < 5$ and the connectivity of G is larger than $k^2 - \alpha k + \alpha - 1$, then any two longest cycles have at least k common vertices for sufficiently large k.* □

I suggest considering also the following function:

*Let $f''(k)$ be defined as $f(k)$, making the additional assumption that the two longest cycles in question are edge-disjoint.*

Another interesting problem is to determine the intersection pattern of two longest odd cycles or of a longest odd cycle and a longest even cycle.

Results on the intersection pattern of two longest paths can be derived from the intersection pattern of two longest cycles in the following way: Let $G$ be a connected graph and let $w$ and $w'$ be two longest paths of $G$ intersecting in a set $W$ of exactly $k$ vertices of $G$, $1 \leqq k \leqq 4$. Then $G + X$ contains the two longest cycles $w \oplus X$ and $w' \oplus X$ intersecting in the set $W \cup \{X\}$ of precisely $k + 1$ vertices ($G + X$ and $w \oplus X$ as defined in section 3.8). Thus the Theorems 3.11.4, 3.11.6 and 3.11.7 can be applied to $G + X$. This implies the following result:

**Theorem 3.11.11.** *Let G be a connected graph and let w and w′ be two longest paths of G intersecting in a set W of exactly k vertices of G, $1 \leqq k \leqq 4$. Then*

(a) *W is a separating set of G.*

(b) *In the case $k = 1$ each longest path of G passes through the unique vertex of W.*

(c) *In the case $k = 2$ each longest path of G passes through at least one vertex of W.* □

## 3.12. On intersections of at least three longest cycles

By Theorem 3.11.1 we know that any two longest cycles of a 2-connected graph have at least two common vertices. M. Grötschel, 1984, showed that if two longest cycles of a 2-connected graph have precisely two common vertices, then all longest cycles pass through them.

It can easily be shown that any two longest paths of a connected graph meet in a common vertex. Theorem 3.11.11 implies that if two longest paths of a 2-connected graph have precisely one common vertex, then all longest paths pass through it.

T. Gallai, 1968, asked: do all longest paths of a connected graph $G$ contain a common vertex $P$? Does there exist a vertex $P$ such that $G - P$ contains no longest path of $G$? H. Sachs, 1968, asked: do all longest cycles of a 3-connected graph pass through at least one of two fixed vertices $P$ and $Q$? Do there exist two vertices $P$, $Q$ such that $G - \{P, Q\}$ contains no longest cycle of $G$?

H. Walther, 1969, was the first to give an answer to Gallai's problem by presenting a (planar) connected graph no vertex of which is contained in all longest paths. The graph of Fig. 3.12.1a found by H. Walther (see H. Walther and H.-J. Voss, 1974) and T. Zamfirescu, 1974, independently, is the smallest known example. The graph of Fig. 3.12.1b, found by W. Schmitz, 1976/7, is the smallest known planar example.

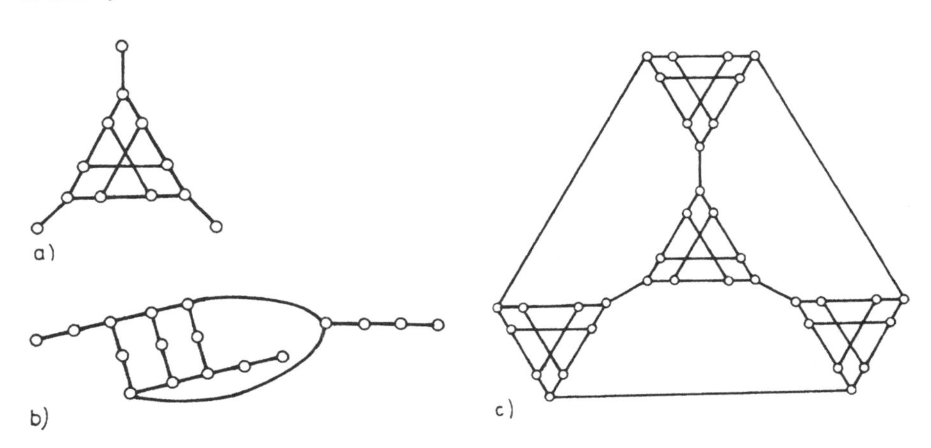


We know that any two longest paths of a connected graph meet in at least one common vertex and that there are connected graphs in which all longest paths have empty intersection. An unsolved problem due to T. Zamfirescu (private communication) is:

Do any three longest paths (cycles) of a connected (2-connected) graph have a common vertex? If yes, what is the largest integer $p$ (respectively $c$) such that any $p$ longest paths ($c$ longest cycles) of any connected (2-connected) graph have a common vertex?

The graph of Fig. 3.12.1b shows that $2 \leqq p \leqq 6$ and the graph of Fig. 3.12.1c found by T. Zamfirescu, 1974, as the smallest known 3-connected example for Gallai's problem shows that $2 \leqq c \leqq 8$. The graphs of Fig. 3.12.1b and 3.12.1c contain seven longest paths and nine longest cycles with an empty intersection, respectively.

In general we ask:

**Question 3.12.1.** *Does an integer $\alpha$ exist with the property that every 2-connected graph $G$ contains a set $A$ of $\alpha$ vertices such that each longest cycle of $G$ intersects $A$?*

**Question 3.12.2.** *Does an integer $\beta$ exist with the property that every connected graph $G$ contains a set $B$ of $\beta$ vertices such that each longest path of $G$ intersects $B$?*

We remark that a similar problem was solved by L. Lovász (unpublished): if every two cycles of a graph $G$ have a common vertex, then $G$ contains a set $A$ of at most three vertices such that each cycle intersects $A$. More generally (P. Erdős and L. Pósa, 1965, M. Simonovits, 1967, and H.-J. Voss, 1968): if among any $k$ cycles at least two intersect, then there is a set $A$ of at most $f(k)$ vertices such that each cycle intersects $A$, where $f(k)$ is an appropriate integer for each fixed $k$. More on these problems can be found in H. Walther and H.-J. Voss, 1974, and B. Bollobás, 1978a.

Now we ask for graphs with minimum vertex number having longest cycles or longest paths with empty intersection set.

T. Zamfirescu, 1972, has defined $C_k^j$ as the smallest integer $n$ such that there exists a $k$-connected graph $G$ of order $n$, with at least one cycle and with the property that for each $j$ of its vertices there is a longest cycle of $G$ avoiding these vertices. If there is no such $G$, we put $C_k^j = \infty$.

The number $P_k^j$ is defined by replacing "cycle" by "path" in the definition above. The analogous numbers for planar graphs are denoted by $\bar{P}_k^j$ and $\bar{C}_k^j$.

The result of H. Walther, 1969, can be formulated as $P_1^1 \leqq 12 \leqq \infty$. B. Grünbaum, 1974, W. Hatzel, 1979, W. Schmitz, 1975, 1976/7, C. Thomassen, 1974b, 1976, H. Walther, 1969, 1970 (see also H. Walther and H.-J. Voss, 1974), and T. Zamfirescu, 1972, 1974, 1975, 1976, subsequently determined and improved bounds for $C_k^j$, $\bar{C}_k^j$, $P_k^j$ and $\bar{P}_k^j$. These bounds are presented in Tables I and II (the capitals in brackets are the initial capitals of the names of the authors).

Table I

| $k$ \ $j$ | $P_k^j$: 1 | $P_k^j$: 2 | $P_k^j$: 3 | $\bar{P}_k^j$: 1 | $\bar{P}_k^j$: 2 | $\bar{P}_k^j$: 3 |
|---|---|---|---|---|---|---|
| 1 | $\leqq 12$ (WZ) | $\leqq 93$ (Z) | ? | $\leqq 17$ (S) | $\leqq 308$ (Z) | ? |
| 2 | $\leqq 32$ (Z) | $\leqq 270$ (Z) | ? | $\leqq 32$ (Z) | $\leqq 914$ (Z) | ? |
| 3 | $\leqq 36$ (Z) | $\leqq 270$ (Z) | ? | $\leqq 224$ (H) | $\leqq 26\,378$ (H) | ? |
| 4 | ? | ? | ? | $\infty$ | $\infty$ | $\infty$ |

Table II

| $k$ \ $j$ | $C_k^j$: 1 | $C_k^j$: 2 | $C_k^j$: 3 | $\bar{C}_k^j$: 1 | $\bar{C}_k^j$: 2 | $\bar{C}_k^j$: 3 |
|---|---|---|---|---|---|---|
| 1 | | $C_1^j = 3j + 3$ (T) | | | $\bar{C}_1^j = 3j + 3$ (T) | |
| 2 | $= 10$ (Z) | $\leqq 75$ (Z) | ? | $\leqq 15$ (T) | $\leqq 135$ (Z) | ? |
| 3 | $= 10$ (Z) | $\leqq 75$ (Z) | ? | $\leqq 57$ (H) | $\leqq 6\,758$ (H) | ? |
| 4 | ? | ? | ? | $\infty$ | $\infty$ | $\infty$ |

T. Zamfirescu, 1976, observed that $C_1^j \leqq \bar{C}_1^j \leqq 3j + 3$. He conjectured that $C_1^j = \bar{C}_1^j = 3j + 3$. C. Thomassen, 1976, proved this conjecture in the affirmative by showing that any graph $G$ with $n$ vertices contains a set $A$ of at most $\lfloor n/3 \rfloor$ vertices such that any longest cycle of $G$ intersects $A$.

W. T. Tutte's theorem, asserting that every 4-vertex-connected graph has a Hamiltonian cycle, implies that $\bar{P}_k^j = \bar{C}_k^j = \infty$ for $j \geqq 1$, $k \geqq 4$.

The non-existence has not been investigated for any other of the numbers under consideration. It would be of great interest to determine which of $P_4^1$, $C_4^2$, $P_1^3$, $C_2^3$, $\bar{P}_1^3$, $\bar{C}_2^3$ is finite. Also it seems particularly interesting to prove or disprove the existence of $P_k^1$ $(k \geqq 4)$.

# Chapter 4

# Isomorphic bridges in 2-connected graphs with circumference $l$

## 4.0. Introduction

It is well known that almost all graphs have the trivial automorphism group. Here we shall deal with the rare case, namely, with classes of graphs with high symmetries, i.e. with graphs which have so-called *isomorphic bridges* (symmetric subgraphs).

In graph classes in which certain parameters are (universally) bounded the graphs of large order have many isomorphic bridges. Here we study this phenomenon for the class $\mathfrak{Z}_l$ of all 2-connected graphs with circumference $l$. As M. Simonovits, 1974, showed the extremal graphs of many extremal graph theorems contain many isomorphic bridges. For instance in the famous theorem of Turán (P. Turán, 1954) on graphs of order $n$ containing no $K_{k+1}$ as a subgraph, the extremal graphs have $\lceil n/k \rceil$ independent vertices all having the same neighbourhood; these independent vertices are the kernels of $\lceil n/k \rceil$ isomorphic bridges.

The length of the longest cycles of the graphs of $\mathfrak{Z}_l$ is equal to a given integer $l$. That means that the lengths of all cycles are bounded by $l$. This restriction of the cycle lengths has the consequence that graphs of $\mathfrak{Z}_l$ of large order have high symmetries, i.e. they have many isomorphic bridges (symmetric subgraphs). This will be stated in Theorem 4.2.3. by H.-J. Voss (see H. Walther and H.-J. Voss, 1974) and Theorem 4.2.5. by M. Simonovits, 1974.

The subclass of all graphs of $\mathfrak{Z}_l$ which have no $p$ isomorphic bridges is finite. As a consequence of this an $n$-reduction can be introduced in $\mathfrak{Z}_l$, which was applied by K. Hauschild and W. Rautenberg, 1971, K. Hauschild, H. Herre and W. Rautenberg, 1972, and B. Dahn, 1974, in proving metamathematical theorems.

Other classes with the same symmetry properties will be investigated in Chapter 6.

At the end of this introduction we mention graphs without two or more isomorphic bridges which have precisely one inner vertex each.

First we make the following definition: two different vertices $X$, $Y$ of a graph $G$ with the same neighbourhood $N$ are called *symmetric*. Since $G$ is simple, $X$ and $Y$ are not adjacent. The vertices $X$ and $Y$ are kernels of two so-called *N-bridges* (for the general definition of isomorphic bridges see the definition in section 4.2).

D. P. Geoffrey and D. P. Sumner, 1978, and others investigated graphs *without* two or more *vertices having the same neighbourhood*, i.e. without two (pairwise) symmetric vertices. These graphs are called *point-determining*. For results the reader is referred to the paper of D. P. Geoffrey and D. P. Sumner, 1978, and to the bibliography of this paper.

## 4.1. The class of all 2-connected graphs with circumference at most 5

This section is devoted to the description of the class $\mathfrak{Z}_l$ for small values of $l$.

**Theorem 4.1.1.** *The classes $\mathfrak{Z}_3$, $\mathfrak{Z}_4$ and $\mathfrak{Z}_5$ consist of all graphs which are obtained from the graphs of Figure 4.1.1 by deleting* 0 *or* 1 *or* 2 *or more edges which are depicted in Fig. 4.1.1 by dotted lines.*

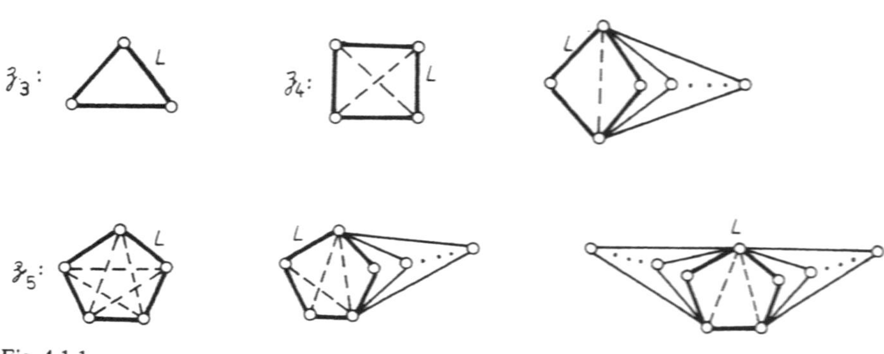


Fig. 4.1.1

*Proof.* Let $G \in \mathfrak{Z}_l$ and let $L$ denote a cycle of $G$ of length $l$. We consider two cases.

Case 1. Assume that $l = 3$. By Theorem 3.3.1 the length $\lambda(B)$ of each bridge $B$ of $L$ in $G$ is at most $\lfloor l/2 \rfloor = 1$. Since $L$ has no chords, the triangle $L$ has no bridges in $G$, and consequently $G = L$.

Case 2. Assume that $4 \leqq l \leqq 5$.

(1) *Each bridge $B$ of $L$ is an $L,L$-path of length* 1 *or* 2,

because by Theorem 3.3.1 the length $\lambda(B)$ of $B$ in $G$ is at most $\lfloor l/2 \rfloor = 2$. □

(2) *If two bridges of $L$ in $G$ overlap, then they are chords of $L$.*

*Proof of (2).* Assume that there are two overlapping bridges $B_1$, $B_2$ of $L$ in $G$. By (1) the bridges $B_1$, $B_2$ are $L,L$-paths of $L$. By Lemma 1.2.1 they are skew. Hence $B_1$ and $B_2$ have no common vertex. Theorem 3.3.4 implies that

$$\lambda(B_1) + \lambda(B_2) \leqq \lfloor l/2 \rfloor = 2 . \square$$

Now it is an easy task to determine all graphs consisting of a circuit $L$ of length 4 or 5 and bridges of $L$ satisfying (1) and (2). □

In H. Walther and H.-J. Voss, 1974, all graphs of $\mathfrak{Z}_6$ and $\mathfrak{Z}_7$ are characterized; in H.-J. Voss, 1973, the characterization of $\mathfrak{Z}_8$ and $\mathfrak{Z}_9$ is given.

## 4.2. Isomorphic bridges and symmetric subgraphs

If we consider the graphs of $\mathfrak{Z}_4$ and $\mathfrak{Z}_5$, then we observe that each graph of order at least 6 contains two bridges (of a longest cycle $L$) of the same structure, namely, two independent vertices with the same neighbourhood in $G$. A similar effect can be observed in $\mathfrak{Z}_6$ and $\mathfrak{Z}_7$ (H. Walther and H.-J. Voss, 1974).

In order to study this problem with respect to $\mathfrak{Z}_l$ for general $l$ we introduce the concept of isomorphic bridges.

**Definition.** Let $G$ be a graph and $H$ a subgraph of it. Two regular bridges $B_1$ and $B_2$ of $H$ with kernels $K_1$ and $K_2$, respectively, are said to be *isomorphic* if and only if $B_1 = B_2$ or there is an isomorphism $\varphi$ from $G - K_2$ onto $G - K_1$ such that $G - K_1 - K_2$ is fixed, i.e. for each vertex $X$ of $G - K_1 - K_2$ we have $\varphi(X) = X$ (see Fig. 4.2.1).

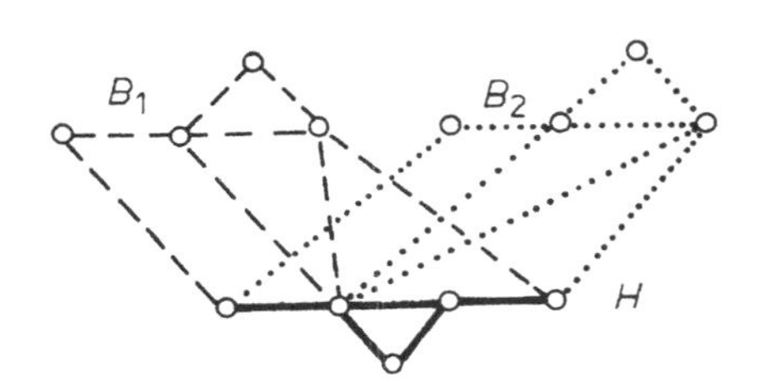


Fig. 4.2.1

Two bridges $B_1$, $B_2$ are called *isomorphic* if and only if they have the same set $H$ of vertices of attachment and they are isomorphic with respect to $H$. The bridges $B_1, B_2, \ldots, B_s$ are said to be *isomorphic* if and only if $B_i$ and $B_j$ are isomorphic for every $1 \leqq i < j \leqq s$. The kernels $K_1$ and $K_2$ of isomorphic bridges $B_1$ and $B_2$ are said to be *symmetric*. This leads to the following definition.

**Definition.** Two connected induced subgraphs $U_1$ and $U_2$ of a graph $G$ are called *symmetric* if and only if they are the kernels of two isomorphic bridges. $U_1, U_2, \ldots, U_s$ are symmetric if and only if the subgraphs $U_i$ and $U_j$ are symmetric for every $1 \leqq i < j \leqq s$.

From this definition we can deduce the following equivalent characterization of symmetric subgraphs.

**Proposition 4.2.1.** *Two connected induced subgraphs $U_1$ and $U_2$ of a graph $G$ are symmetric if and only if $U_1 = U_2$ or $U_1 \neq U_2$ and*

1) *$U_1$, $U_2$ are disjoint,*
2) *no vertex of $U_1$ has a neighbour in $U_2$,*
3) *there is an isomorphism $\varphi$ from $U_1$ onto $U_2$ such that for any vertices $X \in U_1$ and $Y \in G - U_1 - U_2$ the vertex $X$ is joined to $Y$ if and only if $\varphi(X)$ is joined to $Y$.*

In the definition of the concept of isomorphic bridges $B_1$, $B_2$ of $H$, the subgraph $H$ can be chosen as the set of vertices of attachment of $B_1$ and $B_2$.

Since by Proposition 1.1.1 the kernels of bridges are connected, we have the following result.

**Proposition 4.2.2.** *The relation "symmetry" is an equivalence relation in the class of all subgraphs of a graph $G$.*

*The same is true for the relation "isomorphism" in the class of the corresponding bridges in the graph $G$.*

The main result of section 4.2 is the following:

**Theorem 4.2.3.** *Let $p$, $l$ be integers with $p \geqq 2$, $l \geqq 3$. Then there is an integer $n(p, l)$ such that every graph $G \in \mathfrak{Z}_l$ of order greater than $n(p, l)$ contains $p$ isomorphic bridges.*

**Remark 4.2.4.** *The assertion of Theorem 4.2.3 is not true for the class of all connected graphs with circumference $l$.*

For instance for each integer $k$ the chain of $2k + 1$ triangles as depicted in Fig. 4.2.2 has circumference 3 and contains no isomorphic bridges.

In section 4.6 we shall prove a more general theorem. It will be formulated here in terms of symmetric subgraphs.

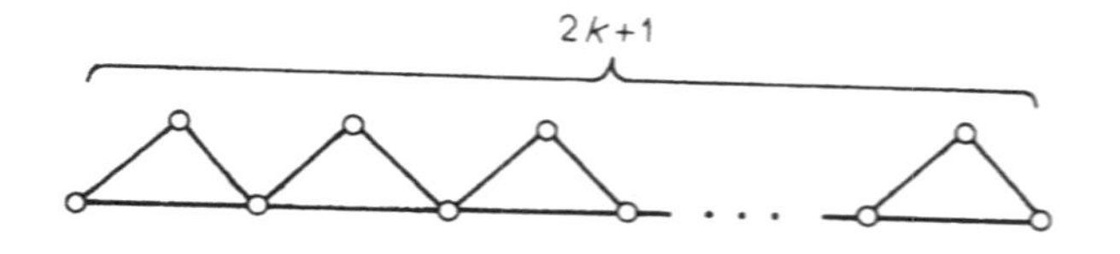


Fig. 4.2.2

**Theorem 4.2.5.** *Let $l$ be a given integer, suppose that $\varepsilon > 0$, and let $\{m_k\}$, $m_k \geqq 2$, be an arbitrary sequence of positive integers. Then there exists an integer $N_0$ such that every graph $G \in \mathfrak{Z}_l$ of order $n > N_0$ contains a family of subgraphs $S_{i,j}$, $i = 1, 2, \ldots$, $j = 1, 2, \ldots$, with the following properties:*

(i) *No two of them have vertices in common.*

(ii) *For every fixed i, the subgraphs $S_{i,j}$ are symmetric in G.*
*If $n(S_{i,j}) = k_i$, then $j = 1, 2, \ldots, m_{k_i}$.*

(iii) $\sum \sum n(S_{i,j}) > n - \varepsilon n$.

Note: $n(S_{i,j}) = n(S_{i',j})$, $i \neq i'$, implies that $k_i = k_{i'}$ and $m_{k_i} = m_{k_{i'}}$. Further, it may happen that these two classes of symmetric subgraphs $S_{i,1}, \ldots, S_{i,m_{k_i}}$ and $S_{i',1}, \ldots, S_{i',m_{k_{i'}}}$ form together a larger class of symmetric graphs.

The proof of Theorem 4.2.5 will be given in section 4.6 by deriving it from the corresponding elegant Theorem 4.6.2 of M. Simonovits, 1974, on graphs containing no path of length greater than $\lambda$.

*Proof of Theorem 4.2.3.* The theorem will be proved by induction on the circumference $l$. For $3 \leqq l \leqq 5$ the validity of Theorem 4.2.3 follows from Theorem 4.1.1.

Suppose that Theorem 4.2.3 is true for all $l \in \{3, \ldots, l_0 - 1\}$. Let $G$ be any graph of $\mathfrak{Z}_{l_0}$ which does not contain $p$ isomorphic bridges. We shall show that there exists an integer $n(p, l_0)$ depending only on $p$ and $l_0$ such that

(*) $\qquad n(G) \leqq n(p, l_0)$.

Since $G$ is an arbitrary graph of $\mathfrak{Z}_{l_0}$ without $p$ isomorphic bridges assertion (*) is true for all $G \in \mathfrak{Z}_{l_0}$ with this property.

Now we start proving (*). Let $L$ denote a largest cycle of $G$. Clearly, $L$ has length $l_0$. Corollary 3.3.3 implies the following two assertions.

(1) *Each bridge $B$ of $L$ consists of at most $l_0/2$ blocks.* □

(2) *Each block of $B$ has at most $l_0/2$ vertices which are cut vertices or vertices of attachment of $B$.* □

By Theorem 3.6.1 each bridge $B$ of $L$ has circumference at most $l_0 - 2$. Hence each block $H$ of $B$ is a $K_2$ or $H \in \mathfrak{Z}_i$ for an index $i$ with $3 \leqq i \leqq l_0 - 2$.

(3) *No block $H$ of $B$ contains at least $p + \frac{l_0}{2}$ isomorphic bridges.*

*Proof of (3).* By (2) the block $H$ of $B$ has at most $l_0/2$ vertices that are cut vertices or vertices of attachment of $B$. Consequently, if $H$ has at least $p + l_0/2$ isomorphic bridges, then at least $p$ of them are also isomorphic in $G$. Contradiction! □

By induction the order of $H$ is

$$(4) \qquad n(H) \leqq n\left(p + \left\lfloor \frac{l_0}{2} \right\rfloor, l_0 - 2\right)$$

($n(p, l)$ is a monotonically increasing function in its two variables $l$ and $p$ with $3 \leqq l \leqq l_0 - 1$ and $p \geqq 2$). By (1) the bridge $B$ consists of at most $l_0/2$ blocks. Therefore, the order of $B$ is

$$(5) \qquad n(B) \leqq \frac{l_0}{2} n\left(p + \left\lfloor \frac{l_0}{2} \right\rfloor, l_0 - 2\right).$$

A bridge $B$ of the cycle $L$ is known if its structure and its intersection with $L$ are known. Therefore, we consider $B \cup L$. By (5) it follows that

$$(6) \qquad n(B \cup L) \leqq l_0 + \frac{l_0}{2} n\left(p + \left\lfloor \frac{l_0}{2} \right\rfloor, l_0 - 2\right).$$

By (6) the number $\alpha(l_0)$ of all possible pairwise not isomorphic bridges of the cycle $L$ (of length $l_0$) is not larger than the number of all graphs on $l_0 + \frac{l_0}{2} n\left(p + \left\lfloor \frac{l_0}{2} \right\rfloor, l_0 - 2\right)$ vertices, i.e.

$$(7) \qquad \alpha(l_0) \leqq 2^{\binom{c}{2}} \quad \text{with} \quad c = l_0 + (l_0/2)\, n(p + \lfloor l_0/2 \rfloor, l_0 - 2).$$

If $\alpha_p(l_0)$ denotes the number of all possible bridges $B$ of $L$ such that no $p$ of them are isomorphic, then

$$(8) \qquad \alpha_p(l_0) = (p-1)\, \alpha(l_0).$$

Each of these bridges $B$ satisfies (5). Consequently, each graph $G \in \mathfrak{Z}_{l_0}$ which does not contain $p$ isomorphic bridges has an order at most

$$n(p, l_0) =_{\text{def}} l_0 + \alpha_p(l_0)\,(l_0/2)\, n(p + \lfloor l_0/2 \rfloor, l_0 - 2). \; \square$$

The proof of Theorem 4.2.5 does not provide a good bound for the graphs of $\mathfrak{Z}_l$ which have no $p$ isomorphic bridges.

## 4.3. A simple property of bridges in 2-connected graphs

In sections 4.3 and 4.4 we shall prove some auxiliary results on bridges which will be needed in the proof of our next main theorem, i.e. Theorem 4.5.1 in section 4.5.

In section 4.3 we shall prove the case $n = 2$ of the following statement: the union of $n$ (not necessarily isomorphic) bridges in an $n$-connected graph is again $n$-connected. We start with Whitney's well-known construction of 2-connected graphs from cycles (H. Whitney, 1932a; see also G. A. Dirac, 1967).

**Whitney's Construction.** *$H$ is a 2-connected graph if and only if $H = H_1 \cup \ldots \cup H_r$, where $H_1$ is a circuit and $H_{i+1}$ is a path having precisely its end vertices in $H_1 \cup \ldots \cup H_i$.* □

We deal with this construction in sections 8.0, 2⁰, and 8.6. Next we prove the following result:

**Proposition 4.3.1.** *Let $B_1$, $B_2$ be two different bridges of a subgraph of a 2-connected graph having the same set of vertices of attachment. Then the graph $B_1 \cup B_2$ is 2-connected.*

*Proof.* Since $B_1$, $B_2$ have the same set of vertices of attachment and $B_1 \neq B_2$, the kernels $K_1$, $K_2$ of $B_1$, $B_2$, respectively, are disjoint and no vertex of $K_1$ is adjacent to a vertex of $K_2$.

Obviously the bridge $B_i$ contains a tree $T_i$ whose end vertices are precisely the vertices of attachment of $B$. With Whitney's construction it can easily be shown that $T_1 \cup T_2$ is 2-connected. Let $B_1$ contain an edge $e \notin T_1 \cup T_2$. Since all vertices of attachment of $B_1$ are in $T_1$, there is a path $p$ of $B_1$ through $e$ having exactly its end vertices in $T_1 \cup T_2$. Then $T_1 \cup T_2 \cup p$ is 2-connected. Introducing recursively all edges of $B_1 \cup B_2$ which are not in $T_1 \cup T_2$ we obtain a sequence of 2-connected subgraphs of $B_1 \cup B_2$ which also includes $B_1 \cup B_2$ itself. □

## 4.4. Properties of bridges in graphs of $\mathfrak{Z}_l$

In this section we shall deal with two properties of large classes of (isomorphic) bridges in graphs of $\mathfrak{Z}_l$. These properties will be applied to the proof of the reduction theorem in section 4.5, but they also have a certain interest of their own.

The first property is true for the wider class of all (connected or disconnected) graphs with circumference $l$.

**Property 4.4.1.** *Let $G$ be a graph with circumference $l$. If $\lfloor l/2 \rfloor + 1$ pairwise distinct bridges have the same set $S$ of vertices of attachment, then*

$$|S| \leqq \lfloor l/2 \rfloor.$$

*Proof.* Assume that there are $s =_{\text{def}} \lfloor l/2 \rfloor + 1 \geqq 2$ pairwise distinct bridges $B_1, B_2, \ldots, B_s$. Let $A_1, A_2, \ldots, A_s$ denote $s$ common vertices of attachment. Let $p_i$ denote an $L,L$-path in $B_i$ with ends $A_i$ and $A_{i+1}$ for all $1 \leqq i \leqq s$ (indices taken modulo $s$). Then $p_1 \cup p_2 \cup \ldots \cup p_s$ is a cycle of length at least $2s = 2\lfloor l/2 \rfloor + 2 \geqq l + 1$. Contradiction! □

**Definition.** A bridge $B$ *contains a bridge* $B'$ if and only if the kernel $K$ of $B$ contains the kernel $K'$ of $B'$.

**Property 4.4.2.** *Let $G \in \mathfrak{Z}_l$ and let $L$ be a longest cycle of $G$. Let $B_1, B_2, \ldots, B_s$ be $s = \lfloor l/2 \rfloor + 1$ pairwise distinct isomorphic bridges. Then either one bridge $B_i$ is a bridge of $L$ or all bridges $B_1, \ldots, B_s$ belong to the same bridge of the cycle $L$.*

*Proof.* First we shall show that

(1) *There exists an index i such that no inner vertex of $B_i$ is on L.*

*Proof of (1).* Assume that for all $1 \leqq i \leqq s$ an inner vertex $X_i$ of $B_i$ belongs to $L$. Since for all $1 \leqq i < j \leqq s$ the kernels of $B_i$ and $B_j$ are disjoint and are not joined by an edge, $\{X_1, \ldots, X_s\} \subseteq L$ is an independent set of vertices in $G$. Therefore, $L$ has length at least $2s = 2\lfloor l/2 \rfloor + 2 \geqq l + 1$. Contradiction! □

Let $B_1$ denote a bridge no inner vertex of which is on $L$. Therefore, the bridge $B_1$ is completely contained in a bridge $B$ of $L$. If $B_1 = B$, then the proof is complete. Suppose that $B_1 \neq B$. Then an inner vertex $P$ of $B$ is a vertex of attachment of $B_1$. Assume that an inner vertex of $B_2$ is on $L$. Obviously, the vertex $P$ of attachment of $B_1$ is also a vertex of attachment of $B_2$.

Let $Q$ denote an inner vertex of $B_2$ which is adjacent to $P$. Since an inner vertex of $B_2$ is on $L$, there is a path $w$ in the kernel of $B_2$ joining $Q$ to a vertex $R$ of $L$ so that $w$ and $L$ have only the vertex $R$ in common. Thus

(2) *$R \in L$ is an inner vertex of $B_2$.* □

By Theorem 3.6.1 the bridge $B_1$ has circumference at most $l - 2$. Since $B_1$ and $B_2$ are isomorphic bridges, $B_2$ also has circumference at most $l - 2$. Therefore, $L$ is not contained in $B_2$. From this and from assertion (2) we deduce that there is an arc $a$ of $L$ with ends $R$ and $S$ such that $a - \{S\}$ belongs to the kernel of $B_2$ and $S$ is a vertex of attachment of $B_2$ (see Fig. 4.4.1). $S$ is also a vertex of attachment of the bridge $B_1$. The path $p =_{\text{def}} (P, Q) \cup w \cup a$ is a $P,S$-path of $B_2$ of which only the end vertices are vertices of attachment of $B_2$.

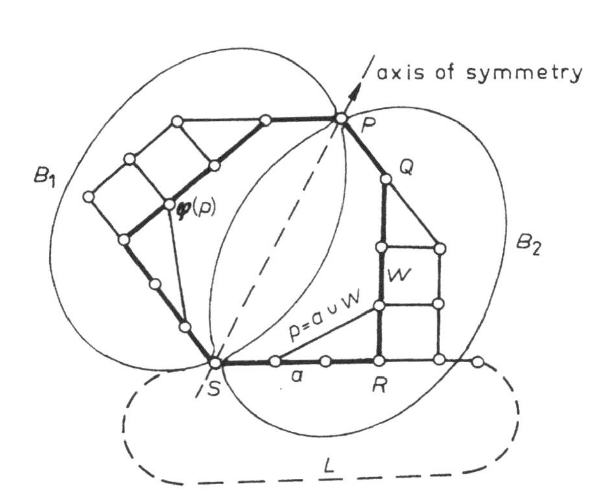


Fig. 4.4.1

Let $\varphi$ denote a bridge isomorphism from $B_2$ onto $B_1$, i.e. an isomorphism from $G - K_1$ onto $G - K_2$ with fixed $G - K_1 - K_2$, where $K_i$ denotes the kernel of $B_i$ for $i = 1, 2$. Then $L \triangle (p \cup \varphi(p))$ is a cycle of length at least $l + 2$. Contradiction!

Thus $B_1, \ldots, B_s$ are contained in $B$. □

## 4.5. $n$-reductions of graphs in the class $\mathfrak{Z}_l$

After defining the concept of $n$-reduction we present a theorem on $n$-reductions of graphs in the class $\mathfrak{Z}_l$ together with its proof.

**Definition.** Let $H$ be a subgraph of a graph $G$ and $\mathfrak{M}$ the set of all bridges of $H$. By Proposition 4.2.2 the relation "isomorphic" is an equivalence relation in $\mathfrak{M}$. Hence $\mathfrak{M}$ can be partitioned into equivalence classes. The following operation on $G$ is called an *elementary n-reduction* of $G$: we label $n$ (isomorphic) bridges in each equivalence class and delete the kernels of all unlabelled bridges of $\mathfrak{M}$ from the graph $G$. An elementary $n$-reduction of $G$ is called *proper* if at least one vertex is deleted from $G$. An $n$-reduction is the recursive application of a sequence of elementary $n$-reductions of a graph $G$. A graph $G$ on which a proper elementary $n$-reduction cannot be applied is called *n-irreducible*, i.e. $G$ does not contain $n+1$ isomorphic bridges.

With these notions we can reformulate Theorem 4.2.3:

*For each pair of integers $p \geqq 2$, $l \geqq 3$ there is an integer $n(p, l)$ such that every p-irreducible graph $G \in \mathfrak{Z}_l$ has order at most $n(p, l)$.*

Now we formulate the main theorem on $n$-reductions of graphs of $\mathfrak{Z}_l$.

**Theorem 4.5.1.** *Let $l$ and $n$ be integers with $l \geqq 3$ and $n \geqq \lfloor l/2 \rfloor + 1$. Then*

(i) *every graph $G \in \mathfrak{Z}_l$ can be transformed into an n-irreducible graph by at most $\lfloor l/2 \rfloor - 1$ elementary n-reductions;*

(ii) *the n-irreducible graph obtained from a graph $G \in \mathfrak{Z}_l$ by n-reductions is (up to isomorphism) uniquely determined and is again a graph of $\mathfrak{Z}_l$;*

(iii) *the number of n-irreducible graphs of $\mathfrak{Z}_l$ is finite.*

The bound $\lfloor l/2 \rfloor - 1$ in assertion (i) cannot be improved. For $l = 6$ this is shown by the following example. Similar examples can be constructed for all $l$.

Let $U_i$ be a graph with vertex set $\{P, X^i, Y_1^i, Y_2^i, \ldots, Y_i^i, Q\}$ and edge set $\{(P, X^i)\} \cup \{(X^i, Y_j^i) \mid 1 \leqq j \leqq i\} \cup \{(Y_j^i, Q) \mid 1 \leqq j \leqq i\}$, where $U_i \cap U_k = \{P, Q\}$ for all $1 \leqq i < k$ (see Fig. 4.5.1). The graph $U_i$ can be interpreted as a bridge of $(\{P, Q\}, \emptyset)$. Let $H_m =_{\text{def}} U_4 \cup U_5 \cup \ldots \cup U_m$ $(m \geqq 4)$.

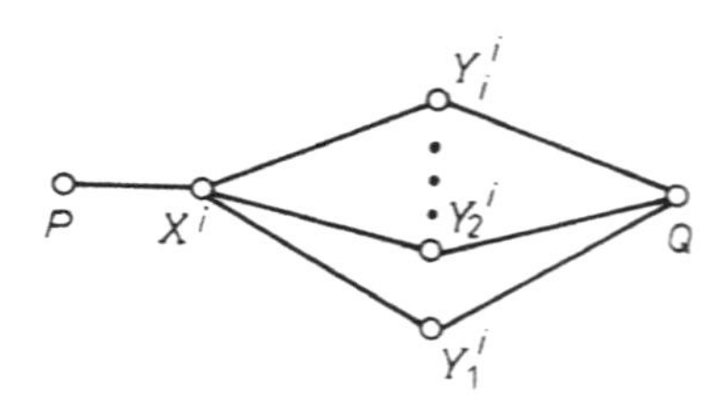


Fig. 4.5.1

Let $U_4^1, U_4^2, U_4^3$ be three distinct bridges of $(\{P, Q\}, \emptyset)$ which are isomorphic to $U_4$, where $U_4^k \neq U_4$ for all $k = 1, 2, 3$. Let $H' =_{\text{def}} U_4 \cup U_4^1 \cup U_4^2 \cup U_4^3$. If $m \geqq 9$, then $H'$ is up to isomorphism the unique 4-irreducible graph which can be obtained by 4-reduction from $H_m$. Obviously, each such 4-reduction can only be accomplished by at least two elementary 4-reductions. The assertion (ii) is not true if $n \leqq \lfloor l/2 \rfloor - 1$. This will be shown by the following example:

Let $l =_{\text{def}} 2s$ be an even integer and let $\Gamma$ denote the graph obtained from the circuit $L =_{\text{def}} [X_1, Y_1, X_2, Y_2, \ldots, X_s, Y_s, X_1]$ by joining each vertex $Y_i$ to all vertices of $L - \{Y_i\}$ for all $1 \leqq i \leqq s$. Then the vertices $X_1, \ldots, X_s$ are kernels of isomorphic bridges of the subgraph induced by $\{Y_1, \ldots, Y_s\}$. Hence by an $n$-reduction of $\Gamma$ with $n \leqq \lfloor l/2 \rfloor - 1 = s - 1$ the largest circuit $L$ will be destroyed. Note that $\Gamma = I_s + K_s$.

*Proof of Theorem 4.5.1.* The theorem will be proved by induction on $l$.

Theorem 4.1.1 implies the validity of Theorem 4.5.1 for $3 \leqq l \leqq 5$. Suppose that Theorem 4.5.1 is true for all $l \in \{3, \ldots, l_0 - 1\}$. We consider an arbitrary graph $G \in \mathfrak{Z}_{l_0}$. Let $L$ denote a longest cycle of $G$. By Property 4.4.2 each elementary $n$-reduction can be considered in the bridges of $L$. By Property 4.4.2 each elementary $n$-reduction reduces bridges of $L$ and/or bridges in blocks of bridges of $L$. Theorem 3.6.1 implies that the blocks of every bridge of $L$ have circumference at most $l - 2$. Consequently, by induction the assertions of Theorem 4.5.1 are valid for all these blocks, where the $n$-reductions in a block $B$ of a bridge $B$ of $L$ are applied only to isomorphic bridges of $B$ which are also isomorphic in $G$. Now the proof of Theorem 4.5.1 can readily be completed. □

A more detailed proof is in H. Walther and H.-J. Voss, 1974; a further proof can be found in R. Halin, 1981.

## 4.6. Graphs without paths of lengths greater than $\lambda$

This section is concerned with the proof of Theorem 4.2.5 and the extension of the results of sections 4.1.-4.5 to the class $\mathfrak{Y}_\lambda$ of all connected graphs with maximal path length $\lambda$. The proof of Theorem 4.2.5 is based on M. Simonovits's Theorem 4.6.2.

Let $X$ be a vertex not in $G$. Let $G + X$ be derived from $G$ by adding $X$ and joining $X$ to all vertices of $G$ by edges (see section 3.8).

If $G \in \mathfrak{Y}_\lambda$, then obviously $G + X \in \mathfrak{Z}_{\lambda+2}$. Thus all the results of sections 4.1-4.5 can be extended to the class $\mathfrak{Y}_\lambda$. For $\mathfrak{Y}_\lambda$ we only formulate those theorems that correspond to the Theorems 4.2.3 and 4.2.5.

Since Theorem 4.2.3 is valid for $\mathfrak{Z}_{\lambda+2}$, for $\mathfrak{Y}_\lambda$ it yields the following result:

**Theorem 4.6.1.** *Let $p$, $\lambda$ be integers with $p \geqq 2$, $\lambda \geqq 1$. Then there is an integer $n(p, \lambda + 2)$ such that every graph $G \in \mathfrak{Y}_\lambda$ of order at least $n(p, \lambda + 2)$ contains $p$ isomorphic bridges.*

In the following we shall show that this assertion is also true for graphs which are not connected. This means that the class of all graphs with maximal path length $\lambda$ has somewhat different properties than the class of all graphs with maximal circuit length $l$.

Next we shall prove the stronger Theorem 4.6.2 of M. Simonovits (as Lemma 3.2.1 in M. Simonovits, 1974).

**Theorem 4.6.2.** *Let $\lambda$ be a given integer, suppose that $0 < \varepsilon < 1$, and let $\{m_k\}$, $m_k \geqq 2$, be an arbitrary sequence of positive integers. Then there exists an integer $N_0$ such that each graph $G$ of order greater than $N_0$ without paths of length greater than $\lambda$ contains a family of subgraphs $S_{i,j}$, $i = 1, 2, \ldots$, $j = 1, 2, \ldots$, with the following properties:*

(i) *No two of them have vertices in common.*

(ii) *For every fixed $i$, the subgraphs $S_{i,j}$ are symmetric in $G$.*
*If $n(S_{i,j}) = k_i$, then $j = 1, 2, \ldots, m_{k_i}$.*

(iii) $\sum \sum n(S_{i,j}) > n - \varepsilon n$.

*Proof.* Here we follow the proof given by M. Simonovits, 1974. The theorem holds trivially for $\lambda = 1$ and $\lambda = 2$. We apply induction on $\lambda$.

Let us suppose that the theorem holds for all $1 \leqq \lambda \leqq \lambda_0 - 1$. For $\lambda_0$ we again write $\lambda$. According to the hypothesis, we determine $\hat{N}_0$ corresponding to $\lambda - 1$, $\varepsilon/2$ and $\hat{m}_k = m_k t 2^{k\lambda}$, where $t$ will be specified later.

Now let $G$ be a graph of order $n$ containing no path of length $\lambda + 1$. We divide the components of $G$ into three classes $C_1$, $C_2$, $C_3$, where $n_k$ denotes the number of vertices of the components in the class $C_k$ $(k = 1, 2, 3)$.

*Class* $C_1$. $C_1$ is the set of all components of order greater than $\hat{N}_0$ containing no path of length $\lambda$.

By the induction hypothesis the graph induced by the components of $C_1$ contains a family of subgraphs $S_{i,j}$ satisfying (i)–(iii), where, of course, $n$ must be replaced by $n_1$, i.e.

(1) $\left(1 - \dfrac{\varepsilon}{2}\right) n_1$ *vertices of components of* $C_1$ *belong to symmetric graphs of the desired type.* □

*Class* $C_2$. $C_2$ is the set of all components of order at most $\hat{N}_0$.

(2) *If* $\mu$ *is the number of all pairwise non-isomorphic graphs of order at most* $\hat{N}_0$, *then at least*

$$n_2 - \hat{N}_0 \mu \max_{k \leqq \hat{N}_0} m_k = n_2 - O(1)$$

*vertices of components of* $C_2$ *belong to symmetric graphs of the desired type.*

*Proof of (2).* We select at least $m_{k_1}$ isomorphic components of $k_1$ vertices from $C_2$ (if we can), then at least $m_{k_2}$ isomorphic components of $k_2$ vertices, and so on. If finally we cannot select at least $m_{k_t}$ isomorphic components of $k_t$ vertices from $C_2$, then there remain fewer than $\mu \max_{k \leqq \hat{N}_0} m_k$ components. Now the only thing we have to notice is that isomorphic components are symmetric subgraphs. □

*Class* $C_3$. $C_3$ is the set of all components of order greater than $\hat{N}_0$ containing a path $p_\lambda$ of length $\lambda$.

(3) $(1 - \varepsilon)\, n_3$ *vertices of components of* $C_3$ *belong to symmetric graphs of the desired type.*

*Proof of (3).* Let $U \in C_3$ and $u =_{\text{def}} n(U)$. Then $U$ contains a path $p_\lambda$ of length $\lambda$. If $U - p_\lambda$ contains another path of length $\lambda$, then $U$ would contain a path of length $\lambda + 1$ (see Theorem 3.8.1). Therefore, $U - p_\lambda$ does not contain a path of length $\lambda$ and we can apply the induction hypothesis. Let $T_{i,j}$ be symmetric subgraphs of $U - p_\lambda$ for every fixed $i$ and $j = 1, \ldots, \hat{m}_{k_i}, \ldots$ Sometimes they are not symmetric in $G$ because they are joined to $p_\lambda$ in different ways. However, among any $2^{k_i\lambda} m_{k_i}$ subgraphs $T_{i,j}$ (for fixed $i$) there exist $m_{k_i}$ subgraphs joined to $p_\lambda$ in the same way. These subgraphs are also symmetric in $G$.

Therefore, one can find at least $2^{k_i\lambda}(t - 1)\, m_{k_i}$ subgraphs among the considered $2^{k_i\lambda} t m_{k_i}$ ones forming $2^{k_i\lambda}(t - 1)$ sets of $m_{k_i}$ symmetric subgraphs of $U$. For $i = 1, 2, \ldots$, the total number of vertices of these graphs will be

$$(1 - \varepsilon/2)(u - \lambda)(1 - 1/t) > (1 - \varepsilon)\, u$$

if $t$ and $N_0$ are large enough. □

(1), (2) and (3) imply that the resulting systems of symmetric graphs contain at least

$$(1-\varepsilon)\,n_1 + n_2 - O(1) + (1-\varepsilon)\,n_3 \geqq (1-\varepsilon)\,n$$

vertices altogether if $n$ is large enough. □

*Proof of Theorem 4.2.5.* In order to prove the validity of Theorem 4.2.5 we apply an improved version of a result of G.A. Dirac, 1952a (H.-J. Voss, 1973), namely Corollary 3.7.3:

**Lemma 4.6.3.** *In every 2-connected graph with circumference $l$ each path has length at most* $\lfloor l^2/4 \rfloor$. □

By this lemma $\mathfrak{Z}_l \subseteq \bigcup_{i=1}^{\lfloor l^2/4 \rfloor} \mathfrak{D}_i$. This implies that the graphs of $\mathfrak{Z}_l$ satisfy the conditions of M. Simonovits's Theorem 4.6.2 and consequently the conditions of Theorem 4.2.5. □

# Chapter 5

# The reconstruction of graphs having many isomorphic bridges

## 5.0. Introduction

There are deep and difficult questions concerning isomorphisms of graphs, the best known being the reconstruction conjecture, first formulated by P. J. Kelly and S. M. Ulam in 1942 and published by P. J. Kelly, 1957. A survey of the reconstruction problem is given by C. St. J. A. Nash-Williams, 1978, and W. T. Tutte, 1984. The notation is taken from W. T. Tutte, 1984.

Let $G$ be a graph with vertices $V_1, \ldots, V_k$. The subgraphs $G_j =_{\text{def}} G - V_j$ are called *primal subgraphs.*

**Reconstruction Conjecture.** *Let the isomorphism classes of the k primal subgraphs of G be given. Then, if $k \geqq 3$, the isomorphism class of G is uniquely determined.* □

Since two primal subgraphs $G_i$, $G_j$, $i \neq j$, can be isomorphic, some isomorphism classes appear with a certain multiplicity.

Up to now the reconstruction conjecture has not been proved or disproved. It has only been solved for certain classes of graphs.

In the following the recognizability and reconstructibility of graphs with a property E will be considered. The graphs with property E are *recognizable* if and only if E can be inferred from the isomorphism classes of $G_j$. The graph $G$ is *reconstructible* if and only if the isomorphism class of $G$ can be inferred from the isomorphism classes of $G_j$. Let us give two examples for these concepts. P. J. Kelly, 1957, proved the following statement:

*Trees are reconstructible.* □

For 2-connected graphs the reconstruction problem could not be solved until now. The following result can easily be proved:

*2-connected graphs are recognizable.* □

Since the reconstruction problem for 2-connected graphs is as yet unsolved, we ask whether 2-connected graphs with certain properties are reconstructible.

Here the reconstruction problem will be investigated for (2-connected) graphs which have many "symmetries", i.e. they have many symmetric subgraphs.

The smallest subgraphs consist of exactly one vertex. Obviously, two such subgraphs are symmetric if and only if the corresponding non-adjacent vertices have the same neighbourhood.

A class of symmetric subgraphs always means an equivalence class. W. Dörfler and W. Imrich proved (W. Dörfler, 1972, W. Dörfler and W. Imrich, 1972) the following result:

*If a graph $G$ contains at least one class of at least $k$ symmetric vertices, $k \geqq 2$, and no class with precisely $k-1$ symmetric vertices, then $G$ is reconstructible.* □

The result of W. Dörfler and W. Imrich is a theorem on symmetric subgraphs of order 1. The main result of Chapter 5 is the extension of this result to arbitrary orders $n$ (Theorem 5.3.2). We include this result in this book because it gives a deeper insight into the structure of graphs with many bridges.

## 5.1. Some theorems concerning the isomorphic reconstruction conjecture

In section 5.1 we present three well-known results relating to the reconstruction conjecture.

**Theorem 5.1.1.** *2-connected graphs are recognizable.*

*Proof.* If a graph $G$ of order at least 3 is 2-connected, then all its primal subgraphs are connected. If $G$ is not 2-connected, then at least one primal subgraph is disconnected. □

Without proof we state the following result (see W. T. Tutte, 1984):

**Theorem 5.1.2.** *Graphs having no spanning block are recognizable. For each such graph the number of blocks in any given isomorphism class is reconstructible.* □

In section 3 we shall apply a lemma of P. J. Kelly, 1957. If $H$ and $K$ are graphs, we write $\gamma(H, K)$ for the number of subgraphs of $H$ isomorphic to $K$. It seems unlikely that there is a good algorithm for determining this number. However, in a theoretical argument, we can suppose it known whenever the isomorphim classes of $H$ and $K$ are given (W. T. Tutte, 1984, p. 119).

**Theorem 5.1.3.** *(Kelly's Lemma). The number $\gamma(G, K)$ is a reconstructible property of $G$ for any graph $K$ such that $|V(K)| \neq |V(G)|$.*

*Proof.* If $K$ has more vertices than $G$, then clearly $\gamma(G, K) = 0$. We may therefore assume that $|V(K)| < |V(G)|$. A subgraph of $G$ isomorphic to $K$ occurs as a subgraph in exactly $|V(G)| - |V(K)|$ of the primal subgraphs $G_j$. Hence $\gamma(G, K)$ is determined as follows:

$$(|V(G)| - |V(K)|)\, \gamma(G, K) = \sum_j \gamma(G_j, K). \quad \square$$

## 5.2. Properties of isomorphic bridges in graphs with circumference $l$ and their primal subgraphs

The concepts of isomorphic bridge and symmetric subgraph have been introduced in section 4.2. In Chapter 5 we shall prefer the concept of symmetric subgraph. All subgraphs $U$ are induced in this section. Since $U$ is the kernel of a bridge, we speak sometimes of inner vertices of $U$ and of vertices of attachment of $U$.

We shall prove some results on classes of at least $\lfloor l/2 \rfloor + 1$ isomorphic bridges in graphs with circumference $l$ and in their primal subgraphs; hence they are also valid for the graphs of $\mathfrak{Z}_l$.

The proof of our main Theorem 5.3.2 is based on these results. Yet they also have a certain interest of their own. Together with the properties of bridges proved in sections 4.3 and 4.4 the results obtained give a deeper insight into the structure of graphs with many isomorphic bridges. For convenience we restate Property 4.4.1 as follows.

**Proposition 5.2.1.** *Let $G$ be a graph with circumference $l$. If $\lfloor l/2 \rfloor + 1$ pairwise distinct bridges have the same set $S$ of vertices of attachment, then $|S| \leqq \lfloor l/2 \rfloor$.* □

Next we state a theorem which has already been proved for 2-connected graphs in H. Walther and H.-J. Voss, 1974.

**Theorem 5.2.2.** *Let $G$ be a (not necessarily 2-connected) graph with circumference $l$. Let $C$ and $C'$ be two classes of symmetric subgraphs of order $n$ of $G$ with*

$$|C|, |C'| \geqq \lfloor l/2 \rfloor + 1.$$

*Then the subgraphs of $C \cup C'$ are pairwise independent or all subgraphs of one class are completely contained in one subgraph of the other class.*

Note that a class of symmetric subgraphs of $G$ is always an equivalence class, i.e. it is a maximal class of symmetric subgraphs.

The assertions $|C|, |C'| \geqq \lfloor l/2 \rfloor + 1$ cannot be weakened. This is shown by the complete bipartite graphs with $s$ and at least $s+1$ vertices in each bipartition class $C$ and $C'$, respectively. Then the bipartition classes $C$ and $C'$ are classes of symmetric vertices of cardinalities $\lfloor l/2 \rfloor$ and at least $\lfloor l/2 \rfloor + 1$, respectively. Yet the vertices of $C \cup C'$ are not pairwise independent.

*Proof of Theorem 5.2.2.* By contradiction. Suppose that the subgraphs of $C \cup C'$ are not pairwise independent. Let $C = \{U_1, \ldots, U_\alpha\}$ and $C' = \{U'_1, \ldots, U'_{\alpha'}\}$ with $\alpha, \alpha' \geqq \lfloor l/2 \rfloor + 1$ such that $|V(U_1)| \geqq |V(U'_1)|$. We consider two cases.

Case 1. Assume that the notation can be chosen so that $U_1, U'_1$ are not independent and $U_1 \setminus U'_1 \neq \emptyset$ and $U'_1 \setminus U_1 \neq \emptyset$. Since $U'_1 \setminus U_1 \neq \emptyset$, the subgraph $U_1$ has a vertex $A_1$ of attachment which is an inner vertex of $U'_1$. Since $C$ is a symmetry class, $A_1$ is a vertex of attachment of all subgraphs $U_1, \ldots, U_\alpha$. Consequently, $U_i$ and $U'_1$ are not independent for all $i$, $1 \leqq i \leqq \alpha$. Since $V(U'_1) \subseteqq V(U_i)$ and $U'_1 \neq U_i$, the subgraph $U'_1$ has a vertex $A'_i$ of attachment which is an inner vertex of $U_i$, $1 \leqq i \leqq \alpha$. Since $U_1, \ldots, U_\alpha$ are independent, the vertices $A'_1, \ldots, A'_\alpha$ are all different. Hence $U'_1$ has $\alpha \geqq \lfloor l/2 \rfloor + 1$ vertices of attachment. This contradicts Proposition 5.2.1! □

Case 2. Assume that

(*) *for all pairs of dependent subgraphs $U_i$, $U'_j$ we have* $U'_j \setminus U_i = \emptyset$, i.e. *$U'_j$ is contained in $U_i$ and $U'_j \neq U_i$.*

Without loss of generality, let $U'_1 \subset U_1$. Then the subgraph $U'_1$ has a vertex $A'_1$ of attachment which is an inner vertex of $U_1$. Since $C'$ is a symmetry class, $A'_1$ is a vertex of attachment of all subgraphs $U'_1, \ldots, U'_{\alpha'}$ of $C'$. Consequently, $U'_i$ and $U_1$ are dependent for all $i$, $1 \leqq i \leqq \alpha'$. Our assumption (*) implies that $U'_i \subset U_1$ for all $i$, $1 \leqq i \leqq \alpha'$. □

**Proposition 5.2.3.** *Let $U_1, \ldots, U_\alpha$ be $\alpha \geqq \lfloor l/2 \rfloor + 2$ symmetric subgraphs of order $n$ of a graph $G$ with circumference $l$. Let $X$ be a vertex of $\bigcup U_j$. If $G - X$ contains $\beta \geqq \lfloor l/2 \rfloor + 1$ symmetric subgraphs $U'_1, \ldots, U'_\beta$ of order at least $n$ with $U_i \neq U'_j$ for all $1 \leqq i \leqq \alpha$, $1 \leqq j \leqq \beta$, then $\beta - 1$ of them are also symmetric in $G$.*

*Proof.* By contradiction. Suppose that the assertion of Proposition 5.2.3 is not true. Then $X$ is a neighbour of at least two of the subgraphs $U'_1, \ldots, U'_\beta$; say $U'_1$ and $U'_2$. Let the notation be chosen so that $X \in U_1$. Since the order of $U'_1$ and $U'_2$ is not less than that of $U_1$ and $X$ is a neighbour of $U'_1$ and $U'_2$, the subgraph $U'_i$ contains a vertex $A_i$ of attachment of $U_1$ $(i = 1, 2)$. Since $U_1, \ldots, U_\alpha$ are symmetric, $A_1$ and $A_2$ are also vertices of attachment of $U_2, \ldots, U_\alpha$. Since in $G - X$ the subgraphs $U_2, \ldots, U_\alpha$ and $U'_1, \ldots, U'_\beta$ are symmetric, by Theorem 5.2.2 the subgraphs $U_2, \ldots, U_\alpha$ are contained in both $U'_1$ and $U'_2$. Therefore, the independent subgraphs $U'_1$, $U'_2$ have a non-empty intersection $U'_1 \cap U'_2 \neq \emptyset$. Contradiction! □

The graph of Fig. 5.2.1 shows that Proposition 5.2.3 cannot be improved: $U'_1 \nsim U'_2 \sim U'_3 \sim \ldots \sim U'_\beta$ in $G$ and $U'_1 \sim U'_2 \sim U'_3 \sim \ldots \sim U'_\beta$ in $G - X$. An improvement is possible if the $U_i$ and $U'_j$ have the same order.

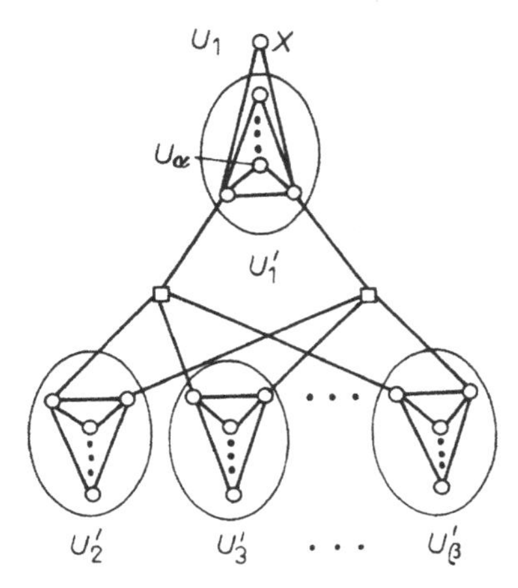


Fig. 5.2.1

**Proposition 5.2.4.** *Let $l \geqq 3$ be an integer and let $G$ be a graph with circumference $l$. Let $\{U_1, \ldots, U_\alpha\}$ be $\alpha \geqq \lfloor l/2 \rfloor + 2$ symmetric subgraphs of $G$ of order $n$. Let $X$ be a vertex of $\bigcup U_j$. If $G - X$ contains $\beta \geqq \lfloor l/2 \rfloor + 1$ symmetric subgraphs $U'_1, \ldots, U'_\beta$ of the same order $n$, then they are also symmetric in $G$.*

*Proof.* Let the notation be chosen so that $X \in U_1$. Then $U_2, \ldots, U_\alpha$ are also symmetric in $G - X$.

(1) *$\{U_2, \ldots, U_\alpha, U'_1, \ldots, U'_\beta\}$ is an independent set of subgraphs in $G - X$.*

It is possible that $U_i = U'_j$ for some $i$ and $j$.

*Proof of (1).* If $U'_1 \sim U_i$ in $G - X$, then assertion (1) is obviously true. If $U'_1 \nsim U_i$ in $G - X$, then assertion (1) follows from Theorem 5.2.2. □

(2) *$\{U_1, U'_1, \ldots, U'_\beta\}$ is an independent set of subgraphs in $G$.*

*Proof of (2).* By contradiction. Suppose that $U_1$ depends on $U'_i$ in $G$. Since $X \in U_1 \setminus U'_i$ and $U_1$ and $U'_i$ have the same order, the subgraph $U'_i$ contains a vertex $A$ of attachment of $U_1$. Since $U_1, \ldots, U_\alpha$ are symmetric subgraphs in $G$, the vertex $A$ is also a vertex of attachment of $U_2, \ldots, U_\alpha$. Consequently, $\{U_2, \ldots, U_\alpha, U'_i\}$ is not an independent set of subgraphs in $G - X$. This contradicts (1)! □

Since $U'_1, \ldots, U'_\beta$ are symmetric in $G-X$ and $X \in U_1$, the assertion (2) implies that $U'_1, \ldots, U'_\beta$ are also symmetric in $G$. □

Proposition 5.2.4 immediately implies the following result:

**Proposition 5.2.5.** *Let $l \geqq 3$ be an integer and let $G$ be a graph with circumference $l$. Let $M$ be the set of all subgraphs of order $n$ of $G$ and let $C_1, \ldots, C_p$ denote all symmetry classes of $M$ of cardinality at least $\lfloor l/2 \rfloor + 2$. By $X$ we denote a vertex of a subgraph $U$ of the class $C_i$, and by $M_X$ the set of all subgraphs of $G-X$ of order $n$. If $|C_i| > \lfloor l/2 \rfloor + 2$ or if $|C_i| = \lfloor l/2 \rfloor + 2$, then all symmetry classes of $M_X$ of cardinality at least $\lfloor l/2 \rfloor + 2$ are $C_1, \ldots, C_{i-1}, C_i \setminus \{U\}, C_{i+1}, \ldots, C_p$ and $C_1, \ldots, C_{i-1}, C_{i+1}, \ldots, C_p$, respectively.*

Now we shall prove the following result:

**Proposition 5.2.6.** *Let $l$, $\alpha$, be integers with $l \geqq 3$, $\alpha \geqq \lfloor l/2 \rfloor + 3$. Let $G$ be a graph with circumference $l$. Suppose that $G$ does not contain $\alpha$ symmetric subgraphs of order $n$. Then there is a primal subgraph without $\alpha$ symmetric subgraphs of order $n$.*

*Proof.* By contradiction. Suppose that

(a) one graph $G$ with circumference $l$ does not contain $\alpha$ symmetric subgraphs of order $n$ but each primal subgraph has $\alpha$ symmetric subgraphs of order $n$.

We consider two cases.

Case 1. Let $G$ contain $\alpha - 1$ symmetric subgraphs. Obviously, $\alpha - 1 \geqq \lfloor l/2 \rfloor + 2$. Let $X$ denote a vertex of one of these subgraphs. Then by Proposition 5.2.5 the graph $G-X$ contains no $\alpha$ symmetric subgraphs. This contradicts (a)!

Case 2. Suppose that $G$ does not contain $\alpha - 1$ symmetric subgraphs. Let $X$ be a vertex of $G$. Then by (a) the graph $G-X$ contains $\alpha$ symmetric subgraphs $U_1, \ldots, U_\alpha$ of order $n$. Let $Y$ be a vertex of $U_1$. Then by (a) the graph $G-Y$ contains $\alpha$ symmetric subgraphs $U'_1, \ldots, U'_\alpha$ of order $n$. If $X \in \bigcup_j U'_j$, then let the notation be chosen so that $X \in U'_1$. Obviously, $U_2, \ldots, U_\alpha$ are symmetric in $G-X-Y$, and $U'_2, \ldots, U'_\alpha$ are also symmetric in $G-X-Y$.

(1) *$\{U_2, \ldots, U_\alpha, U'_2, \ldots, U'_\alpha\}$ is an independent set of subgraphs in $G-X-Y$.*

It is possible that $U_i = U'_j$ for some $i$ and $j$.

*Proof of (1).* If $U_i \sim U'_j$ in $G-X-Y$ for some $2 \leqq i, j \leqq \alpha$, then assertion (1) is obvious. If $U_i \nsim U'_j$ in $G-X-Y$, then assertion (1) follows from Theorem 5.2.2. □

(2) *$U_1, U'_2, \ldots, U'_\alpha$ are independent in $G$.*

*Proof of (2).* By contradiction. Suppose that $U_1$ depends on $U'_i$ in $G$ for some $2 \leqq i \leqq \alpha$. Since $Y \in U_1 \setminus U'_i$ and $U_1$ and $U'_i$ have the same order, the subgraph $U'_i$ contains a vertex $A$ of attachment of $U_1$. Since $U_1, \ldots, U_\alpha$ are symmetric in $G-X$, the vertex $A$ is also a vertex of attachment of $U_2, \ldots, U_\alpha$. Consequently, $\{U_2, \ldots, U_\alpha, U'_i\}$ is not an independent set in $G-X-Y$. This contradicts (1)!

Since $U'_2, \ldots, U'_\alpha$ are symmetric in $G-Y$ and $Y \in U_1$, (2) implies that $U'_2, \ldots, U'_\alpha$ are also symmetric in $G$. This contradicts our assumption at the beginning of case 2! □

The assertion (2) of the proof of Proposition 5.2.6 cannot be improved. This is shown by the graph of Fig. 5.2.2; in this graph $U_1, U'_1$ are not independent.

Fig. 5.2.3 shows that in the proof of Proposition 5.2.6 the vertex $Y$ has to be taken from $U_1$ because $G \setminus X$ and $G \setminus Y$ ($G - X \sim G - Y$) contain three symmetric subgraphs which are not symmetric in $G$.

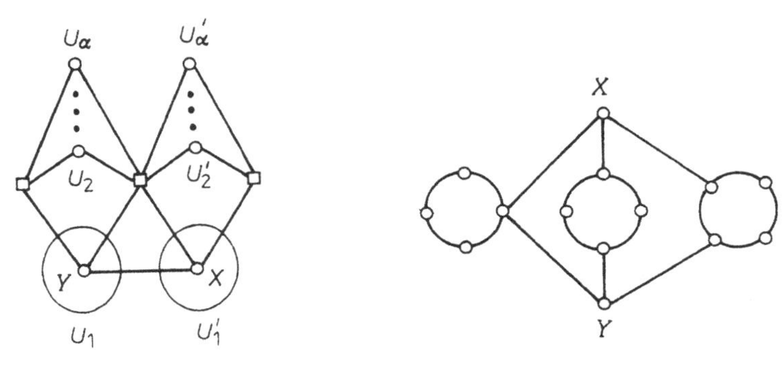


Fig. 5.2.2

Fig. 5.2.3

## 5.3. Symmetric subgraphs and the reconstruction conjecture

In the Introduction we mentioned that the reconstruction problem could not be solved for 2-connected graphs until now. It has been solved only for some classes of 2-connected graphs.

In Chapter 5 we study the reconstruction conjecture in graphs with many symmetric subgraphs of a given order $n$. The results are also valid in $\mathfrak{Z}_l$, and $\mathfrak{Z}_l$ is a class of 2-connected graphs. The simplest case is $n = 1$. Obviously, two subgraphs of order 1 are symmetric if and only if the corresponding vertices have the same neighbourhood. Such vertices are said to be symmetric (equivalent). With respect to symmetric vertices W. Dörfler and W. Imrich (W. Dörfler, 1972, and W. Dörfler and W. Imrich, 1972) proved the following result:

**Theorem 5.3.1.** *Let $k \geqq 2$ be an arbitrary integer. If a graph $G$ contains at least one class of at least $k$ symmetric vertices and no class with precisely $k - 1$ symmetric vertices, then $G$ is reconstructible.* □

Our intention is to extend Theorem 5.3.1 to graphs with many symmetric subgraphs of order $n \geqq 2$. In order to ensure that the elements of two different large classes of symmetric subgraphs are independent, we have to apply Theorem 5.2.2. Therefore we restrict our investigations to graphs with circumference $l$ containing at least $\lfloor l/2 \rfloor + 4$ symmetric subgraphs (isomorphic bridges). In a graph $G$ with circumference $l$ let $C_1, \ldots, C_s$ denote the classes of symmetric subgraphs of given order $n$ having cardinalities $|C_i| \geqq \lfloor l/2 \rfloor + 4$ $(1 \leqq i \leqq s)$. By Theorem 5.2.2 the union $\bigcup C_j$ is a set of pairwise independent subgraphs in $G$. Let $K$ denote the union of all bridges of $G$ which have their kernels in $\bigcup C_j$.

With $\bigcup C_j$ this subgraph $K$ is uniquely determined, i.e. $\gamma(G, K) = 1$ (for the definition of $\gamma(G, K)$, see section 5.1). Thus by P. J. Kelly's lemma (Theorem 5.1.3) the subgraph $K$ can be inferred from the isomorphism classes of the primal subgraphs of $G$ if (perhaps by other arguments) it is known that the order of $K$ is less than the order of $G$. Yet in this case also all possible arrangements $H$ of classes of at least $\lfloor l/2 \rfloor + 4$ isomorphic bridges have to be checked to see whether they are subgraphs of $G$ with $\gamma(G, H) = 1$ or not. Such a maximal subgraph $H$ would be the solution $K$. In the proof of the following theorem the structure of $K$ will be directly inferred from the isomorphism classes of the primal subgraphs of $G$.

**Theorem 5.3.2.** *Let $H_1, \ldots, H_k$ denote the isomorphism classes of the primal subgraphs of $G$. From $H_1, \ldots, H_k$ up to isomorphism all equivalence classes $C_1, \ldots, C_s$ of symmetric subgraphs of order $n$ of $G$ with cardinalities $|C_1| \geqq |C_2| \geqq \ldots \geqq |C_s| \geqq \lfloor c(G)/2 \rfloor + 4$ can be inferred. The union of their corresponding classes of isomorphic bridges can also be inferred.*

*Moreover, if $|C_i| - |C_{i+1}| \geqq 2$ for some index $i$ or if $|C_s| > \lfloor l/2 \rfloor + 4$, then $G$ is reconstructible.*

By Theorem 5.3.2 the subgraph $K$ formed by all bridges of $G$ with kernels in $\bigcup C_j$ can be inferred from the isomorphism classes of the primal subgraphs.

Comparing Theorems 5.3.1 and 5.3.2 we discover a common feature: $G$ is reconstructible if the sequence $|C_1|, \ldots, |C_s|$ has a gap. This means that $G$ can be inferred from the isomorphism classes of their primal subgraphs if there is an integer $k$, $k \geqq 2$ in Theorem 5.3.1 and $k \geqq \lfloor l/2 \rfloor + 5$ in Theorem 5.3.2, with the following property: $G$ contains a class of symmetric subgraphs of cardinality $k$ and no such class (with the same order of the subgraphs) of cardinality $k-1$.

From Theorem 5.3.2 the following two corollaries can be deduced:

**Corollary 5.3.3.** *The graphs with circumference $l$ containing at least $\lfloor l/2 \rfloor + 4$ symmetric subgraphs are recognizable.*

**Corollary 5.3.4.** *Let $l$; $m$ be integers with $l \geqq 3$ and $k \geqq \lfloor 1/2 \rfloor + 4$. Let $G$ contain classes of at least $k$ symmetric subgraphs of order $n$ and no class with precisely $k-1$ symmetric subgraphs of order $n$. Then $G$ is reconstructible.*

## 5.4. Proof of Theorem 5.3.2

In this section our main Theorem 5.3.2 will be proved. The proof includes a description of "how to infer information relating to $G$ from the isomorphism classes of the primal subgraphs".

First we continue the investigations of classes of at least $\lfloor l/2 \rfloor + 2$ symmetric subgraphs. Lemmas 5.4.1-3 present results on graphs of isomorphism classes of the primal subgraphs of $G$. The proof of Theorem 5.3.2 is based on these lemmas.

Let $G$ be a graph with circumference $l$ and let $C_1, \ldots, C_p$ be all the equivalence classes of its symmetric subgraphs of order $n$ with cardinalities $|C_1| \geqq \ldots \geqq |C_p| \geqq \lfloor l/2 \rfloor + 2$. The non-increasing sequence $S(G) = (|C_1|, \ldots, |C_p|)$ is called the *s-sequence* of $G$. Let $\gamma(G) =_{\text{def}} \sum |C_j|$.

Correspondingly, for each graph $H_i$ of the isomorphism class of $G_i$ the $s$-sequence $S(H_i)$ and $\gamma(H_i)$ are defined.

**Lemma 5.4.1.** *Let $G$ be a graph of circumference $l$ and $C_1, \ldots, C_p$ its equivalence classes of sym-*

*metric subgraphs of order n with cardinalities* $|C_1| \geqq \ldots \geqq |C_p| \geqq \lfloor l/2 \rfloor + 2$. *Let H be a graph of the isomorphism class of the primal subgraph* $G - X$. *Then*

(i) $\gamma(H) \geqq \gamma(G)$ *if and only if* $X \notin \bigcup C_j$,

(ii) $\gamma(H) = \gamma(G) - 1$ *if and only if* $X \in C_j$ *and* $|C_j| > \lfloor l/2 \rfloor + 2$,

(iii) $\gamma(H) = \gamma(G) - (\lfloor l/2 \rfloor + 2)$ *if and only if* $X \in C_j$ *and* $|C_j| = \lfloor l/2 \rfloor + 2$.

*Proof.* The lemma has to be proved in both directions. First the direction "$\leftarrow$" of the lemma will be proved. If $X \notin \bigcup C_j$, then $G - X \supseteqq \bigcup C_j$. Consequently,

(1) *if* $X \notin \bigcup C_j$, *then* $\gamma(H) \geqq \gamma(G)$. □

If $X \in C_j$, then by Proposition 5.2.5 the $s$-sequence of the primal subgraph $G - X$ has the following property:

(2) *if* $X \in C_j$ *and* $|C_j| > \lfloor l/2 \rfloor + 2$, *then* $\gamma(H) = \gamma(G) - 1$,

and

(3) *if* $X \in C_j$ *and* $|C_j| = \lfloor l/2 \rfloor + 2$, *then* $\gamma(H) = \gamma(G) - (\lfloor l/2 \rfloor + 2)$.

(1), (2) and (3) imply the assertions of Lemma 5.4.1 in the case "$\leftarrow$".

The validity of direction "$\rightarrow$" of the lemma also follows from (1), (2) and (3) (for instance by applying Hauber's theorem (see J. Naas-H. L. Schmid, Mathematisches Wörterbuch, Berlin – Stuttgart 1961) to (1), (2), (3)). □

The usual order "$<$" in the set of integers $\{\gamma(H_i) \mid 1 \leqq i \leqq k\}$ induces a partical order in the set $\{S(H_i) \mid 1 \leqq i \leqq k\}$. In this set a second partial order "$<$" is introduced. For $S(H_i) = (|C_1^i|, \ldots, |C_{p_i}^i|)$ and $S(H_h) = (|C_1^h|, \ldots, |C_{p_h}^h|)$ we define:

$$S(H_i) < S(H_h) \quad \text{if and only if } |C_j^i| \leqq |C_j^h| \quad \text{for all } j, \quad 1 \leqq j \leqq \min\{p_i, p_h\}, \quad \text{and} \quad |C_j^i| < |C_j^h| \quad \text{for at least one } j.$$

We introduce

$$N =_{\text{def}} \left\{ H_i \mid \gamma(H_i) = \min_r \gamma(H_r) \right\}$$

and observe the following property of $N$.

**Lemma 5.4.2.** *Let G and* $C_1, \ldots, C_p$ *be defined as in Lemma* 5.4.1.

(i) *Let* $|C_p| > \lfloor l/2 \rfloor + 2$. *Then* $H \in N$ *if and only if* $\gamma(H) = \gamma(G) - 1$.

(ii) *Let* $|C_p| = \lfloor l/2 \rfloor + 2$. *Then* $H \in N$ *if and only if* $\gamma(H) = \gamma(G) - (\lfloor l/2 \rfloor + 2)$.

*Proof.* By Lemma 5.4.1, $\min_r \gamma(H_r) = \gamma(G) - 1$ if and only if $|C_p| > \lfloor l/2 \rfloor + 2$ and $\min_r \gamma(H_r) = \gamma(G) - (\lfloor l/2 \rfloor + 2)$ if and only if $|C_p| = \lfloor l/2 \rfloor + 2$. □

Next from the isomorphism classes of the primal subgraphs of $G$ we shall infer whether

$$|C_1| > |C_p| > \lfloor l/2 \rfloor + 2$$

or

$$|C_1| > |C_p| = \lfloor l/2 \rfloor + 2$$

or

$$|C_1| = \ldots = |C_p| \geqq \lfloor l/2 \rfloor + 3.$$

**Lemma 5.4.3.** *Let G and* $C_1, \ldots, C_p$ *be defined as in Lemma 5.4.1. Let* $|C_1| \geqq \lfloor l/2 \rfloor + 3$. *Then*

(i) *$N$ contains two graphs $H_1$, $H_2$ with different s-sequences $S(H_1) \neq S(H_2)$ if and only if $p \geqq 2$ and $|C_1| > |C_p| > \lfloor l/2 \rfloor + 2$.*

(ii) *All graphs $H_i \in N$ have the same s-sequence $S(H_i) = S$ and one graph $H_h \notin N$, $H_h \cong G - X_h$, has an s-sequence $S(H_h)$ with $S(H_h) < S$ if and only if $p \geqq 2$ and $|C_1| > |C_p| = \lfloor l/2 \rfloor + 2$.*

(iii) *All graphs $H_i \in N$ have the same s-sequence $S(H_i) = S$ and all graphs $H_h \notin N$, $H_h \cong G - X_h$, have an s-sequence $S(H_h)$ with $S(H_h) > S$ if and only if $p \geqq 1$ and $|C_1| = \ldots = |C_p| \geqq \lfloor l/2 \rfloor + 3$.*

*Proof.* Let $H_i \cong G - X_i$, $X_i \in G$, $1 \leqq i \leqq k$. The lemma has to be proved in both directions.

First the direction "$\Leftarrow$" will be proved. Since $|C_1| \geqq \lfloor l/2 \rfloor + 3$, precisely one of the following three cases occurs:

Case 1. Let $|C_1| > |C_p| > \lfloor l/2 \rfloor + 2$. We choose $H_1 \cong G - X_1$, $X_1 \in C_1$, and $H_2 \cong G - X_2$, $X_2 \in C_p$. By Proposition 5.2.5 the s-sequences of $H_1$ and $H_2$ are

$$S(H_1) = S(G - X_1) = (|C_1| - 1, |C_2| \ldots, |C_p|)$$

and

$$S(H_2) = S(G - X_2) = (|C_1|, |C_2| \ldots, |C_p| - 1)$$

(note: if $|C_1| = |C_2|$, the order of the integers in $S(H_1)$ has to be changed). Therefore, $S(H_1) \neq S(H_2)$ and $\gamma(H_1) = \gamma(H_2) = \sum |C_j| - 1 = \gamma(G) - 1$. By Lemma 5.4.2 the graphs $H_1$, $H_2$ belong to $N$. Consequently,

(1) *if $|C_1| > |C_p| > \lfloor l/2 \rfloor + 2$, then $N$ contains two graphs $H_1$, $H_2$ with different s-sequences $S(H_1) \neq S(H_2)$.* □

Case 2. Let $|C_1| > |C_p| = \lfloor l/2 \rfloor + 2$. Let $H \cong G - X$, $X \in G$. By Lemma 5.4.2 the graph $H$ belongs to $N$ if and only if $\gamma(H) = \gamma(G) - (\lfloor l/2 \rfloor + 2)$. Lemma 5.4.1 implies that $H \in N$ if and only if $X \in C_j$ and $|C_j| = \lfloor l/2 \rfloor + 2$. By Proposition 5.2.5 the s-sequence of $H$ is

$$S(H) = (|C_1|, \ldots, |C_{j-1}|, |C_{j+1}|, \ldots, |C_p|),$$

which is obtained from $S(G)$ by deleting the integer $\lfloor l/2 \rfloor + 2$. Hence all $H_i \in N$ have the same s-sequence

$$S(H_i) = (|C_1|, \ldots, |C_j|, \ldots, |C_{p-1}|).$$

We choose $H_h \cong G - X_h$ with $X_h \in C_1$. Since $|C_1| \geqq \lfloor l/2 \rfloor + 3$, by Proposition 5.2.5 the s-sequence of $H_h$ is

$$S(H_h) = (|C_1| - 1, |C_2|, \ldots, |C_p|).$$

Hence $S(H_h) < S(H_i)$. Consequently,

(2) *if $|C_1| > |C_p| = \lfloor l/2 \rfloor + 2$, then all graphs $H_i \in N$ have the same s-sequence $S(H_i) = S$ and one graph $H_h \notin N$, $H_h \cong G - X_h$, has an s-sequence $S(H_h)$ with $S(H_h) < S$.* □

Case 3. Let $|C_1| = \ldots = |C_p| \geqq \lfloor l/2 \rfloor + 3$. Then the Lemmas 5.4.2 and 5.4.1 imply that $H_i \in N$ if and only if $\gamma(H_i) = \gamma(G) - 1$ and $X_i \in C_j$ for some $j$, $1 \leqq j \leqq p$. Therefore, by Proposition 5.2.5 all graphs $H_i \in N$ have the same s-sequence

$$S =_{\text{def}} S(H_i) = (|C_1|, \ldots, |C_1|, |C_1| - 1).$$

Hence for all graphs $H_h \notin N, H_h \cong G - X_h$, the graph $G - X_h \supseteqq \bigcup C_j$. Thus

$$S(H_h) = S(G - X_h) \geqq (|C_1|, \ldots, |C_1|) > S(H_i).$$

Consequently,

(3) *if* $|C_1| = \ldots = |C_p| \geqq \lfloor l/2 \rfloor + 3$, *then all graphs* $H_i \in N$ *have the same s-sequence* $S(H_i) = S$ *and every graph* $H_h \notin N$, $H_h \cong G - X_h$, *has an s-sequence* $S(H_h) > S(H_i)$.

(1), (2) and (3) imply the validity of the assertions of Lemma 5.4.3 in the case "$\leftarrow$".
The validity of the direction "$\rightarrow$" of the lemma also follows from (1), (2) and (3) (for instance by applying Hauber's theorem, see p. 98, to (1), (2), (3)). □

Finally we shall prove Theorem 5.3.2. For convenience we restate it.

**Theorem 5.4.4.** *Let G be a graph and* $C_1, \ldots, C_s$ *its equivalence classes of symmetric subgraphs of order n with cardinalities*

$$|C_1| \geqq |C_2| \geqq \ldots \geqq |C_s| \geqq \lfloor c(G)/2 \rfloor + 4.$$

*Then up to isomorphism the classes* $C_1, \ldots, C_s$ *and the subgraph induced by the bridges having their kernels in* $\bigcup C_j$ *can be inferred from isomorphism classes* $H_1, \ldots, H_k$ *of the primal subgraphs of G.*
*Moreover, if* $|C_i| - |C_{i+1}| \geqq 2$ *for some index i or if* $|C_s| > \lfloor l/2 \rfloor + 4$, *then G is even reconstructible.*

Note: $s = 0$ if $G$ has no equivalence class of cardinality at least $\lfloor c(G)/2 \rfloor + 4$.

*Proof.* Let $G$ be a graph and let $C_1, \ldots, C_s, \ldots, C_p$ denote all equivalence classes of symmetric subgraphs of order $n$ with cardinalities

$$|C_1| \geqq \ldots \geqq |C_s| \geqq \lfloor l/2 \rfloor + 4 > |C_{s+1}| \geqq \ldots \geqq |C_p| \geqq \lfloor l/2 \rfloor + 2.$$

The reconstruction of $\bigcup C_j$ and of $G$ will be done in three steps.

Step 1. From the graphs $H_1, \ldots, H_k$ we shall infer that $G$ does not have $\lfloor l/2 \rfloor + 4$ symmetric subgraphs, i.e. $s = 0$, or $p \geqq 1$ and $|C_1| \geqq \lfloor l/2 \rfloor + 3$. This will be done by considering two questions ($Q_1$) and ($Q_2$).

($Q_1$) *Does an integer* $\alpha$ *exist such that one* $H_j$ $(1 \leqq j \leqq k)$ *contains* $\alpha + 3$ *symmetric subgraphs of order n with at most* $\alpha$ *vertices of attachment?*

(1) *If the answer to* ($Q_1$) *is "No", then* $s = 0$.

*Proof of (1).* By contradiction. Suppose that $s \geqq 1$, i.e. $G$ contains $\lfloor l/2 \rfloor + 4$ symmetric subgraphs of order $n$. Then by Proposition 5.2.5 for each vertex $X \in V(G)$ the graph $G - X$ contains $\lfloor l/2 \rfloor + 3$ symmetric subgraphs of order $n$. Therefore, the answer to ($Q_1$) is "Yes" with $\alpha = \lfloor l/2 \rfloor$. Contradiction! □

Let the answer to ($Q_1$) be "Yes". Then $H_j = G - X_j$ contains $\alpha + 3$ symmetric subgraphs $U_1, \ldots, U_{\alpha+3}$ with $\alpha' \leqq \alpha$ vertices of attachment $A_1, \ldots, A_{\alpha'}$. Hence the only neighbours of $U_1, \ldots, U_{\alpha+3}$ are $A_1, \ldots, A_{\alpha'}, X_j$. Since $U_1, \ldots, U_{\alpha+3}$ are independent, there is no cycle containing at least one vertex of each $U_i$, $1 \leqq i \leqq \alpha + 3$. Therefore $G$ is not Hamiltonian and the circumference of $G$ can be determined.

(2) *The circumference of G is* $l = \max_j c(H_j)$. □

($Q_2$) *Does each $H_j$, $1 \leqq j \leqq k$, contain $\lfloor l/2 \rfloor + 3$ symmetric subgraphs of order $n$?*

(3) a) *If the answer to* ($Q_2$) *is "No", then $s = 0$.*
b) *If the answer is "Yes", then $p \geqq 1$ and $|C_1| \geqq \lfloor l/2 \rfloor + 3$.*

*Proof of (3), a).* By contradiction. Suppose that $s \geqq 1$, i.e. $G$ contains $\lfloor l/2 \rfloor + 4$ symmetric subgraphs of order $n$. Then by Proposition 5.2.5 for each vertex $X_i \in V(G)$ the graphs $G - X_i$ and $H_i$, $H_i \cong G - X_i$, contain at least $\lfloor l/2 \rfloor + 3$ symmetric subgraphs of order $n$. Contradiction! □

*Proof of (3), b).* Since each primal subgraph of $G$ contains at least $\lfloor l/2 \rfloor + 3$ symmetric subgraphs of order $n$, by Proposition 5.2.6 the graph $G$ has at least $\lfloor l/2 \rfloor + 3$ symmetric subgraphs of the same order $n$. □

Step 2. By step 1 (assertion (3) b) only graphs with $p \geqq 1$ and $|C_1| \geqq \lfloor l/2 \rfloor + 3$ will be investigated further. From the graphs $H_i$, $1 \leqq i \leqq k$, we shall infer the isomorphism classes of $C_1, \ldots, C_p$ and of the subgraph of $G$ induced by all bridges having their kernels in $\bigcup_{j=1}^{s} C_j$. By Lemma 5.4.3 three cases have to be investigated.

Case 1. Let all graphs $H_i \in N$ have the same $s$-sequence $S(H_i) = S$ and one graph $H_h \notin N$ has an $s$-sequence $S(H_h)$ with $S(H_h) < S$. Lemma 5.4.3(ii) implies that

$$p \geqq 2 \quad \text{and} \quad |C_1| > |C_p| = \lfloor l/2 \rfloor + 2.$$

Then by Proposition 5.2.5 the $s$-sequence of $H_i$ is

$$S = S(H_i) = (|C_1|, \ldots, |C_{p-1}|).$$

Hence $p = p_i + 1$ and $C_r \cong C_r^i$ for all $1 \leqq r \leqq p - 1$. Since $|C_p| = \lfloor l/2 \rfloor + 2$, the integer $s \leqq p - 1$, so the isomorphism classes of $\bigcup_{j=1}^{s} C_j$, and the isomorphism class of the subgraph of $G$ induced by all bridges with kernels in $\bigcup_{j=1}^{s} C_j$ have been reconstructed.

Case 2. Let $N$ contain two subgraphs $H_1$, $H_2$ with different $s$-sequences. Lemma 5.4.3(i) implies that

$$p \geqq 2 \quad \text{and} \quad |C_1| > |C_p| > \lfloor l/2 \rfloor + 2.$$

We choose $H_i \in N$ with $|C_{p_i}^i| = \min_{H_j \in N} |C_{p_j}^j|$. Then by Proposition 5.2.5 the $s$-sequence of $H_i$ is

$$S(H_i) = (|C_1|, \ldots, |C_{p-1}|, |C_p| - 1)$$

with $|C_{p_i-1}^i| = |C_{p-1}| > |C_p| - 1 = |C_{p_i}^i|$. Hence $C_{p_i}^i$ is the unique equivalence class of $H_i$ with cardinality $|C_{p_i}^i| = |C_p| - 1$, $|C_p| - 1 \geqq \lfloor l/2 \rfloor + 2$. We add a symmetric subgraph to $C_{p_i}^i$ and denote the new class by $\hat{C}_{p_i}^i$. The corresponding graph is $\hat{H}_i$. Consequently, $p = p_i$ and $C_r \cong C_r^i$ for all $1 \leqq r \leqq p - 1$ and $C_p \cong \hat{C}_{p_i}^i$. Thus the isomorphism classes of $\bigcup_{j=1}^{p} C_j$ and the isomorphism class of the subgraph of $G$ induced by all bridges with kernels in $\bigcup_{j=1}^{p} C_j$ have been reconstructed.

Case 3. Let all graphs $H_i \in N$ have the same s-sequence $S(H_i) = S$ and all graphs $H_h \notin N$ have an s-sequence $S(H_h)$ with $S(H_h) > S$. Lemma 5.4.3(iii) and $|C_1| \geqq \lfloor l/2 \rfloor + 3$ imply that

$$p \geqq 1 \quad \text{and} \quad |C_1| = \ldots = |C_p| > \lfloor l/2 \rfloor + 2.$$

Then by Proposition 5.2.5 the graph $H_i$ has an s-sequence

$$S = S(H_i) = (|C_1|, |C_2|, \ldots, |C_{p-1}|, |C_p| - 1) = (|C_1|, |C_1|, \ldots, |C_1|, |C_1| - 1).$$

Hence $C^i_{p_i}$ is the unique equivalence class of $H_i$ with cardinality $|C_1| - 1$, where $|C_1| - 1 \geqq \lfloor l/2 \rfloor + 2$. We add a symmetric subgraph to $C^i_{p_i}$ and denote the new class by $\hat{C}^i_{p_i}$. The corresponding graph is $\hat{H}_i$. Consequently, $p = p_i$ and $C_r \cong C^i_r$ for all $1 \leqq r \leqq p-1$ and $C_p \cong \hat{C}^i_{p_i}$. Thus the isomorphism classes of $\bigcup_{j=1}^{p} C_j$ and of the subgraph of $G$ induced by all bridges with kernels in $\bigcup_{j=1}^{p} C_j$ have been reconstructed.

Step 3. By step 1 (assertion (3)b) only graphs with $p \geqq 1$ and $|C_1| \geqq \lfloor l/2 \rfloor + 3$ will be investigated further. From the graphs $H_i$, $1 \leqq i \leqq k$, we shall infer the isomorphism class of $G$ in the case when $|C_p| > \lfloor l/2 \rfloor + 2$ or $|C_i| - |C_{i+1}| \geqq 2$ for some $i$. We consider two cases.

Case 1. Let $|C_p| > \lfloor l/2 \rfloor + 2$. In this case we continue the investigations of cases 2 and 3 of step 2. In step 2 the isomorphism class of $\bigcup_{j=1}^{p} C_j$ and the isomorphism class of the subgraph of all bridges with kernels in $\bigcup_{j=1}^{p} C_j$ have been reconstructed.

Moreover, $G$ is isomorphic to a subgraph $\bar{G}$ of $\hat{H}_i$. Of course, the integers $p$ and $s$ are also known.

If $n = 1$, then $G \cong \hat{H}_i$ and $G$ is reconstructed. Now suppose that $n \geqq 2$.

The aim of the following investigations is to find the vertices of $\hat{H}_i - \bar{G}$, $\bar{G} \subseteqq \hat{H}_i$, where $G$ and $\bar{G}$ are isomorphic. We define an integer $t$ as follows: if $|C_p| = \ldots = |C_1|$, then let $t = 0$; otherwise let $t$ be defined so that $|C_p| = \ldots = |C_{t+1}| < |C_t|$. Let $c_p =_{\text{def}} |C_p|$. By Proposition 5.2.5 and step 2

(4) *the graph* $H_i \in N$ *and* $|C^i_{p_i}| = c_p - 1$ *if and only if* $X_i \in U \in \bigcup_{j=t+1}^{p} C_j$. □

The problem now is to find an $H_i \in N$ so that the subgraph $D$ corresponding to $U - X_i$ in $G_i$ can be recognized. Then $G$ would be isomorphic to $\hat{H}_i - D$.

In order to find such a subgraph $H_i$ we consider certain classes of symmetric subgraphs of order $n-1$. Let $D_1, D_2, \ldots, D_r$ denote the equivalence classes of the connected induced subgraphs of order $n-1$ with $|D_1| \geqq |D_2| \geqq \ldots \geqq |D_r|$ such that for each class $D_k$ there is a subgraph $U \in \bigcup_{j=t+1}^{p} C_j$ and a vertex $X_i \in U$ with the property that $U - X_i$ is equivalent to the elements of $D_k$. Obviously, the primal subgraph $G_i$ contains all subgraphs of $\bigcup_j D_j$ with all vertices of attachment.

Let $H_i \in N$ be such that $|C^i_{p_i}| = c_p - 1$. Let $D^i_1, D^i_2, \ldots, D^i_r$ denote the equivalence classes of the connected subgraphs of order $n-1$ with $|D_1^i| \geqq |D_2^i| \geqq \ldots \geqq |D_{r_i}^i|$ such that for each class $D_k^i$ there is a subgraph $U \in \hat{C}_{p_i}^i$ and a vertex $X_i \in U$ with the property that $U - X_i$ is symmetric to the elements of $D_k^i$. Let $d_1 =_{\text{def}} \max |D_1^i|$, where the maximum is to be taken over all graphs $H_i \in N$ with $|C_{p_i}^i| = c_p - 1$. We define an integer $\varrho$ as follows: if $|D_1| = \ldots = |D_r|$, then let $\varrho = r$; otherwise let $\varrho$ be defined so that $|D_1| = \ldots = |D_\varrho| \geqq |D_{\varrho+1}|$. Proposition 5.2.5 implies that

(5) *the graph* $G_i \in N$ *has the properties* $|C_{p_i}^i| = c_p - 1$ *and* $|D_1^i| = d_1$ *if and only if* $X_i \in U \in \bigcup_{j=t+1}^{p} C_j$ *and* $U - X_i$ *is symmetric to a subgraph of* $\bigcup_{k=1}^{\varrho} D_k$; i.e. *the equivalence classes of symmetric subgraphs of order n with cardinality at least* $\lfloor l/2 \rfloor + 2$ *are*

$$C_1, \ldots, C_t, C_{t+1}, \ldots, C_j - U, \ldots, C_p$$

*and* $U - X_i \in D_1^i$, *i.e.*

$$|D_1^i| = |D_1| + 1 > |D_2^i| = |D_2| \geqq \ldots \geqq |D_r^i| = |D_r|.$$

With (5) the reconstruction of $G$ can be completed: in $H_i \in N$ with $|C_{p_i}^i| = c_p - 1$ and $|D_1^i| = d_1$ we add a new subgraph $U'$ symmetric to the members of $C_{p_i}^i$ and delete a member of $D_1^i$. The resulting graph is isomorphic to $G$.

Since all information on $G$ has been derived from $H_1, \ldots, H_k$, the graph $G$ is uniquely determined by $H_1, \ldots, H_k$. Thus the proof of Theorem 5.4.4 is complete in case 1.

Case 2. Let $|C_p| = \lfloor l/2 \rfloor + 2$ and $|C_i| - |C_{i+1}| \geqq 2$ for some index $i$. Then put $\mu =_{\text{def}} |C_{i+1}|$. The reconstructibility of $G$ can be proved by repeating the proof of case 1 with the inequality $|C_p| > \mu + 1$ instead of $|C_p| > \lfloor l/2 \rfloor + 2$. □

# Chapter 6

# New results concerning isomorphic bridges and their applications to critical graphs

## 6.0. Introduction

Here the investigations of Chapter 4 are continued. By Theorem 4.2.3 in the class $\mathfrak{Z}_l$ of all 2-connected graphs with circumference $l$ each "large" graph has a subgraph with $p$ isomorphic bridges. This and other theorems of Chapter 4 will be extended to some new classes of graphs. There are interesting applications of these results to graphs which are critical with respect to the chromatic number, or to the independence number. For instance if such a graph $G$ has a "large" order and contains an odd cycle, then it also contains a "large" odd cycle; or if $G$ has a "large" order and contains a subgraph "contractible to $K_\eta$", then it also contains a "large" subgraph "contractible to $K_\eta$". The latter graph will be referred to as a Hadwiger graph with Hadwiger number $\eta$ (see section 6.1).

Similar results are true for the classes of all graphs with a given chromatic index, a given achromatic number or a given size of the maximum matchings, respectively.

All these results can be derived from theorems on isomorphic bridges because the graphs considered have no isomorphic bridges or only a "small" number of isomorphic bridges.

### 1° Critically $k$-chromatic graphs

The critically $k$-chromatic graphs have been very well investigated by many authors (see B.Toft, 1986). First we repeat the well-known concepts of colouring theory (see J.A.Bondy and U.S.R. Murty, 1976).

A $k$-vertex colouring of a graph $G$ is an assignment of $k$ colours 1, 2, ..., $k$ to the vertices of $G$; the colouring is proper if adjacent vertices have different colours. Thus a proper $k$-vertex colouring of $G$ is a partition $(V_1, V_2, \ldots, V_k)$ of the vertex set of $G$ into $k$ (possibly empty) independent sets.

Throughout this paper only proper $k$-vertex colourings will be considered. Therefore it will be convenient to refer to a "proper $k$-vertex colouring" simply as a "$k$-colouring".

A graph $G$ is said to be *$k$-colourable* if it has a $k$-colouring. The chromatic number, $\chi(G)$, of $G$ is the minimum $k$ for which $G$ is $k$-colourable; if $\chi(G) = k$, the graph $G$ is said to be $k$-chromatic.

A graph $G$ is said to be *critically $k$-chromatic* if $G$ is $k$-chromatic and by deletion of any edge or vertex the resulting graph is $(k-1)$-colourable. In other words, $G$ is critically $k$-chromatic if $\chi(H) < \chi(G)$ for every proper subgraph $H$ of $G$. Such graphs were first introduced by G.A. Dirac, 1952b.

Since 1952 critically $k$-chromatic graphs have been well investigated (for a survey see B. Toft, 1986). Here we only mention the remarkable papers of T. Gallai, 1963 and 1964. These critically $k$-chromatic graphs are extremal (minimal) in the class of all $k$-chromatic graphs. As a consequence of this, many results on the class of the critically $k$-chromatic graphs can be extended to the class of all $k$-chromatic graphs.

## 2° The Hadwiger Conjecture

Let $e$ be an edge of a graph $G$. If $e$ is deleted and its end vertices are identified, then we say that the resulting graph is obtained from $G$ by contracting the edge $e$.

A graph $H$ is a contraction of $G$ if $H$ can be obtained by contracting some edges of $G$. A graph $H$ is a subcontraction of $G$ if $H$ is a subgraph of a contraction of $G$.

In 1943 H. Hadwiger published his famous conjecture (H. Hadwiger, 1943 and 1958):

**Conjecture.** *Every $k$-chromatic graph has $K_k$ as a subcontraction.*

In Chapter 6 we also deal with subgraphs $H$ of $k$-chromatic graphs contractible to $K_\pi$. Unfortunately we are not able to prove Hadwiger's conjecture; we shall only show that if a "large" critically $k$-chromatic graph contains a subgraph contractible to $K_\pi$, then it also contains a "large" one.

Hadwiger's conjecture can easily be verified for $k = 2$ and $k = 3$. G. A. Dirac, 1952b, proved its validity for $k = 4$ by showing that every critically 4-chromatic graph contains a subdivision of $K_4$.

B. Zeidl, 1958, has shown that every critically 4-chromatic graph $G$ has a subdivision of $K_4$ containing a cycle of odd length. Here we shall show that such a graph of "large" order contains a subdivision of $K_4$ with a large odd cycle. This result generalizes results of J. B. Kelly and L. M. Kelly, 1954, and G. A. Dirac, 1955, on large cycles in critically $k$-chromatic graphs (see also R. C. Read, 1957, and T. Gallai, 1963).

Problems relating to Hadwiger's conjecture are also considered in section 8.0, 4°.

## 3° Graphs which are edge-critical with respect to the independence number

First some definitions. A set $I$ of vertices of a graph $G$ is said to be *independent* if any two vertices are not adjacent, i.e. the subgraph of $G$ induced by $I$ has no edge. The *independence number* $\alpha(G)$ of $G$ is the cardinality of a maximum independent (vertex) set.

Obviously, every independent set induces a complete graph in the complement $\bar{G}$ of $G$.

A graph $G$ is said to be *edge-critical* with respect to the independence number if deleting any edge $e$ of $G$ results in a graph with a larger independence number, i.e. $\alpha(G - e) = \alpha(G) + 1$.

If $\alpha(G) = \alpha$, then $G$ is simply called *$\alpha$-critical*. The graphs $G_1$ and $G_3$ of Fig. 6.0.1 are 2-critical and $G_2$ is not critical with respect to the independence number.

The concept of an $\alpha$-critical graph was introduced by P. Erdős and T. Gallai, 1961, and these graphs have been investigated by many authors (B. Andrásfai, 1967, M. D. Plummer, 1967, L. Lovász, 1976 und 1979, L. Surányi, 1976, and W. Wessel, 1976 and 1981, and others).

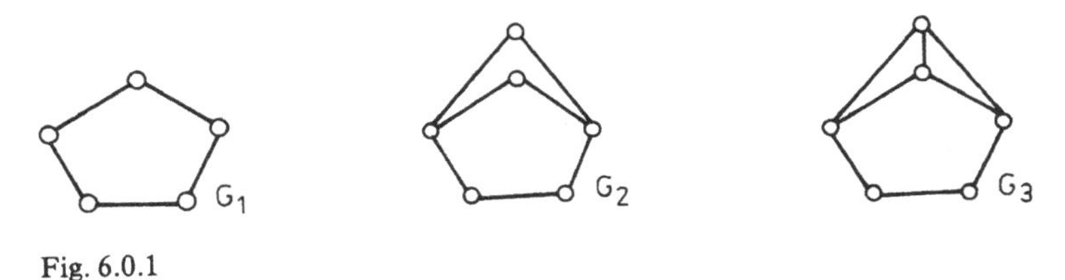


Fig. 6.0.1

Here we only show that if an $\alpha$-critical graph of "large" order contains a subgraph of a certain well-defined class of graphs, then it also contains a "large" one from the same class.

## 6.1. Two special classes of graphs

In sections 6.2–6.7 we need the concepts of Hadwiger graph and prism graph.

**Definition.** Let $H_\eta$ be a graph with minimum degree at least 2 which is obtained from $\eta$ disjoint trees by adding $\binom{\eta}{2}$ edges so that each pair of distinct trees is joined by precisely one of these edges. Each such graph is said to be a *Hadwiger graph.* A graph $G$ has *Hadwiger number* $\eta(G) = \eta$ if $G$ contains a Hadwiger graph $H_\eta$ but no Hadwiger graph $H_{\eta+1}$.

Hadwiger graphs with Hadwiger numbers $\eta = 3, 4, 5$ are depicted in Fig. 6.1.1.

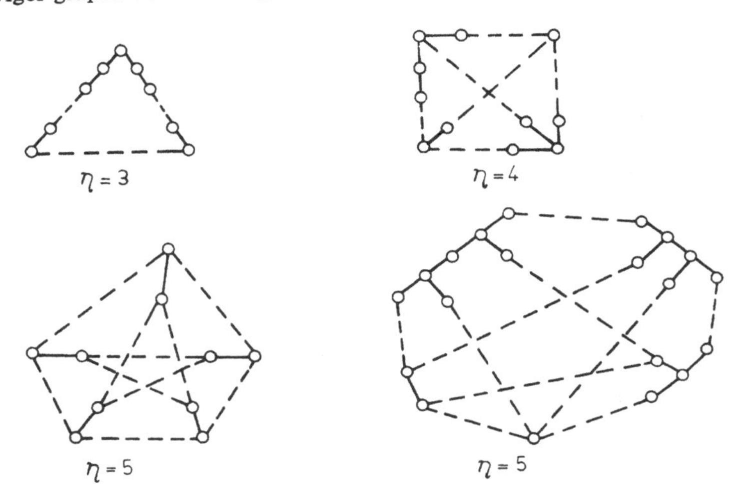


Fig. 6.1.1

Obviously, $H_\eta$ is 2-connected and has Hadwiger number $\eta$. Moreover, a graph $H_\eta$ is a Hadwiger graph with Hadwiger number $\eta$ if and only if $K_\eta$ is a contraction of $H_\eta$.

Now the famous conjecture of H. Hadwiger, 1943 and 1958, can be formulated as follows:

**Conjecture 6.1.1.** *For any graph $G$ we have $\chi(G) \leqq \eta(G)$.*

In section 6.3 it will be proved that every critically $k$-chromatic graph of large order with $\eta(G) = \eta$ contains a large Hadwiger graph $H_\eta$.

**Definition.** Let $C_1$ and $C_2$ be two cycles which have at most one vertex in common. We link $C_1$ and $C_2$ by three disjoint paths, where one of these paths has length 0 if $C_1$ and $C_2$ have a common vertex. The resulting graph is said to be a *prism graph* (see Fig. 6.1.2).

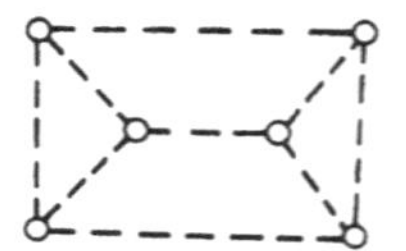
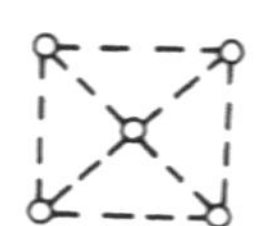

Fig. 6.1.2

## 6.2. Extensions of the results of Chapter 4 to other classes of graphs

First we shall introduce some new classes of graphs which have similar properties to the class $\mathfrak{Z}_l$.

**Definition.** A class $\mathfrak{K}$ of finite graphs is said to have *property* E if all graphs $G \in \mathfrak{K}$ satisfy the following condition: if $t$ is a path with ends in $G$, all inner vertices of which are not in $G$, then $G \cup t$ contains a subgraph $G' \in \mathfrak{K}$ so that $t \subseteqq G'$.

**Proposition 6.2.1.** *The following classes of graphs have property* E:

(i) *the class $\mathfrak{W}$ of all paths;*

(ii) *the class $\mathfrak{W}_{A,B}$ of all paths with fixed ends $A$ and $B$;*

(iii) *the class $\mathfrak{C}$ of all circuits;*

(iv) *the class $\mathfrak{O}$ of all odd circuits;*

(v) *the class $\mathfrak{H}_\eta$ of all Hadwiger graphs with Hadwiger number $\eta$ ($\eta \geqq 3$);*

(vi) *the class $\mathfrak{P}$ of all prism graphs;*

(vii) *the class $\mathfrak{T}(X_1, \ldots, X_s)$ of all trees with fixed ends $X_1, \ldots, X_s$.*

The class of all even circuits does not have property E. For all classes of graphs with property E the following theorem is true:

**Theorem 6.2.2.** *Let $\mathfrak{K}$ denote a class of finite graphs with property* E, *and $N$ a positive integer. Let $G$ be a 2-connected (finite or infinite) graph which contains a subgraph $H \in \mathfrak{K}$ of order $N$ but no subgraph from $\mathfrak{K}$ of order greater than $N$. Then $G$ has circumference $l \leqq N^2$.*

*Proof.* Let $L$ denote a largest cycle of length $l$ in the graph $G$. We consider two cases: Case 1. Assume that $L$ contains at most one vertex of the subgraph $H$. Since $G$ is 2-connected, there are two disjoint paths $p_1$ and $p_2$ joining $L$ and $H$ (if $L$ and $H$ have a common vertex, then $p_1$ has length 0). The ends of $p_1$ and $p_2$ on $L$ split $L$ into two arcs $b_1$ and $b_2$. The path $t_i =_{\text{def}} p_1 \cup b_i \cup p_2$ is an $H,H$-path, i.e. only the ends of $t_i$ belong to $H (i = 1, 2)$. Hence $H \cup t_i$ contains a subgraph $H_i$ with $t_i \subseteqq H_i \in \mathfrak{K}$. The graph $H$ is the largest subgraph of $G$ belonging to $\mathfrak{K}$. Therefore,

$$\begin{aligned}\lambda(L) &\leqq \lambda(b_1) + \lambda(b_2) \leqq \lambda(t_1) - 1 + \lambda(t_2) - 1\\ &\leqq n(t_1) - 2 + n(t_2) - 2\\ &\leqq n(H_1) + n(H_2) - 4 \leqq 2N - 4 \leqq N^2.\end{aligned}$$

Case 2. Assume that $L$ contains at least two vertices of the subgraph $H$. The subgraph $H$ has $N$ vertices. These vertices split $L$ into the arcs $b_1, b_2, \ldots, b_s$ $(s \leqq N)$. Hence $H \cup b_i$ contains a subgraph $H_i$ with $b_i \subseteqq H_i \in \mathfrak{K}$. By the maximality of $H$ it follows that

$$\lambda(b_i) \leqq n(b_i) - 1 \leqq n(H_i) - 1 \leqq n(H) - 1 \leqq N - 1.$$

Consequently, $\lambda(L) = \sum_{i=1}^{s} \lambda(b_i) \leqq N(N-1)$. □

If $\mathfrak{K} = \mathfrak{W}$ and if $\mathfrak{K} = \mathfrak{O}$, then a better bound can easily be obtained.

**Theorem 6.2.3.** *If* $\mathfrak{K} = \mathfrak{W}$, *then* $l \leqq N$. *If* $\mathfrak{K} = \mathfrak{O}$, *then* $l \leqq 2N - 2$.

*Proof.* Obviously, for $\mathfrak{K} = \mathfrak{W}$ the assertion is true. In order to prove the stronger theorem for $\mathfrak{K} = \mathfrak{O}$ we shall more carefully examine the proof of Theorem 6.2.2. Let $H \in \mathfrak{O}$ and let $L$ be defined as in the proof of Theorem 6.2.2. If $L$ is an odd cycle, then $L \in \mathfrak{O}$ and $\lambda(L) \leqq N$, which is smaller than $2N - 2$.

Now let $L$ be an even cycle. If this even cycle $L$ contains at most one vertex of the subgraph $H$, then the proof of case 1 of Theorem 6.2.2 provides the bound $2N - 4$ which is smaller than $2N - 2$. Now let the even cycle contain at least two vertices of the subgraph $H$. Let $P$ and $Q$ denote two arbitrary vertices of $H \cap L$. Since $H \in \mathfrak{O}$ and $L$ is an even circuit, the parity of one of the two arcs of $H$ between $P$ and $Q$ is different from the parity of the two arcs of $L$ between $P$ and $Q$. From this it can easily be deduced that there exists an arc $\bar{H}$ of $H$ with $\| \bar{H} \cap L \| = 2$ and the parity of $\bar{H}$ is different from the parity of the two arcs $L_1, L_2$ of $L$ connecting the two vertices of $\bar{H} \cap L$ in $L$. Hence $L_i \triangle \bar{H} \in \mathfrak{O}$ and

$$\lambda(L) + 2\lambda(\bar{H}) = \lambda(L_1 \triangle \bar{H}) + \lambda(L_2 \triangle \bar{H}) \leqq 2N. \ \square$$

In Theorem 6.2.2 the bound $N^2$ is not the best possible one. The bound $2N - 2$ in Theorem 6.2.3 is best possible. This is shown by the graph which consists of two vertices of valency 3 which are linked by an edge and by two topological edges of length $N - 1$ ($N$ odd).

By Theorem 6.2.3 in every 2-connected non-bipartite graph $G$ we have

$$l_o \leqq l \leqq 2(l_o - 1),$$

where $l$ and $l_o$ denote the maximum circuit length and the maximum odd circuit length, respectively.

A similar assertion for the maximum even circuit length $l_e$ does not exist. This is shown by the graphs of the class $\mathfrak{O}(r)$ in Chapter 7 at page 139.

Let $\mathfrak{Z}_N(\mathfrak{K})$ denote the class of all 2-connected graphs $G$ having the following property: $G$ contains a subgraph $H \in \mathfrak{K}$ of order $N$ but no subgraph from $\mathfrak{K}$ of order greater than $N$. Obviously, $\mathfrak{Z}_l(\mathfrak{C}) = \mathfrak{Z}_l$. Theorem 6.2.2 implies the following result:

**Lemma 6.2.4.** $\mathfrak{Z}_N(\mathfrak{K}) \subseteqq \bigcup_{l=3}^{N^2} \mathfrak{Z}_l(\mathfrak{C}) = \bigcup_{l=3}^{N^2} \mathfrak{Z}_l$.

Hence the assertion of Theorem 4.2.3 is also true for the graphs of $\mathfrak{Z}_N(\mathfrak{K})$. This result will be formulated as follows:

**Theorem 6.2.5.** *Let $\mathfrak{K}$ denote a class of finite graphs with property* E. *Let $N_{\mathfrak{K}}$ be the order of a smallest graph of $\mathfrak{K}$. Let $p$, $N$ be integers with $p \geqq 2$, $N \geqq N_{\mathfrak{K}}$. Then there is an integer $n(\mathfrak{K}, p, N)$ such that every graph $G \in \mathfrak{Z}_N(\mathfrak{K})$ of order greater than $n(\mathfrak{K}, p, N)$ contains $p$ isomorphic bridges.*

We also refer to this result as "every large graph of $\mathfrak{Z}_N(\mathfrak{K})$ contains $p$ isomorphic bridges".

With the aid of Lemma 6.2.4 we conclude that Theorem 4.2.3, Simonovits's Theorem 4.2.5 (with respect to cycles), and the reduction Theorem 4.5.1 also hold for the class $\mathfrak{Z}_N(\mathfrak{K})$. The formulation of the corresponding theorems is left to the reader (see H.-J. Voss, 1977d).

## 6.3. Critically $k$-chromatic graphs

First we state a well-known simple property.

**Proposition 6.3.1.** *Every critically $k$-chromatic graph has minimum degree at least $k-1$.* □

In the following we shall study properties of bridges of subgraphs in critically $k$-chromatic graphs.

We consider a proper subgraph $H$ of a critically $k$-chromatic graph $G$. Then $\chi(H) < \chi(G) = k$ and $H$ has some $(k-1)$-colourings. Obviously, each bridge $B$ of $H$ has chromatic number $\chi(B) < k$ and also has some $(k-1)$-colourings.

Two colourings are said to be *distinct* if their partitions into independent sets are distinct.

**Definition.** Let $B$ be a bridge of a subgraph $H$ of a critically $k$-chromatic graph with $\emptyset \neq H \neq G$. We say that *$B$ forbids the $(k-1)$-colouring $C_H$ of $H$* if there is no $(k-1)$-colouring $C_B$ of $B$ such that $H \cup B$ is properly $(k-1)$-coloured by $C_H$ and $C_B$. □

In Fig. 6.3.1 an example is depicted: $G$ is critically 4-chromatic; $H$ consists of the two vertices $X$ and $Y$. There is no 3-colouring of $B_1$ or $B_2$ such that $X$ and $Y$ have the same colour or distinct colours, respectively. Consequently, $B_1$ forbids a colouring in which $X$ and $Y$ have the same colours, and $B_2$ forbids a colouring in which $X$ and $Y$ have distinct colours.

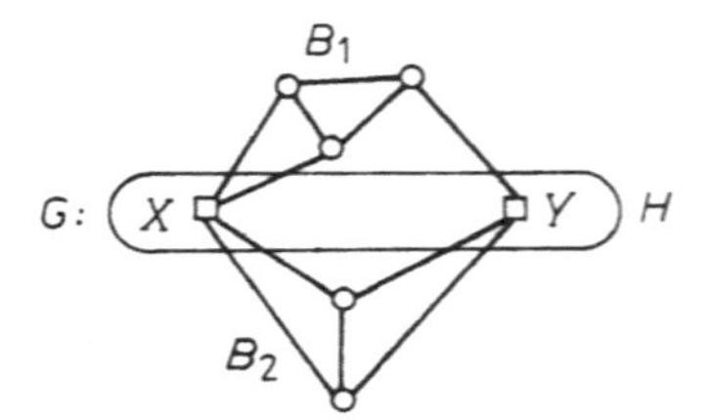


Fig. 6.3.1

**Theorem 6.3.2.** *Let $B$ be a bridge of a proper subgraph $H$ of a critically $k$-chromatic graph with $\emptyset \neq H \neq G$. Then a $(k-1)$-colouring $C_H$ of $H$ exists so that $B$ and no other bridge of $H$ forbids $C_H$.*

*Proof.* If $B$ has inner vertices, then we delete all inner vertices of $B$, and if $B$ is a $K_2$, then we delete the unique edge of $B$. The resulting graph has a $(k-1)$-colouring $C$. Since $G$ is $k$-chromatic, there is no $(k-1)$-colouring $C_B$ of $B$ such that $G$ is properly $(k-1)$-coloured by $C$ and $C_B$. Consequently, $B$ forbids the $(k-1)$-colouring $C_H$ of $H$ induced by $C$, and no other bridge of $H$ forbids $C_H$! □

**Corollary 6.3.3.** *Every critically k-chromatic graph G of order at least* 3 *is* 2*-connected.*

*Proof.* By contradiction. Assume that $G$ has at least two components or a cut vertex. In the first and in the second case let $A$ denote any vertex or a cut vertex of $G$, respectively. Then $G$ contains at least two bridges of $A$, but $A$ has only one $(k-1)$-colouring. This contradicts Theorem 6.3.2. □

**Corollary 6.3.4.** *Let H denote an induced subgraph of a critically k-chromatic graph G which has at least two bridges. Then H is not complete.*

*Proof.* By contradiction. Assume that $H$ is a complete graph. $G$ contains at least two bridges of $H$, but $H$ has only one $(k-1)$-colouring. This contradicts Theorem 6.3.2. □

**Corollary 6.3.5.** *No critically k-chromatic graph G contains two isomorphic bridges.*

*Proof.* By contradiction. Let $B_1$ and $B_2$ be two isomorphic bridges of a certain subgraph $H$ of $G$. Then $B_1$ and $B_2$ forbid the same set of $(k-1)$-colouring of $H$. This contradicts Theorem 6.3.2. □

It is also very easy to give a direct proof of Corollary 6.3.5. In the following a critical graph is always a critically $k$-chromatic graph for some $k$.

From Theorem 6.2.5 and Corollaries 6.3.3 and 6.3.5 we shall derive the following result:

**Theorem 6.3.6.** *Let $\mathfrak{K}$ be a class of finite graphs with property* E. *Then for each integer N there is an integer $n(\mathfrak{K}, N)$ such that every critical graph of order greater than $n(\mathfrak{K}, N)$ with a subgraph in $\mathfrak{K}$ also has a subgraph in $\mathfrak{K}$ of order greater than N.*

*Proof.* Let $G$ be a critical graph, a subgraph of which belongs to $\mathfrak{K}$. Let $G$ have more than $n(\mathfrak{K}, N) =_{\text{def}} n(\mathfrak{K}, 2, N)$ vertices, where $n(\mathfrak{K}, 2, N)$ is the integer defined in Theorem 6.2.5. By Corollary 6.3.3 the graph $G$ is 2-connected.

(1) *G contains a subgraph $H \in \mathfrak{K}$ of order greater than N.*

*Proof of (1).* By contradiction. Assume that $G$ does not contain a graph of $\mathfrak{K}$ of order greater than $N$ as a subgraph. Then $G \in \mathfrak{Z}_i(\mathfrak{K})$ for some index $i \leqq N$. Since $G$ has order greater than $n(\mathfrak{K}, 2, N) \geqq n(\mathfrak{K}, 2, i)$, by Theorem 6.2.5 the graph $G$ contains two isomorphic bridges. This contradicts Corollary 6.3.5. □

With (1) the proof of Theorem 6.3.6 is complete. □

In the following we refer to this result as "every large critical graph, a subgraph of which belongs to $\mathfrak{K}$, also contains a large graph from $\mathfrak{K}$".

Theorem 6.3.6 states that every critically $k$-chromatic graph of large order which has an element of $\mathfrak{K}$ as a subgraph also contains a large graph of $\mathfrak{K}$. If $F(\mathfrak{K}, n, k)$ is the largest integer such that every critically $k$-chromatic graph of order $n$ which has an element of $\mathfrak{K}$ as a subgraph contains a subgraph of $\mathfrak{K}$ of order at least $F(\mathfrak{K}, n, k)$, then we have the following result:

**Proposition 6.3.7.** $\lim_{n \to \infty} F(\mathfrak{K}, n, k) = +\infty$. □

Obviously, for $k \geqq 3$ every critically $k$-chromatic graph contains a subgraph $H' \in \mathfrak{C}$ and a subgraph $H'' \in \mathfrak{D}$. Consequently, by Theorem 6.3.6, we have the following result:

**Corollary 6.3.8.** *Every large critical graph contains a large circuit and also a large odd circuit.* □

The first assertion was proved by J. B. Kelly and L. M. Kelly, 1954; the second assertion gives an answer to the case $\varkappa = 3$ of the question posed by J. Nešetřil and V. Rödl at the International Colloquium on Finite and Infinite Sets held at Keszthely in Hungary in 1973 (oral communication, see also B. Toft, 1986).

**Problem 6.3.1.** *Let $\varkappa$, $k$, $N$ be arbitrary positive integers with $\varkappa < k$. Does there exist a positive integer $n$ such that every critically $k$-chromatic graph $G$ with at least $n$ vertices contains a critically $\varkappa$-chromatic subgraph $G'$ with at least $N$ vertices?*

The order of the magnitude of $F(\mathfrak{C}, n, k)$ was investigated by J. B. Kelly and L. M. Kelly, 1954, G. A. Dirac, 1955, and R. C. Read, 1957. T. Gallai, 1963, has obtained a stronger result by showing that

*for an infinite set of different positive integers $n$ there exist critically $k$-chromatic graphs of order $n$ of circumference at most $c_k \log n$, where $c_k$ is an appropriate constant.* □

By Theorem 6.2.3 it follows that $F(\mathfrak{C}, n, k)$ and $F(\mathfrak{D}, n, k)$ have the same magnitude. It also follows that the result "every large critical graph contains a large odd circuit" can be derived from the result of Kelly and Kelly that "every large critical graph contains a large circuit".

In the following for a critically $k$-chromatic graph we briefly write $k$-critical graph.

G. A. Dirac, 1952b, has proved that every 4-critical graph contains a subdivision of $K_4$. B. Zeidl, 1958, has shown that every 4-critical graph has a subdivision of $K_4$ containing a circuit of odd length. An improvement of this result is the Theorem 8.4.1 of C. Thomassen and B. Toft, 1981 (see Chapter 8): every 4-critical graph $G$, $G \neq K_4$, contains a subdivision of $K_4$ in which a 4-cycle becomes an odd cycle and one of the two remaining edges is left undivided.

G. A. Dirac, 1952b, has proved the following result:

**Theorem 6.3.9.** *Each circuit of a $k$-critical graph $G$, $k \geqq 4$, is contained in a subdivision of $K_4$.* □

If we apply this result to the largest circuits and to the largest odd circuits, then we obtain the following result from Theorem 6.3.6 with respect to $\mathfrak{K} = \mathfrak{C}$ and $\mathfrak{K} = \mathfrak{D}$:

**Theorem 6.3.10.** *For $k \geqq 4$ every large $k$-critical graph has a large subdivision of $K_4$ and also a subdivision of $K_4$ having a large odd circuit.* □

The first statement is directly implied by Theorem 6.3.6. The second statement cannot be derived from Theorem 6.3.6, because the class of all subdivisions of $K_4$ having an odd circuit does not have property E.

Proposition 6.3.1 implies:

**Proposition 6.3.11.** *Every $k$-critical graph $G$ of order $n$ has at least $\frac{(k-1)}{2} n$ edges.*

*Proof.* By Proposition 6.3.1 the graph $G$ has minimum degree at least $k-1$. A well-known result is that the sum of the degrees of all vertices of $G$ is twice the number $e$ of edges of $G$. Consequently, $2e \geqq (k-1)\,n$. □

This lower bound was improved by T. Gallai, 1963, and G. A. Dirac, 1974. For $k \geqq$ 6every $k$-critical graph contains at least $\frac{5}{2}n$ edges. A result of G. A. Dirac, 1964, is that

*every simple graph of order $n \geqq 5$ with at least $\frac{5}{2}n - 3$ edges contains a subdivision of $K_5^-$, where $K_5^-$ is obtained from $K_5$ by deleting one edge.* □

Since this graph contains a special prism graph, we have the following result:

**Theorem 6.3.12.** *For $k \geqq 6$ every $k$-critical graph contains a prism graph.* □

It can easily be proved that for integers $n \geqq 3$ the class of all 2-connected graphs contractible to $K_n^-$ has property E. Applying Theorem 6.3.6 we obtain the following result:

**Theorem 6.3.13.** *For $k \geqq 6$ every large $k$-critical graph contains a large prism graph and also a large 2-connected subgraph contractible to $K_5^-$.*

Next we state the following result:

**Theorem 6.3.14.** *For each positive integer $r$ there exist integers $k_r$ and $k'_r$ such that for all positive integers $k \geqq k_r$ or $k \geqq k'_r$ every $k$-critical graph contains a Hadwiger graph $H_r$ (with Hadwiger number $r$) or a subdivision of $K_r$, respectively.* □

The first result is due to K. Wagner, 1964, and the second result is due to H. A. Jung, 1965 (see also W. Mader, 1967). By Theorem 6.3.6 we obtain the following result:

**Theorem 6.3.15.** *For all $k \geqq k_r$ every large $k$-critical graph contains a large Hadwiger graph $H_r$.* □

By our methods a similar proposition with respect to subdivisions of $K_r$ cannot be proved. Moreover, I do not know whether such a proposition is true.

By Theorem 6.3.6 the set of all critical graphs which contain an element of $\mathfrak{K}$ of order $N$ but which do not contain an element of $\mathfrak{K}$ of order greater than $N$ is finite. Therefore, there is an integer $k(\mathfrak{K}, N)$ such that the class of all graphs containing an element of $\mathfrak{K}$ of order $N$ but none of order greater than $N$ is empty. In other words:

**Theorem 6.3.16.** *Every graph containing an element of $\mathfrak{K}$ of order $N$ but none of order greater than $N$ is properly colourable by $k(\mathfrak{K}, N)$ colours.* □

By a result of P. Erdős and H. Hajnal, 1966, we can take $k(\mathfrak{D}, N) = N+2$ because they showed that

*every graph which does not contain circuits of lengths $2j+1$ for all $j \geqq i$ is properly colourable by $2i$ colours.* □

Some results on colouring graphs of $\mathfrak{Z}_l$ are presented in H.-J. Voss, 1973, and in H. Walther and H.-J. Voss, 1974.

## 6.4. Graphs which are edge-critical with respect to the independence number

In section 6.4 we deal with properties of graphs which are edge-critical with respect to the independence number. These results are needed in section 6.5.

For convenience we refer to "graphs which are edge-critical with respect to the independence number" as, simply, edge-critical graphs.

There are infinite sequences of 2-connected edge-critical graphs. Simple families are the odd cycles and the complete graphs. Other edge-critical graphs are obtained by replacing some vertices of an odd cycle by complete graphs (see Fig. 6.4.1).

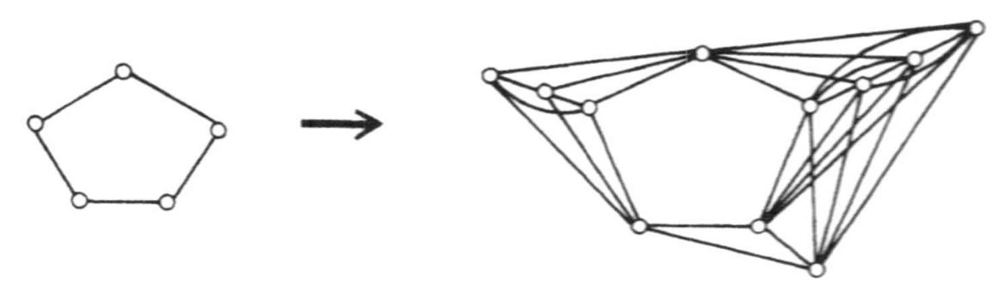

Fig. 6.4.1

From a series of papers giving other families we mention A. Hajnal, 1965, B. Andrasfai, 1967, M. D. Plummer, 1967, M. M. Krieger, 1970, L. Surányi, 1973 and 1976, L. Lovász, 1976, and W. Wessel, 1973 and 1976.

Next we formulate a proposition.

**Proposition 6.4.1.** *A bipartite graph G is edge-critical if and only if the components of G are disjoint $K_1$'s and $K_2$'s.*

This proposition follows from the next theorem:

**Theorem 6.4.2.** *Each pair of adjacent edges of an edge-critical graph is contained in an odd circuit without chords.*

*Proof.* Let $G$ be an edge-critical graph with $\alpha(G) = \alpha$ and $e_1$, $e_2$ any two adjacent edges of $G$. Let $Y_i$, $X$ denote the end vertices of $e_i\ (i = 1, 2)$. Since $G$ is edge-critical, $\alpha(G - e_i) = \alpha + 1$. Obviously, each maximum independent vertex set of $G - e_i$ contains both vertices $Y_i$ and $X$. Let $I_i \cup \{X\}$, $X \notin I_i$, denote such a maximum independent set of $G - e_i$. Obviously,

(1) $\qquad Y_i \in I_i \quad (i = 1, 2).$

We consider the subgraph $H$ of $G$ induced by $(I_1 \cup I_2) \cup \{X\}$. By construction the vertex $X$ has two neighbours in $H$, namely, $Y_1$ and $Y_2$. Hence the graph $H - (I_1 \cap I_2) - (X, Y_2)$ is bipartite with the two vertex classes $I_1 \setminus I_2$ and $(I_2 \setminus I_1) \cup \{X\}$; all vertices of $I_1 \cap I_2$ consist of isolated vertices (see Fig. 6.4.2).

Since for $i = 1, 2$ the set $I_i \cup \{X\}$ is independent in $G - e_i$ and $(X, Y_{i+1})$, $Y_3 =_{\text{def}} Y_1$, is an edge of $G - e_i$, it follows that $Y_2 \notin I_1$ and $Y_1 \notin I_2$. Therefore, by (1):

(2) $\qquad Y_1 \in I_1 \setminus I_2 \quad$ *and* $\quad Y_2 \in I_2 \setminus I_1.$

Let $K$ denote the component of $H - X$ containing $Y_1$.

(3) $\qquad$ *$K$ contains $Y_2$.*

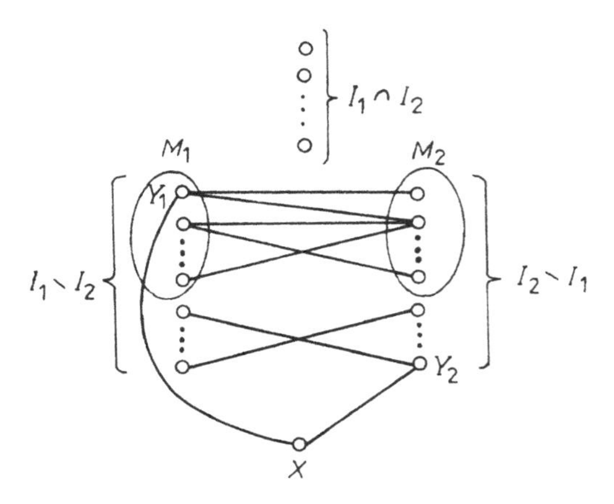


Fig. 6.4.2

*Proof of (3).* By contradiction. Assume that $Y_2 \notin K$. $K$ is bipartite. We define two vertex classes of $K$:

$$M_1 =_{\text{def}} K \cap (I_1 \setminus I_2) \quad \text{and} \quad M_2 =_{\text{def}} K \cap (I_2 \setminus I_1).$$

By construction all neighbours of $M_1$ are in $M_2 \cup \{X\}$ and all neighbours of $M_2$ are in $M_1$. In $H$ we exchange $M_1$ and $M_2 \cup \{X\}$; we obtain the sets $(I_1 \setminus M_1) \cup M_2 \cup \{X\}$ and $(I_2 \setminus M_2) \cup M_1$.

The definitions of $K$, $M_1$, $M_2$ and the assumption that $Y_2 \notin K$ imply that $Y_1 \in M_1$ and $Y_2 \notin M_2$. We conclude that $X \in (I_1 \setminus M_1) \cup M_2 \cup \{X\}$ and $Y_1, Y_2 \in (I_2 \setminus M_2) \cup M_1$. Consequently, both $(I_1 \setminus M_1) \cup M_2 \cup \{X\}$ and $(I_2 \setminus M_2) \cup M_1$ are independent vertex sets of $G$, where either $(I_1 \setminus M_1) \cup M_2 \cup \{X\}$ or $(I_2 \setminus M_2) \cup M_1$ has cardinality at least $\alpha(G) + 1$. Contradiction! □

By (2) and (3) we have $Y_1 \in M_1$ and $Y_2 \in M_2$. We choose a shortest $Y_1, Y_2$-path $p$ in $K$. Obviously, this path has odd length and no chord. Consequently, $p \cup [Y_2, X, Y_1]$ is an odd chordless cycle containing the two adjacent edges $e_1$ and $e_2$. □

Theorem 6.4.2 implies the following result:

**Proposition 6.4.3.** *Each component of an edge-critical graph is either an isolated vertex or a $K_2$ or a 2-connected graph.*

For an example see the graph of Fig. 6.4.3.

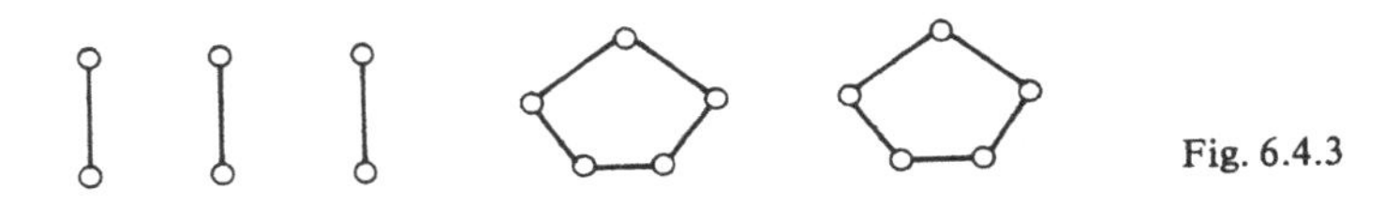

Fig. 6.4.3

*Proof.* Let $K$ be a component of order at least 3 of an edge-critical graph. Hence $K$ has at least two edges. Since by Theorem 6.4.2 every two adjacent edges are in an (odd) cycle, $K$ has neither cut edges nor cut vertices. Consequently, $K$ is 2-connected. □

## 6.5. Properties of graphs which are edge-critical with respect to the independence number

In section 6.5 the results of section 6.2 will be applied to graphs which are edge-critical with respect to the independence number. First a lemma will be proved.

**Lemma 6.5.1.** *Let $G$ be a graph which has two distinct isomorphic bridges $B_1$ and $B_2$ with kernels $K_1$ and $K_2$, respectively. Let $I$ denote a maximum independent vertex set of $G$. Then $I \cap B_1$ and $I \cap B_2$ have equal cardinality,* i.e. $|I \cap B_1| = |I \cap B_2|$.

*Proof.* By contradiction. Assume that $|I \cap B_1| > |I \cap B_2|$ and, consequently, $|I \cap K_1| > |I \cap K_2|$ for a graph $G$, two isomorphic bridges $B_1$, $B_2$ in $G$ and a maximum independent set $I$ of $G$.

Since $B_1$ and $B_2$ are isomorphic bridges, there is an isomorphism $\varphi\colon G - K_2 \to G - K_1$ such that $\varphi(X) = X$ for all vertices $X$ of $G - K_2 - K_1$. Then $I' = (I \setminus (I \cap K_2)) \cup \varphi(I \cap K_1)$ is a new independent vertex set with $|I'| > |I|$. Contradiction! □

**Proposition 6.5.2.** *No connected edge-critical graph has two distinct isomorphic bridges.*

For disconnected edge-critical graphs, Proposition 6.5.2 is not true (see Fig. 6.4.3).

*Proof.* By contradiction. Assume that there exists a connected edge-critical graph which has two distinct isomorphic bridges $B_1$ and $B_2$. Since $G$ is connected, $B_1$ and $B_2$ have a common vertex of attachment $R$.

Let $X$ be an inner vertex of $B_1$ adjacent to $R$. Since $G$ is edge-critical, the graph $G - (X, R)$ has a maximum independent vertex set $I$ of order $\alpha(G - (X, R)) = \alpha(G) + 1$. Obviously, $I$ contains $X$ and $R$. Therefore, $I - X$ and $I - R$ are two maximum independent sets of $G$. Then one of the two sets does not satisfy the equality condition of Lemma 6.5.1. Contradiction! □

Now we formulate the main theorem of this section.

**Theorem 6.5.3.** *Let $\mathfrak{K}$ be a class of finite graphs with property E. Then for each integer $N$ there is an integer $n(\mathfrak{K}, N)$ such that every connected edge-critical graph of order greater than $n(\mathfrak{K}, N)$ with a subgraph in $\mathfrak{K}$ also has a subgraph from $\mathfrak{K}$ of order greater than $N$.*

*Proof.* Let $G$ be a connected edge-critical graph, a subgraph of which belongs to $\mathfrak{K}$. Let $G$ have more than $n(\mathfrak{K}, N) =_{\text{def}} \max\{3, n(\mathfrak{K}, 2, N)\}$ vertices, where $n(\mathfrak{K}, 2, N)$ is the integer of Theorem 6.2.5. By Proposition 6.4.3 the graph $G$ is 2-connected.

(1) *$G$ contains a subgraph $H \in \mathfrak{K}$ of order greater than $N$.*

*Proof of (1).* By contradiction. Assume that $G$ does not contain a subgraph of $\mathfrak{K}$ of order greater than $N$ as a subgraph. Then $G \in \mathfrak{Z}_i(\mathfrak{K})$ for some index $i \leqq N$. Since $G$ has order greater than $n(\mathfrak{K}, 2, N) \geqq n(\mathfrak{K}, 2, i)$, by Theorem 6.2.5 the graph $G$ contains two isomorphic bridges. This contradicts Proposition 6.5.2! □

By Theorem 6.5.3 we have the following result:

**Theorem 6.5.4.** *Every large connected edge-critical graph containing a cycle, an odd cycle, a prism graph or a Hadwiger graph $H$, respectively, also has a large one.* □

# Chapter 7

# Long cycles in graphs with given minimum degree

## 7.0. Introduction

The concept of cycle plays a fundamental role in the theory of undirected graphs and there are numerous papers dealing with cycles. The literature on cycles in graphs is extensive. For surveys on cycles in undirected graphs, the reader is referred to the monographs of H. Walther and H.-J. Voss, 1974, and G. Schaar, M. Sonntag, and H.-M. Teichert, 1988, and the survey articles by J. C. Bermond, 1978, J. A. Bondy, 1978, and L. Lesniak-Foster, 1977. The last book and the three articles partially deal with Hamiltonian cycles.

A survey article on cycles in digraphs is due to J. C. Bermond and C. Thomassen, 1981. More references on cycles in digraphs can be found there.

In this section we restrict ourselves to the investigation of valency conditions for the existence of long cycles having certain properties. We present a survey on these problems in $1° - 5°$.

### 1° Large cycles in graphs with given minimum degree

The well-known theorem of G. A. Dirac, 1952a, stands at the very beginning of a discussion of large cycles and, especially, of Hamiltonian cycles (see Theorem 7.2.1).

(1) *Every 2-connected graph $G$ with minimum degree at least $r \geqq 3$ has a Hamiltonian cycle or contains a cycle of length at least $2r$.*

In a graph with $n$ vertices the assumption that $G$ has minimum degree at least $n/2$ implies that $G$ is 2-connected. Hence in (1) the 2-connectedness of $G$ can be dropped.

(2) *Every graph $G$ of order $n$ with minimum degree at least $n/2$ has a Hamiltonian cycle.*

This has been strengthened by O. Ore, 1960.

(3) *Let $G$ be a graph of order $n$ with $v(G:U) + v(G:V) \geqq n$ for each non-adjacent pair of distinct vertices $U$ and $V$. Then $G$ has a Hamiltonian cycle.*

This was later generalized by J. A. Bondy, 1971, J. C. Bermond, 1976, N. Linial, 1976, and G. A. Dirac, 1978. An elegant proof appears in L. Lovász, 1979 (Exercise 10.27, pp. 68, 393).

(4) *Let $G$ be a 2-connected graph with $v(G:U) + v(G:V) \geqq m$ for each non-adjacent pair of distinct vertices $U$ and $V$. Then $G$ has a Hamiltonian cycle or a cycle of length at least $m$.*

Later P. Erdős and T. Gallai, L. Pósa, 1963, J. A. Bondy, 1970, 1971, J. C. Bermond, 1975, 1976, N. Linial, 1976, M. Grötschel, 1977, H.-J. Voss, 1976a, 1977a, H.-J. Voss and C. Zuluaga, 1977, H. A. Jung, 1981, 1986, 1990, H. Enomoto, 1984, G. Fan, 1984, 1990a, b, S. C. Locke, 1985a, B. Jackson, 1985, I. Fournier and P. Fraisse, 1985, I. Fournier, 1985, P. Fraisse, 1986, Y. Egawa, R. Glas and S. C. Locke, 1987, P. Fraisse and H. A. Jung, 1989, and Min Aung, 1989a, b, proved various extensions of Dirac's result.

O. Ore, 1967b, determined all graphs of (1) with circumference $2r$ and H.-J. Voss, 1976a, has done this for all graphs with circumference $2r+1$. H. Enomoto, 1984, has characterized all graphs of (4) with circumference $m$. Later D. Bauer, H. A. Jung and E. Schmeichel, 1989, characterized all graphs of (4) with circumference $m$ and $m+1$.

With respect to the parity of the cycles H.-J. Voss and C. Zuluaga, 1977, have proved that

(5) *every non-bipartite 2-connected graph G of order at least $2r$ with minimum degree at least $r \geqq 3$ has both an odd cycle of length at least $2r-1$ and an even cycle of length at least $2r$.*

All graphs having only longest odd cycles of length $2r-1$ or longest even cycles of length $2r$ have been determined.

For bipartite graphs H.-J. Voss and C. Zuluaga, 1977, proved that

(6) *every bipartite 2-connected graph G with minimum degree at least $r \geqq 3$ having at least $2r-2$ vertices in each bipartition set contains a cycle of length at least $4r-4$.*

In sections 7.2, 7.4 and 7.6 we present the bridges of longest cycles in graphs which attain the lower bound in (1), (4) and (5). These graphs will be referred to as extremal graphs of (1), (4) and (5), respectively. The extremal graphs will be determined.

In a similar way longest cycles through given vertices and edges will be treated in sections 7.4 and 7.5.

We conclude 1° by presenting a result on Dirac-type conditions on the degrees of the vertices, due to P. Erdős and T. Gallai, 1959. This result stands at the very beginning of investigations of longest cycles through given sets of vertices and edges.

(7) *Each pair P, Q of distinct vertices of a 2-connected graph with minimum degree at least $r \geqq 3$ is joined by a P,Q-path of length at least r.*

## 2° Large cycles in regular graphs

Regular graphs enjoy many properties that one might not expect. For example, it follows from the theorem of P. Hall, 1935, that every non-empty regular bipartite graph has a 1-factor; and a theorem of J. Petersen, 1891, asserts that every non-empty regular graph of even degree has a 2-factor. So it is with cycles.

There are non-Hamiltonian graphs with $\delta(G) \geqq r \geqq 3$ and order $n(G) \geqq r+2$.

For regular graphs B. Jackson, 1980a, has proved the following result:

*If the 2-connected r-regular graph G has order $n(G) \leqq 3r$, then G contains a Hamiltonian cycle.*

G. Fan, 1985, R.-Y. Chen (see G. Fan, 1985) and H.-J. Voss, 1989b, independently proved the following result:

*Except for the Petersen graph all 2-connected r-regular graphs of order at least* $2r+4$ *have circumference at least* $2r+4$.

A similar result relating to odd and even cycles in 2-connected $r$-regular graphs can be derived from a result of H.-J. Voss and C. Zuluaga, 1977.

These theorems and related results will be discussed in section 7.7.

## 3° Cycles with prescribed bridges

The simplest bridges of cycles are the diagonals. Hence cycles having only singular bridges, namely, diagonals, are the Hamiltonian cycles. There is a vast literature on this subject and we refer the reader to the three articles mentioned above. More complicated bridges of cycles are the bridges with precisely one inner vertex. A cycle $C$ of a graph having only bridges with at most one inner vertex is said to be *dominating*. Obviously, a cycle $C$ of $G$ is dominating if and only if $G-C$ has no edge. In other words, $C$ is dominating if and only if every edge of $G$ is adjacent to one vertex of $C$. Such a cycle is called a *D-cycle* by some authors.

The *line graph* $L(G)$ of a graph $G$ has vertex set $E(G)$, and two vertices $e_1, e_2 \in E(G)$ are joined by an edge if $e_1, e_2$ are adjacent. The dominating cycles are closely related to the Hamiltonian cycles. So F. Harary and C. St. J. A. Nash-Williams, 1965, proved the following result: Let $G$ be a graph having at least four vertices. Then the line graph of $G$ is Hamiltonian if and only if $G$ has a dominating closed trail or $G$ is a $K_{1,s}$, $s \geqq 3$.

The concept of dominating cycle can be generalized in two directions. First a cycle $C$ is called *vertex-dominating* if each vertex of $G$ is incident or adjacent to a vertex of $C$ (some authors call these cycles *dominating* and our dominating cycles *D-cycles*).

The second generalization is due to H. J. Veldman, 1983b: $C$ is called a *$D_\lambda$-cycle* if each component of $G-C$ has fewer than $\lambda$ vertices. Thus a $D_1$-cycle is Hamiltonian, a $D_2$-cycle is dominating.

In sections 7.8 and 7.9 we shall present conditions on the degrees of the vertices and other conditions for the existence of dominating, vertex-dominating and $D_\lambda$-cycles.

The well-known theorem of C. St. J. A. Nash-Williams, 1971a, stands at the very beginning of a discussion of dominating cycles.

(1) *Every 2-connected graph $G$ with minimum degree at least* $r \geqq 3$ *on at most* $3r-2$ *vertices has a dominating cycle.*

A result without restrictions on the number of vertices is due to H.-J. Voss, 1977a, and H. A. Jung, 1981.

(2) *Every 3-connected graph $G$ with minimum degree at least* $r \geqq 3$ *has a dominating cycle or a cycle of length at least* $3r-3$.

There are further results on dominating, vertex-dominating and $D_\lambda$-cycles, due to H.-J. Voss, 1977a, H.-J. Voss and C. Zuluaga, 1977, J. A. Bondy, 1980b, H. A. Jung, 1981, 1986, 1990, H. J. Veldman, 1983a and b, 1986, P. Ash and B. Jackson, 1984, B. N. Clark, Ch. J. Colbourn and P. Erdős, 1985, D. Bauer, 1985, P. Fraisse, 1986, J. A. Bondy and G. Fan, 1987.

We conclude the introduction with two problems which are not dealt with in sections 7.1-7.9, but which are closely related to the problems treated in Chapter 7.

1) According to 1° we shall consider long cycles through given vertices and edges in sections 7.4 and 7.5. In 4° we shall amplify this with a survey on the most important results with respect to the following problem: which sufficient conditions ensure the existence of a cycle through any set of vertices and edges of given cardinality?

2) In 5° we present sufficient conditions ensuring that the long cycles of a graph generate the cycle space.

### 4° Long cycles through given vertices and edges

There are many papers which deal with conditions that a cycle or path contains a given set of vertices and edges. In the introduction of the paper of D. A. Holton, B. D. McKay, M. D. Plummer and C. Thomassen, 1982, a good survey on the most important problems is given. We present the survey here.

As proved by G. A. Dirac, 1960, it is an immediate consequence of Menger's theorem that a $k$-connected graph contains a cycle through any $k$ prescribed vertices. M. E. Watkins and D. M. Mesner, 1967, characterized the graphs which show this to be best possible. These are precisely the $k$-connected graphs which contain $k$ vertices whose deletion results in a graph with more than $k$ components. J. A. Bondy and L. Lovász, 1981, proved that the set of cycles through $k$ specified vertices in a $(k+1)$-connected graph generates the cycle space of the graph. They also proved that if $G$ is not bipartite, then any $k-1$ vertices lie on an odd cycle and in general, any $k$ vertices lie on an even cycle.

The Dirac bound can be improved for $k$-connected $k$-regular graphs. It was proved by D. A. Holton, 1982, and independently by A. K. Kelmans and M. V. Lomonossov, 1982a, that a $k$-connected $k$-regular graph contains a cycle through any $k+4$ prescribed vertices. This lower bound is not believed to be best possible. For example, D. A. Holton, B. D. McKay, M. D. Plummer and C. Thomassen, 1982, proved that any 3-connected cubic graph contains a cycle through any nine vertices. This result was independently obtained by A. K. Kelmans and M. V. Lomonossov, 1982b. A. K. Kelmans and M. V. Lomonossov, 1982b, and M. N. Ellingham, D. A. Holton and C. H. C. Little, 1984, independently characterized all graphs which show this to be best possible, i.e., they proved that all 3-connected cubic graphs containing ten vertices not lying on a common cycle can be contracted to a Petersen graph. If $G$ is a planar 3-connected cubic graph, then any set of at most 17 vertices is contained in a cycle (J.-L. Fouquet and H. Thuillier, 1985).

The edge analogue of Dirac's result has also been studied. L. Lovász, 1974, conjectured that a $k$-connected graph has a cycle through any $k$ prescribed independent edges unless these edges form an odd cut set; he verified this for $k=3$. R. Häggkvist and C. Thomassen, 1982, proved a weaker conjecture made by D. R. Woodall, 1977, that a $(k+1)$-connected graph has a cycle through any $k$ prescribed independent edges.

Degree conditions for the existence of a cycle through a given set $M$ of independent edges have been investigated by R. Häggkvist, 1979, K. A. Berman, 1983, A. P. Wojda, 1988, A. Benhocine and A. P. Wojda, 1990, and B. Jackson and N. C. Wormald, 1990.

W. D. McCuaig and M. Rosenfeld, 1985, proved that in a 3-connected graph $G$ any pair of edges is in both an even and an odd cycle unless $G$ has a structure which is completely described by the authors.

N. Homobone and C. Peyrat, 1989, characterized all graphs such that every two edges are contained in a shortest cycle. In sections 7.4 and 7.5 we shall deal with longest cycles through given edges.

### 5° Cycle spaces

It is very interesting that in some cases the long cycles generate the cycle space. The cycle space of a graph is the vector space of all edge sets of Eulerian subgraphs of $G$ over GF(2). If $K$ and $L$ are sets of edges of two Eulerian subgraphs of $G$, their symmetric difference $K \triangledown L$ is defined as the set of edges of $G$ which belong to one of $K$ and $L$, but not to both. It is well known that the cycle space has a basis which consists only of cycles, where a cycle is regarded as the set of its edges (a definition of cycle space can also be found in 2.2.0, 2°).

A set $\mathfrak{C}$ of cycles in a graph $G$ is said to generate the cycle space if every cycle $C$ of $G$ can be written as a symmetric difference of cycles $C_i$ in $\mathfrak{C}$.

G. A. Dirac, 1952a, proved that

(1) *every graph $G$ with minimum degree at least $r \geqq 3$ has a cycle of length at least $r+1$.* □

J. Ben-Arroyo Hartman, 1983, proved that if $G$ is also 2-connected, then there are sufficiently many long cycles to generate the cycle space.

In the complete graph $K_{r+1}$ with $r$ odd and $r \geqq 3$, the cycles of length $r+1$ are all of even length, and hence cannot generate the cycle space, because no odd cycle can be expressed as a symmetric difference of even cycles.

The result of J. Ben-Arroyo Hartman, 1983, is:

(2) *Let $G$ be a 2-connected graph with $\delta(G) \geqq r \geqq 3$. Then the cycles of length at least $r+1$ generate the cycle space of $G$, unless $G \cong K_{r+1}$, where $r$ is odd, in which case the cycles of length at least $r$ generate the cycle space.* □

A short proof of Hartman's theorem is due to S. C. Locke, 1985c.

If the graphs in question are even 3-connected, then a stronger assertion can be proved. G. A. Dirac, 1952a, showed that every 2-connected graph $G$ of order at least $2r$ with minimum degree at least $r \geqq 3$ has a cycle of length at least $2r$.

J. A. Bondy made the following conjecture (see S. C. Locke, 1985b):

(3) **Conjecture.** *Let $G$ be a 3-connected graph of order at least $2r$ with $\delta(G) \geqq r \geqq 3$. Then the cycles of length at least $2r-1$ generate the cycle space of $G$.*

The conjecture would imply the result (5) in 1° of H.-J. Voss and C. Zuluaga, 1977.

The conjecture (3) has been proved by S. C. Locke, 1985b, in the following two cases: 1) $G$ is non-Hamiltonian, 2) $G$ has order at least $4r-5$.

We conclude this section with a result of J. A. Bondy and L. Lovász, 1981: If $S$ is a set of $k$ vertices in a $(k+1)$-connected graph, then the set of all cycles of $G$ through $S$ generates the cycle space of $G$.

## 7.1. Preliminaries

In this chapter the investigations of Chapter 3 will be continued for graphs with given minimum degree. First we repeat some notation.

Let $H$ be a subgraph of a graph $G$. A path $p$ of $G$ is said to be an *H-path* if and only if one end vertex and no other vertices of $p$ belongs to $H$. A path $v$ of $G$ is an *H,H-path* if and only if the two end vertices and no other vertices of $p$ belong to $H$. The end vertex $E$ of an $H$-path $p$, $E \in H$ (or $E \notin H$), is called the *first* (or the *second*) *end vertex* of $p$.

We shall apply the results obtained in Chapter 3, particularly Theorems 3.3.1, 3.3.4 and 3.3.6, on the lengths of bridges of longest cycles. For convenience we reformulate these results:

**Theorem 7.1.1.** *Let L be a longest circuit of length l of a 2-connected graph G. Then*

(i) $\lambda(B) \leqq \lfloor l/2 \rfloor$ *if B is a bridge of L;*

(ii) $\lambda(B_1) + \lambda(B_2) \leqq \lfloor l/2 \rfloor$ *if $B_1$ and $B_2$ are overlapping bridges of L having no common vertices of attachment;*

(iii) $\lambda(B_1) + \lambda(B_2) + \lambda(B_3) \leqq \lfloor l/2 \rfloor$ *if $B_1$, $B_2$, $B_3$ are bridges of L such that no two of them have a common vertex of attachment and $B_1$ does not overlap $B_2$ but $B_3$ overlaps both $B_1$ and $B_2$.* □

We remark that Theorem 7.1.1 remains true if the length $\lambda(B)$ of an $L$-bridge $B$ is replaced by a second measure $\tilde{\lambda}(B)$, introduced at the end of section 3.3. Theorem 7.1.1 is also true for bridges of which some vertices of attachment are deleted. Such bridges are sometimes called *reduced* bridges.

## 7.2. Properties of bridges of longest cycles in graphs with minimum degree at least $r$

First we introduce some notation. Let $G_1 = (V_1, E_1)$ and $G_2 = (V_2, E_2)$ be two graphs. Then the union of $G_1$ and $G_2$ is $G_1 \cup G_2 =_{\text{def}} (V_1 \cup V_2, E_1 \cup E_2)$. If $G_1$ and $G_2$ are not subgraphs of the same graph, then $G_1 \cup G_2$ always denotes the disjoint union of $G_1$ and $G_2$, i.e., $V_1 \cap V_2 = \emptyset$.

If $V_1 \cap V_2 = \emptyset$, then the join $G_1 + G_2$ is obtained from $G_1 \cup G_2$ by linking each vertex of $G_1$ with each vertex of $G_2$ by an edge. The union of $h$ disjoint copies of $G_1$ is denoted by $hG_1$. We again denote the complete graph on $n$ vertices by $K_n$ and the edgeless graph on $n$ vertices by $I_n =_{\text{def}} \bar{K}_n$. If $G_1$ is isomorphic to a subgraph of $G_2$, we shall write $G_1 \subseteqq G_2$.

Let $c(G)$, $n(G)$ and $\delta(G)$ again denote the circumference, the order and the minimum degree of a graph $G$, respectively. A graph $G$ is said to have *property* $E_r$ if and only if $G$ is 2-connected, non-Hamiltonian and $\delta(G) \geqq r$, $n(G) \geqq 2r + 1$.

We start with the well-known theorem of G. A. Dirac, 1952a.

**Theorem 7.2.1.** *If G has* $E_r$, $r \geqq 3$, *then* $c(G) \geqq 2r$.

There are several proofs of Dirac's theorem. One is due to J. A. Bondy, I. B.-A. Hartman and S.C. Locke, 1981. For bipartite graphs a larger lower bound for $c(G)$ can be obtained. H.-J. Voss and C. Zuluaga, 1977, proved that if $G$ is a 2-connected bipartite graph with bipartition $(A, B)$ having minimum degree at least $r$, then $c(G) \geqq 2 \min \{|A|, |B|, 2r-2\}$. This result and an improvement due to B. Jackson, 1985, will be discussed at the end of section 7.6. It is based on a bridge theorem presented in section 7.6.

Notation. Let $G_1 = (V_1, E_1)$ and $G_2 = (V_2, E_2)$ be two graphs and $I$ an independent set of vertices such that

$$V_1 \cap V_2 = I \cap V_1 = I \cap V_2 = \emptyset.$$

Then $B = (\underline{I} \cup G_1) + G_2$ denotes a bridge of $I$, where $\underline{I}$ is the set of vertices of attachment of $B$ and $G_1 + G_2$ the kernel of $B$ (see Fig. 7.2.1).

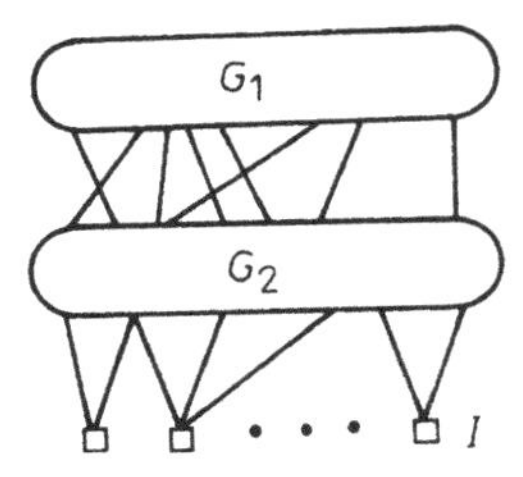


Fig. 7.2.1

In the figures the vertices of attachment and the inner vertices of a bridge $B$ will always be depicted by small rectangles and small circles, respectively.

O. Ore, 1967b, determined all graphs $G$ of Theorem 7.2.1 with $c(G) = 2r$. We shall refer to these graphs as the extremal graphs of Theorem 7.2.1.

First we present all possible bridges of a longest cycle in a graph $G$ with $E_r$, $r \geqq 3$, having circumference $2r$ or $2r+1$.

**Theorem 7.2.2.** *Let $G$ be a graph with property $E_r$, $r \geqq 3$, and $c(G) \leqq 2r+1$. Let $L$ denote a longest cycle of $G$ and $B$ a regular $L$-bridge of $G$. Then $B \cong \underline{I}_r + K_1$ or $B \cong \underline{I}_2 + K_{r-1}$ or, if $r = 4$, $l = 9$, also $B \cong \underline{I}_3 + K_2$* (see Fig. 7.2.2).

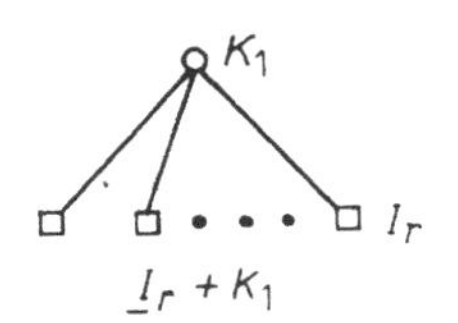


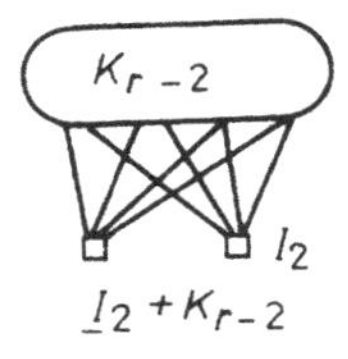


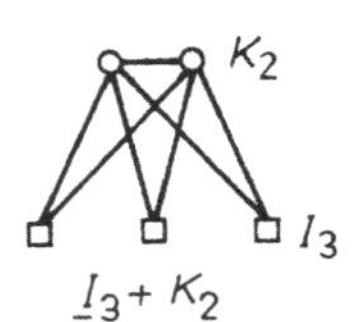


Fig. 7.2.2

We shall refer to $\underline{I}_s + K_1$, $s \geqq r$, as the one-vertex bridges, because their kernels contain precisely one vertex. The bridge $\underline{I}_2 + K_{r-1}$ is a complete graph on $r+1$ vertices with one edge missing, where the two vertices of attachment are not joined by an edge.

By Theorem 7.2.2 we know all bridges of a longest cycle of the graphs with property $E_r$, $r \geqq 3$, and circumference $2r$. Hence we can construct all extremal graphs of Theorem 7.2.2 and formulate Ore's theorem (O. Ore, 1967b).

**Theorem 7.2.3.** *A graph $G$ satisfying property $E_r$, $r \geqq 3$, has circumference $c(G) = 2r$ if and only if*

$$I_r + I_s \subseteqq G \subseteqq K_r + I_s, \qquad s \geqq r+1,$$

*or*

$$I_2 + hK_{r-1} \subseteqq G \subseteqq K_2 + hK_{r-1}, \qquad h \geqq 3. \ \square$$

By Theorem 7.2.2 we know all possible bridges of longest cycles in graphs with $E_r$, $r \geqq 3$, and circumference at most $2r+1$. From a longest cycle $L$ and these bridges all possible graphs with $E_r$ and $c(G) \leqq 2r+1$ can be constructed. Before formulating the result obtained we define some classes of graphs.

**Definition 7.2.4.** Let $r \geqq 3$ be an integer. Five classes of graphs $\mathfrak{C}_1^r$, $\mathfrak{C}_2^r$, $\mathfrak{C}_3^r$, $\mathfrak{C}_4$ and $\mathfrak{C}_5$ will be introduced.

(i) $\mathfrak{C}_1^r =_{\text{def}} \{K_r + I_s \mid s \geqq r+1\} \cup \{K_r + (K_2 \cup I_{s-1}) \mid s \geqq r+1\}$. For $r=3$ the elements of $\mathfrak{C}_1^r$ are depicted in Fig. 7.2.3a and b.

(ii) $\mathfrak{C}_2^r =_{\text{def}} \{K_2 + hK_{r-1} \mid h \geqq 3\} \cup \{K_2 + (K_r \cup (h-1)K_{r-1}) \mid h \geqq 3\}$. For $r=3$ the elements of $\mathfrak{C}_2^r$ are depicted in Fig. 7.2.3c and d.

(iii) Let $H_i =_{\text{def}} K_2 + iK_{r-1}$, $i \geqq 1$, and let $A_i$, $B_i$ denote the two vertices of $K_2$. All graphs of $\mathfrak{C}_3^r$ are obtained from disjoint unions $H_s \cup H_h$, $s, h \geqq 1$, $s+h \geqq 3$, by identifying $A_s$ and $A_h$ and joining $B_s$ and $B_h$ by an edge $(B_s, B_h)$ (see Fig. 7.2.3e).

(iv) $\mathfrak{C}_4 =_{\text{def}} \{K_3 + hK_2 \mid h \geqq 4\}$ (see Fig. 7.2.3f).

(v) All graphs of $\mathfrak{C}_5$ are obtained from disjoint unions $(sI_1 + K_3) \cup (I_2 + hK_2)$, $s \geqq 2$, $h \geqq 1$, $h+s \geqq 4$, by identifying the two vertices of $I_2$ with two distinct vertices of $K_3$ (see Fig. 7.2.3g).

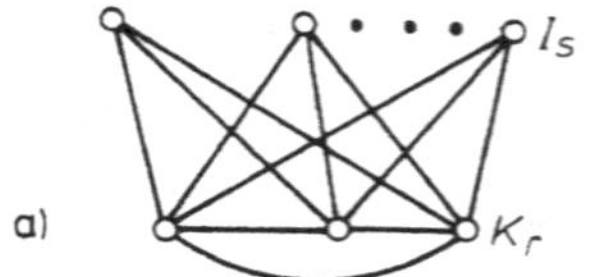


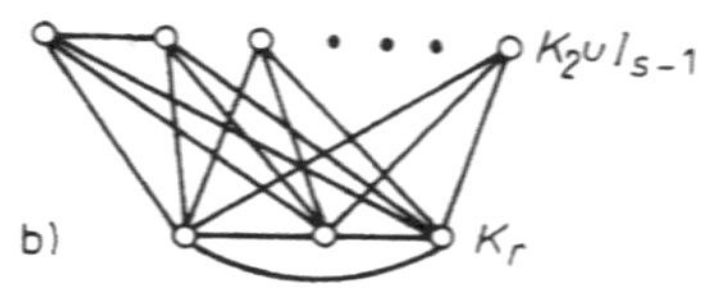


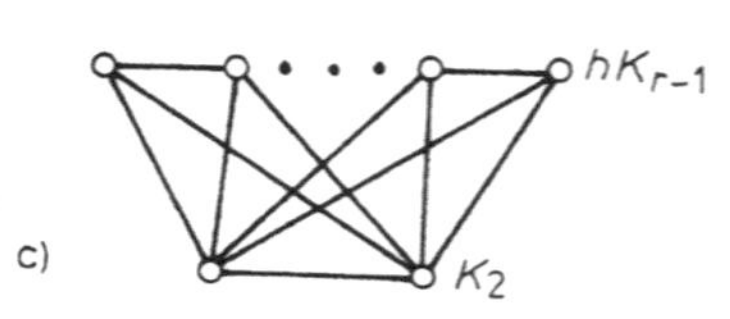


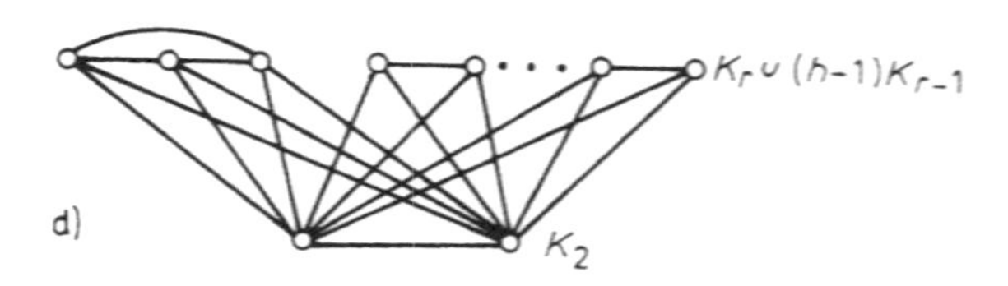


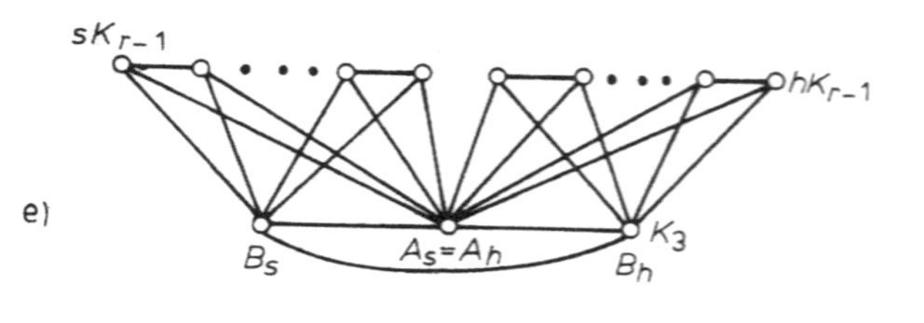

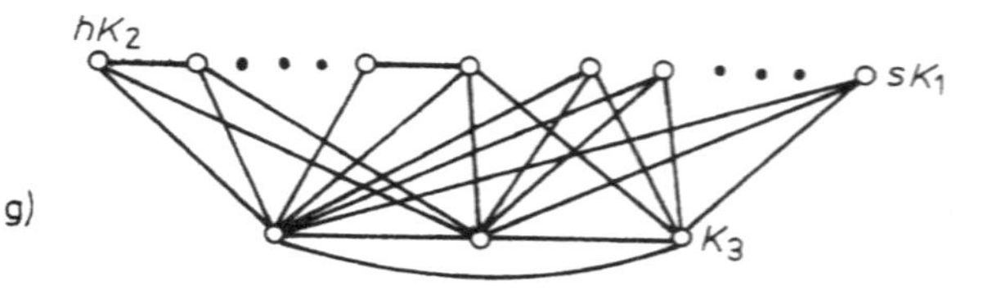


Fig. 7.2.3

Next we state the following result.

**Theorem 7.2.5.** *Let $r \geqq 3$ be an integer and G a graph with circumference $c(G) \leqq 2r+1$ and property $E_r$. Then G is a 2-connected subgraph of minimum degree at least r of a graph H, where*

$$H \in \mathfrak{C}_1^r \cup \mathfrak{C}_2^r \cup \mathfrak{C}_3^r \qquad \text{if } r \geqq 5,$$
$$H \in \mathfrak{C}_1^4 \cup \mathfrak{C}_2^4 \cup \mathfrak{C}_3^4 \cup \mathfrak{C}_4 \qquad \text{if } r = 4,$$
$$H \in \mathfrak{C}_1^3 \cup \mathfrak{C}_2^3 \cup \mathfrak{C}_3^3 \cup \mathfrak{C}_5 \qquad \text{if } r = 3. \quad \square$$

An extension of Dirac's Theorem 7.2.1 has been obtained by J.A.Bondy, 1971, J.C.Bermond, 1976, and N.Linial, 1976, who generalized Ore's condition for a graph to be Hamiltonian (O.Ore, 1960) as follows:

**Theorem 7.2.6.** *Let G be a 2-connected graph of order greater than m, and suppose that $v(G:U) + v(G:V) \geqq m$ if U, V are non-adjacent vertices. Then $c(G) \geqq m$.* $\square$

It is useful to make the following definition: if $G$ is not a complete graph, then

$$\sigma(G) =_{\text{def}} \min \{v(G:U) + v(G:V) \mid U \neq V \text{ and } (U, V) \notin E(G)\}.$$

$\sigma(K_n)$ is undefined.

Obviously, $\sigma(G) \geqq 2\delta(G)$.

H.Enomoto, 1984, determined the extremal graphs of Theorem 7.2.6. These graphs are the same as in Theorem 7.2.3 with $r = \frac{m}{2}$ (i.e., $m$ has to be taken as an even integer). From this result the following assertion of O. Ore, 1960, H. A. Jung, 1978, and C. Nara, 1980, on Hamiltonian cycles can be derived.

**Theorem 7.2.7.** *Let G be a graph of order n with $\sigma(G) \geqq n-1$. If G is not Hamiltonian, then either*

$$I_{(n-1)/2} + I_{(n+1)/2} \subseteqq G \subseteqq K_{(n-1)/2} + I_{(n+1)/2}$$

*or*

$$G \cong K_1 + (K_k \cup K_{n-k-1})$$

*for some integer k between 1 and $n-2$.* $\square$

The characterization of the case $\sigma(G) = n-2$ has been recently accomplished by a number of authors: A.Ainouche and N.Christofides, 1985, E.Schmeichel and D.Hayes, 1985, Z.Skupień, 1985, and D.Bauer and E.Schmeichel, 1987. D.Bauer, H.A.Jung and E.Schmeichel, 1989, classified all 2-connected graphs with $c(G) \leqq \sigma(G) + 2$.

**Theorem 7.2.8.** *Let $\sigma(G) \geqq 6$ and let G be a 2-connected non-Hamiltonian graph on n vertices with*

$$c(G) \leqq \sigma(G) + 1.$$

*Then G is a 2-connected spanning subgraph of a graph H with $\sigma(H) = \sigma(G)$, where either*

(i) *H is one of the graphs of Theorem 7.2.5 with for some r or*

(ii) $H \cong K_2 + (tK_h \cup K_{h-1})$ *for some* $t \geqq 2$, $h \geqq 2$. $\square$

It is interesting that a comparison of Theorem 7.2.5 of Voss with Theorem 7.2.8 shows that the only class of exceptional graphs differing from those found in Theorem 7.2.5 is (ii) (this is the only case with an odd $\sigma(G)$).

The main obstacle to a further improvement is given by the graphs with the component problem, where a graph $G$ is said to have a *component problem* if there exists a non-empty set $X \subseteqq V(G)$ such that $G - X$ has at least $|X| + 1$ components. H. A. Jung, 1978, proved the following result:

**Theorem 7.2.9.** *Let $G$ be a 2-connected graph on $n \geqq 11$ vertices with $\sigma(G) \geqq n - 4$ having no component problem. Then $G$ is Hamiltonian.* □

We remark that every graph of order at least 3 without a component problem is necessarily 2-connected.

H. A. Jung, 1978, pointed out that there exists an infinite family of 2-connected non-Hamiltonian graphs $G$ on $n$ vertices with $\sigma(G) = n - 5$ having no component problem. One such family (due to Karel Zikan; see E. Schmeichel and D. Hayes, 1984) can be obtained from $I_{p-2} + (K_3 \cup I_p)$ by joining three independent edges so that they link each vertex of $K_3$ with a vertex of $I_p$. C. Nara, 1980, and H. A. Jung and C. Nara, 1982, dealt with properties of the exceptional graphs of Theorem 7.2.9. Problems related to Theorem 7.2.9 have been investigated by N. Köhler, 1981a–c. Since all the graphs of Theorem 7.2.8 have the component problem, it implies the following result first established by D. Bauer and E. Schmeichel, 1986, and independently by Y. Tian and L. Zhao, 1988:

*Let $G$ be a graph on $n \geqq 3$ vertices without a component problem. Then*

$$c(G) \geqq \min(n, \sigma(G) + 2).$$

This result is best possible (see D. Bauer, H. A. Jung and E. Schmeichel, 1990).

Solving a conjecture of J. A. Bondy, 1980b, I. Fournier and P. Fraisse, 1985, proved the following generalization of Theorem 7.2.6.

**Theorem 7.2.10.** *Let $G$ be a $k$-connected graph, $k \geqq 2$, such that the degree sum of any $k + 1$ independent vertices is at least $m$. Then $G$ has a Hamiltonian cycle or there is a cycle of length at least $\frac{2m}{k+1}$.* □

J. A. Bondy, 1980b, proved Theorem 7.2.10 for $k = 2$, $m \geqq n + 2$, and $m > (k + 1)(n - 1)/2$, where $n$ is the order of $G$.

Let $\sigma_{k+1}$ be the maximum $m$ such that any $k + 1$ independent vertices of $G$ have degree sum at least $m$. Then Theorem 7.2.10 means for $k = 3$: if $G$ is 3-connected, then $G$ has a Hamiltonian cycle or $c(G) \geqq (2\sigma_3)/3$. In a forthcoming paper H. A. Jung presents an improvement of this result: $c(G) \geqq (2\sigma_3)/3 + c$ for an appropriate constant $c$.

P. Fraisse and H. A. Jung, 1989, proved a similar result: if $G$ is 3-connected, then $G$ has a dominating cycle or $c(G) \geqq \sigma_3 - 3$ (this is Theorem 7.9.7 for $k = 3$).

We conclude section 7.2 with extensions of the theorems of Dirac and Pósa/Bondy. D. R. Woodall, 1975, conjectured that Dirac's Theorem 7.2.1 can be improved as follows:

**Conjecture.** *If $G$ is a 2-connected graph of order $n$ with at least $n/2 + r$ vertices of degree at least $r$, then $G$ has a cycle of length at least $2r$.*

This conjecture has been proved by R. Häggkvist and B. Jackson, 1985, for some special cases. For example, they proved that if a 2-connected graph on at most $3r - 2$ vertices has at least $2r$ vertices of degree at least $r$, then $G$ has a cycle of length at least $2r$.

Another extension of Dirac's theorem is presented by N. Alon, 1986. He proved that every graph $G$ on $n$ vertices with $\delta(G) \geqq n/k$, $k \geqq 2$, contains a cycle of length at least $\lfloor n/(k-1) \rfloor$. When $k = 2$, this result reduces to the classical theorem of Dirac, which asserts that if $\delta(G) \geqq n/2$, then $G$ is Hamiltonian. Y. Egawa and T. Miyamoto, 1989, proved an Ore-type version of this theorem: if $\sigma(G) \geqq 2n/k$, then $c(G) \geqq \lceil n/(k-1) \rceil$.

R. Y. Zhu, 1983, proved an Ore-type theorem: every 2-connected graph has a Hamiltonian cycle or a cycle of length at least $2(\sigma(G) - \delta(G))$, provided that $\sigma(G) \geqq 4$.

G. Fan, 1984, proved a strengthening of the Ore-type theorem of Pósa/Bondy (a shorter proof is due to T. Feng, 1988).

**Theorem 7.2.11.** *If $G$ is a 2-connected graph of order $n$, and for any two vertices $U$, $V$*

$$d(U, V) = 2 \Rightarrow \max(v(G:U), v(G:V)) \geqq m/2,$$

*then $c(G) \geqq m$, where $3 \leqq m \leqq n$, and $d(U, V)$ is the distance between $U$ and $V$ in $G$.* □

If $G$ is also bipartite, a stronger result was obtained by K.-Q. Dang, 1989a.

A further generalization of Ore's theorem is due to G. Fan, 1990a, b.

Some authors established a number of results concerning long cycles in graphs $G$ with large degree having small toughness. A graph $G$ is called *t-tough* if for each subgraph $H$ of $G$ having at least two regular $H$-bridges the number of regular $H$-bridges is at most $|V(H)|/t$. Thus a graph is 1-tough if and only if it has no component problem. The toughness of $G$ is the maximum $t$ for which $G$ is $t$-tough. For a survey on such problems and results I refer the reader to D. Bauer, H. J. Veldman, A. Morgana and E. Schmeichel, 1989/90.

## 7.3. Proofs of the theorems of section 7.2

We shall prove Theorems 7.2.2 and 7.2.5. First we prove a lemma.

**Lemma 7.3.1.** *Let $G$ be a 2-connected graph and $C$ a cycle of $G$. Let $\pi$ denote a $C$-path of length at least 1 with the second end vertex $E$. Let $E$ have $s \geqq 2$ neighbours in $\pi$. Then there is a $C,C$-path $w$ of length $\lambda(w) \geqq s + 1$.*

*Proof.* Let $Z_1, \ldots, Z_s$ denote the neighbours of $E$ on $\pi$ such that $Z_{i+1} \in \pi[Z_i, E]$ and let $X$ be the first end vertex of $\pi$.

(1) *$C$ and $\pi[Z_1, E]$ are linked by two disjoint paths $w_1$ and $w_2$ so that $Z_1$ is an end vertex of $w_1$ or $w_2$.*

*Proof of (1).* Since $G$ is 2-connected, there are two disjoint paths $\bar{w}_1$ and $\bar{w}_2$ joining $C$ and $\pi[Z_1, E]$. Let

$$\{\bar{X}_i\} =_{\text{def}} \bar{w}_i \cap C \quad \text{and} \quad \{\bar{V}_i\} =_{\text{def}} \bar{w}_i \cap \pi[Z_1, Y] \quad \text{for } i = 1, 2.$$

If $Z_1 \in \{\bar{V}_1, \bar{V}_2\}$, then

$$w_i =_{\text{def}} \bar{w}_i \quad \text{for} \quad i = 1, 2.$$

If $Z_1 \notin \{\bar{V}_1, \bar{V}_2\}$ and $\pi[X, Z_1] \cap (\bar{w}_1 \cup \bar{w}_2) = \emptyset$, then

$$w_1 =_{\text{def}} \bar{w}_1 \quad \text{and} \quad w_2 =_{\text{def}} \pi[X, Z_1].$$

If $Z_1 \notin \{\bar{V}_1, \bar{V}_2\}$ and $\pi[X, Z_1] \cap (\bar{w}_1 \cup \bar{w}_2) \neq \emptyset$, then let $Q$ denote a vertex of $\pi[X, Z_1] \cap (\bar{w}_1 \cup \bar{w}_2)$ so that

$$\pi[Z_1, Q] \cap (\bar{w}_1 \cup \bar{w}_2) = \{Q\}.$$

Without loss of generality, let $Q \in \bar{w}_2$. Then

$$w_1 =_{\text{def}} \bar{w}_1 \quad \text{and} \quad w_2 =_{\text{def}} \bar{w}_2[\bar{X}_2, Q] \cup \pi[Q, Z_1]. \ \square$$

Now let $Z_1$ and $V$ denote the end vertices of $w_1$ and $w_2$ on $\pi[Z_1, E]$. Then $V \in \pi[Z_i, Z_{i+1}] - \{Z_i\}$ for some index $i \in \{1, 2, \ldots, s\}$, where $Z_{s+1} =_{\text{def}} E$. Then

$$w =_{\text{def}} w_1 \cup \pi[Z_1, Z_i] \cup [Z_i, E] \cup \pi[E, V] \cup w_2$$

is a $C,C$-path of length at least $s + 1$. □

***Proof of Theorem 7.2.2.*** Let $G$ be a 2-connected graph having minimum degree $\delta(G) \geqq r \geqq 3$, circumference $l =_{\text{def}} c(G) \leqq 2r + 1$ and order $n(G) > c(G) = l$. Let $L$ be a largest cycle of $G$; the cycle $L$ has length $l$. Since $n(G) > l$, the graph $G$ contains a regular bridge $B$ of $L$. In the set of all $L$-paths in $B$ let $p$ be one with the following property: the second end vertex of $p$, say $E$, has the largest number of neighbours on $L$. Let $A_1$ denote the first end vertex of $p$ and $A_2, \ldots, A_\alpha$ the neighbours of $E$ on $L - A_1$.

Case 1. Assume that $\alpha \geqq 2$. Then $T =_{\text{def}} p \cup \bigcup_{i=2}^{\alpha} [E, A_i]$ is a tree whose end vertices are $A_1, A_2, \ldots, A_\alpha$, $\alpha \geqq 2$, where $A_1, \ldots, A_\alpha$ are vertices of attachment of $B$.

We shall prove that

(1) *all neighbours of $E$ are on $L \cup p$.*

*Proof of (1).* Assume that there is a neighbour $Y$ of $E$, $Y \notin L \cup p$. Then $p \cup [E, Y]$ is an $L$-path longer than $p$. Contradiction! □

The length $\lambda(T)$ of a tree $T$ is the number of edges of $T$, i.e., $\lambda(T) = e(T)$.

Since $v(G: E) \geqq r$, assertion (1) and Theorem 7.1.1(i) imply that $r \leqq \lambda(T) \leqq \lambda(B) \leqq r$. Hence

(2) $\lambda(T) = \lambda(B) = r$. □

Hence $T$ has precisely $r + 1$ vertices, namely, the vertex $E$ and the $r$ neighbours of $E$. Consequently, all vertices of $p - E$ are neighbours of $E$.

(3) *Each vertex $X$ of $p - A_1$ is an end vertex of a longest $L$-path $p_X$ in $B$ with $p_X \subseteqq G\langle p \rangle$.*

*Proof of (3).* Let $X \in p - \{A_1, E\}$. Then the predecessor $X'$ of $X$ on $p$ is a neighbour of $E$. Therefore, $p[A_1, X'] \cup [X', E] \cup p[E, X]$ has the desired properties. □

(4) *Each vertex $X$ of $p - A_1$ has all its neighbours on $p \cup \{A_1, \ldots, A_\alpha\}$, i.e., on $T$.*

*Proof of (4).* By contradiction. Assume that $X \in p - A_1$ has a neighbour $A'$ which does not belong to $L - \{A_1, \ldots, A_\alpha\}$. Assertion (2) implies that $\lambda(T \cup [X, A']) = r + 1$. All end vertices of $T \cup [X, A']$ are again vertices of attachment of $B$. By Theorem 7.1.1 the length $\lambda(T \cup [X, A']) \leqq r$. Contradiction! □

Since by (2) the tree $T$ has order $r + 1$, each vertex of $T - L$ is joined to each vertex of $T$ by an edge (note that $\delta(G) \geqq r$). This and (4) imply that

(5) $B \cong \underline{I}_\alpha + K_{r+1-\alpha}$. □

Hence

(6) *every two vertices* $A_i$, $A_j$, $1 \leqq i < j \leqq \alpha$, *are joined by an* $L,L$*-path of* $B$ *of length* $r - \alpha + 2$. □

By the definition of $\tilde{\lambda}(B)$ (see our remark at the end of Section 3.3) we have

$$2\tilde{\lambda}(B) = \alpha(r - \alpha + 2).$$

Theorem 7.1.1(i) implies that $\alpha(r - \alpha + 2) = 2\tilde{\lambda}(B) \leqq 2r + 1$. Therefore,

(7) $(\alpha - 2)(r - \alpha) \leqq 1$. □

This implies that

(8) $\alpha = r$ *or* $\alpha = 2$ *or* $\alpha = 3, r = 4$. □

With (8) and (5) in Case 1 the proof of Theorem 7.2.2 is complete.

Case 2. Assume that $\alpha = 1$. Then the vertex $E$ has at least $r$ neighbours on $p$. By Lemma 7.3.1 there is an $L,L$-path of length at least $r + 1$ contradicting Theorem 7.1.1(i)! □

***Proof of Theorem** 7.2.5.* Here we shall only determine the graphs with a maximal number of edges.

Let $G$ be a 2-connected graph having minimum degree $\delta(G) \geqq r \geqq 3$, circumference $l =_{\text{def}} c(G) \leqq 2r + 1$ and order $n(G) > c(G) = l$. Let $L$ be a longest cycle of $G$; the cycle $L$ has length $l$. Since $n(G) > l$, the graph $G$ contains at least one regular bridge of $L$. By Theorem 7.2.2 all regular bridges are of type $\underline{I}_r + K_1$ or $\underline{I}_2 + K_{r-1}$ and in the case $r = 4$, $l = 9$ they can also be of type $\underline{I}_3 + K_2$. We consider three cases.

Case 1. Assume that $L$ has a bridge $B$ of type $\underline{I}_3 + K_2$. Then $r = 4$ and $l = 9$ and $L$ has only bridges of types $\underline{I}_3 + K_2$, $\underline{I}_4 + K_1$ and $\underline{I}_2 + K_3$.

(1) *If $L$ has a bridge $B^*$ of type $\underline{I}_2 + K_3$ or $\underline{I}_4 + K_1$, then the bridge $B$ avoids $B^*$.*

*Proof of (1).* By contradiction. Assume that $L$ has a bridge $B^*$ of type $\underline{I}_2 + K_3$ or $\underline{I}_4 + K_1$ which overlaps $B^*$. Since $\underline{I}_2 + K_3$ and $\underline{I}_4 + K_1$ have two and four vertices of attachment, respectively, the Lemma 1.2.1 of W.T.Tutte implies that $B^*$ and $B$ are skew. By the definition of skew bridges an $L,L$-path $v$ of $B$ overlaps an $L,L$-path $v^*$ or $B^*$. Obviously, $\lambda(v) = 3$ and $\lambda(v^*) \geqq 2$. Therefore, $\lambda(v) + \lambda(v^*) \geqq 3 + 2 = 5$. This contradicts assertion (ii) of Theorem 7.1.1 that $\lambda(v) + \lambda(v^*) \leqq \lfloor l/2 \rfloor = 4$. □

(2) *If $L$ has a regular bridge $B^*$, then $B^*$ is also of type $\underline{I}_3 + K_2$, and $B$, $B^*$ have the same set of vertices of attachment.*

*Proof of (2).* If $B^*$ is of type $\underline{I}_2 + K_3$ or $\underline{I}_4 + K_1$, then by (1) the bridge $B^*$ avoids $B$. All segments of $L$ with respect to $B$ have length 3. But the vertices of attachment of $B^*$ cannot belong to a path of length 3. Thus $B^*$ is of type $\underline{I}_3 + K_2$. If $B^*$ has a vertex of attachment not in $B$, then $B$ and $B^*$ have disjoint sets of vertices of attachment and $B$ overlaps $B^*$. This contradicts assertion (ii) of Theorem 7.1.1 because

$$\lambda(B) + \lambda(B^*) = 8 > \lfloor l/2 \rfloor = 4 .$$

Consequently, $B^*$ is of type $\underline{I}_3 + K_2$ and the sets of vertices of attachment of $B$ and $B^*$ coincide. □

By (2) an investigation of the possible diagonals of $L$ leads to the conclusion that $G \in \mathfrak{C}_4$ (see Fig. 7.2.3f).

Case 2. Assume that $L$ has a bridge $B$ of type $\underline{I}_2 + K_{r-1}$. By (2) the circuit $L$ has no bridge of type $\underline{I}_3 + K_2$.

(3) *If $L$ has a bridge $B^*$ of type $\underline{I}_r + K_1$, then $r = 3$, $l = 7$, the bridges $B$ and $B^*$ do not overlap, and they are in the position depicted in Fig. 7.3.1.*

*Proof of (3).* Since $\lambda(B) = \lambda(\underline{I}_2 + K_{r-1}) = r$, the bridges $B$ and $B^*$ do not overlap. The segments of $L$ with respect to $B^*$ have length at most 3. Hence $\lambda(B) = r \leqq 3$. Consequently, $r = 3$ and a segment of $L$ with respect to $B^*$ has length 3. Therefore, $l = 7$ and the two bridges of $L$ are in the position depicted in Fig. 7.3.1.

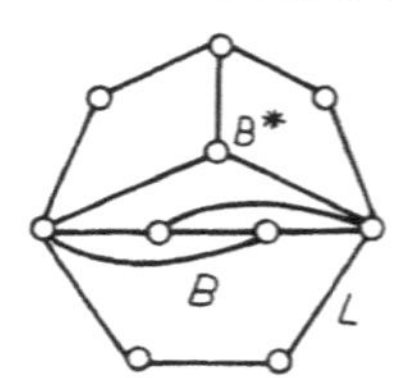


Fig. 7.3.1

We consider two subcases.

Case 2.1. Assume that $L$ has only bridges of type $\underline{I}_2 + K_{r-1}$. Since these bridges have length $r$, they do not overlap (see Theorem 7.1.1(ii)). This and an investigation of the possible diagonals of $L$ leads to the conclusion that $G \in \mathfrak{C}_2' \cup \mathfrak{C}_3'$ (see Fig. 7.2.3c,d,e).

Case 2.2. Assume that $r = 3$, $l = 7$ and $L$ has at least one bridge $B$ of type $\underline{I}_2 + K_2$ and a bridge $B^*$ of type $\underline{I}_3 + K_1$. By (3) the two bridges are in the position depicted in Fig. 7.3.1. By (3) for each further bridge $\bar{B}$ the set of vertices of attachment of $\bar{B}$ coincides with the vertices of attachment of $B$ if $\bar{B}$ is of type $\underline{I}_2 + K_2$, and of $B^*$ if $\bar{B}$ is of type $\underline{I}_3 + K_1$. This and an investigation of the possible diagonals of $L$ lead to the conclusion that $G \in \mathfrak{C}_5$ (see Fig. 7.2.3g).

Case 3. Assume that $L$ has only bridges of type $\underline{I}_r + K_1$.

(4) *Every two bridges $B$, $B^*$ of $L$ of type $\underline{I}_r + K_1$ have the same set of vertices of attachment.*

*Proof of (4).* By contradiction. Assume that there is a vertex $A_1$ of attachment of $B$ which is not in $B^*$. Let $s$ denote the segment of $L$ with respect to $B^*$ containing $A_1$, and let $A_1^*, A_2^* \in B^*$ denote the ends of $s$. Obviously, $\lambda(s) \leqq 3$.

Since $B$ has $r \geqq 3$ vertices of attachment and each segment of $L$ with respect to $B$ has length at least 2, there is a vertex $A_2$ of attachment of $B$ with $A_2 \notin s$. Thus the $L,L$-path $v$ of $B$ with ends $A_1$, $A_2$ overlaps the $L,L$-path of $B^*$ with ends $A_1^*$, $A_2^*$. Therefore, $v$ overlaps

$B^* - A_2$, where $v$ and $B^* - A_2$ have no common vertices. Therefore, for "reduced" bridges the assertion (ii) of Theorem 7.1.1 is valid. Hence $\lambda(v) + \lambda(B^* - A_2) \leqq r$. But the number of edges in $v \cup (B^* - A_2)$ is

$$\lambda(v) + \lambda(B^* - A_2) = 2 + (r - 1) = r + 1.$$

Contradiction! □

Assertion (4) and an investigation of the possible diagonals of $L$ lead to the conclusion that $G \in \mathfrak{C}_1^r$. □

## 7.4. Longest cycles through given vertices and properties of their bridges

The investigations of graphs with property $E_r$, $r \geqq 3$, i.e. of 2-connected non-Hamiltonian graphs with minimum degree $\delta(G) \geqq r$, are continued.

For each vertex $V$ the length of longest cycles through $V$ will be considered. Let $L_V$ denote a longest cycle through $V$ (H. Walther refers to $L_V$ as a "Großkreis" in H. Walther and H.-J. Voss, 1974). Let $p$ be any $L_V$,$L_V$-path with ends $P$, $Q$. It is not obvious whether the following assertion of Lemma 3.2.1 is also true for $L_V$: do both arcs of $L_V$ with ends $P$, $Q$ have length at least $\lambda(p)$? The main problem of the proofs is to answer this question. I have actually succeeded in proving this assertion with one exception only (H.-J. Voss, 1976a). This exception causes the existence of a new class of bridges in our theorems.

First we define this new class $\{\mathfrak{B}(2; r-1, i) \mid r \geqq 3, i \geqq 1\}$: The bridge $\mathfrak{B}(2; r-1, i)$ is obtained from $K_2 + (K_1 \cup iK_{r-1})$ by deleting an edge $e$ which joins the $K_2$ to the $K_1$ and choosing the end vertices of that edge $e$ as vertices of attachment (for $r = 3$, $i \geqq 2$ the bridge $\mathfrak{B}(2; 3, i)$ is depicted in Fig. 7.4.1).

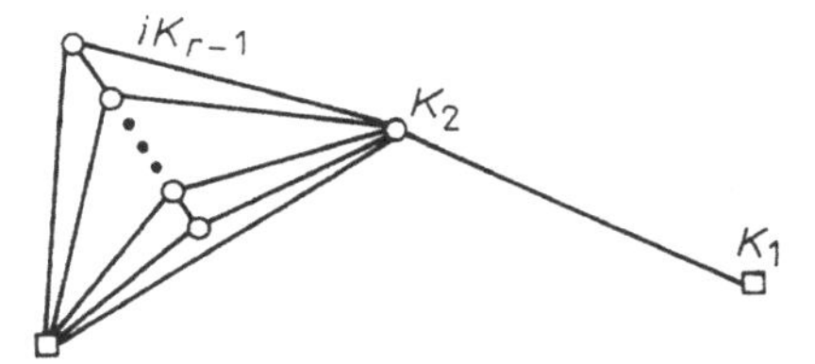


Fig. 7.4.1

I have proved (H.-J. Voss, 1976) the following result:

**Theorem 7.4.1.** *Let $r \geqq 3$ be an integer and $G$ be a graph with property $E_r$ which contains a vertex $V$ so that each cycle through $V$ has length at most $2r + 1$. Let $L_V$ denote a longest cycle of $G$ through $V$. Then $L_V$ has only bridges of types $\underline{I}_r + K_1$, $\underline{I}_2 + K_{r-1}$, $\mathfrak{B}(2; r-1, i)$ with $i \geqq 2$, $\underline{I}_3 + K_2$ and $(\underline{I}_2 + K_1) + jK_1$ with $j \geqq 2$. Bridges of type $\underline{I}_3 + K_2$ only occur in the case $r = 4$, $l = 9$ and bridges of types $(\underline{I}_2 + K_1) + jK_1$ with $j \geqq 2$ occur only in the case $r = 3$, $l = 7$.* □

For all graphs of Theorem 7.4.1 with circumference $c(G) \leqq 2r$ we have the following result:

**Corollary 7.4.2.** *Let the graph $G$ of Theorem 7.4.1 have circumference at most $2r$, then $L_V$ has only bridges of types $\underline{I}_r + K_1$ and $\underline{I}_2 + K_{r-1}$.* □

Theorem 7.4.1 (Corollary 7.4.2) immediately implies a theorem of P. Erdős and T. Gallai, 1959 (see also M. Grötschel, 1977).

**Theorem 7.4.3.** *Let $r \geqq 3$ be an integer and let be $G$ a graph with property $E_r$. Then each vertex $V$ of $G$ is contained in a cycle of length at least $2r$.* □

By Theorem 7.4.1 it is possible to determine all graphs satisfying the hypotheses of Theorem 7.4.1. Before formulating the corresponding theorem we introduce a new class of graphs.

**Definition 7.4.4.** Let $r \geqq 3$ be an integer. The graphs of $C_6^r$ consist of a subgraph $H$ of order 2 and $\alpha$ bridges $\underline{I}_2 + K_{r-1}$, $\beta$ bridges $\underline{I}_2 + K_r$, $\gamma$ bridges $\mathfrak{B}(2; r-1, i)$, $i \geqq 2$, and in the case $r = 3$ also $\delta$ bridges $(\underline{I}_2 + K_1) + jK_1$, $j \geqq 3$, where $\alpha \geqq 1$, $\alpha + \beta + \gamma + \delta \geqq 3$ and $\delta = 0$ for all $r \geqq 4$.

Thus only in the case $r = 3$ the integer $\delta$ can be positive.

For $r = 3$ some graphs $G_1 \in C_6^3$ are depicted in Fig. 7.4.2 with $H = (\{X^*, Y^*\}, (X^*, Y^*))$ containing a bridge $\underline{I}_2 + K_{r-1}$, a bridge $\underline{I}_2 + K_r$ and two bridges $\mathfrak{B}(2; r-1, i)$ with different choice of the vertices of attachment $X^*$ and $Y^*$. The bridge $\underline{I}_2 + K_{r-1}$ which is necessarily in $G_1$ (note that $\alpha \geqq 1$) is drawn with thick lines.

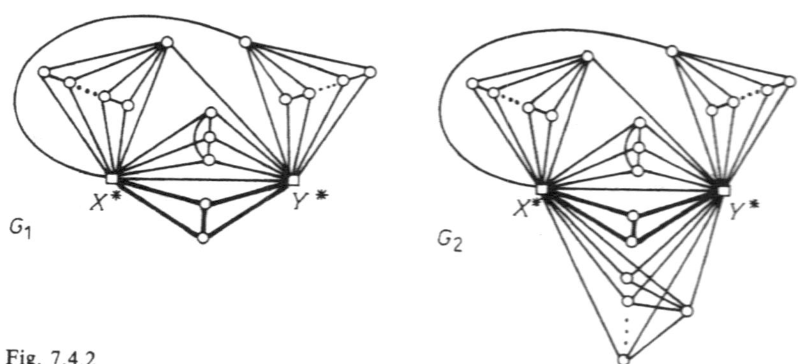


Fig. 7.4.2

Adding a bridge $(\underline{I}_2 + K_1) + jK_1$, $j \geqq 3$, to some $G_1$ we obtain a graph $G_2 \in \mathfrak{C}_6^3$ depicted in Fig. 7.4.2. Note that the graphs $G_1$ are example graphs of $\mathfrak{C}_6^3$ which can be generalized to arbitrary $r$, but the $G_2$'s contain a bridge $(\underline{I}_2 + K_1) + jK_1$, $j \geqq 3$, which can not be generalized to arbitrary $r$.

It can easily be checked that if $H$ has at least two bridges of types $\underline{I}_2 + K_r$, $\mathfrak{B}(2; r-1, i)$, $i \geqq 2$, or $(\underline{I}_2 + K_1) + jK_1$, $j \geqq 3$, then $c(G) = 2r + 2$, but each inner vertex of a bridge $\underline{I}_2 + K_{r-1}$ is contained in cycles of length at most $2r + 1$ only. For instance this is true for the inner vertices of the thickly drawn bridge of the graphs $G_1$ and $G_2$ of Fig. 7.4.2.

Theorem 7.4.1 implies (H.-J. Voss, 1976a) the following result:

**Theorem 7.4.5.** *Let $G$ satisfy the hypotheses of Theorem 7.4.1. Then $G$ is either one of the graphs listed in Theorem 7.2.5 or a 2-connected spanning subgraph of minimum degree at least $r$ of a graph of $\mathfrak{C}_6^r$.* □

A similar theorem involving Ore-type conditions has been proved by H. Enomoto, 1984, extending a result of J. A. Bondy, 1971, J. C. Bermond, 1976, and N. Linial, 1976.

**Theorem 7.4.6.** *Let $G$ be a 2-connected graph of order greater than $m$ with $\sigma(G) \geqq m$. Then* *each vertex V of G is contained in a cycle of length at least m. If some vertex of G is contained in no cycle of length greater than m, then either*

$$I_{m/2} + I_s \subseteqq G \subseteqq K_{m/2} + I_s, \qquad s \geqq m/2 + 1,$$

*or*

$$I_2 + (K_{k-1} \cup l K_{m-k-1}) \subseteqq G \subseteqq K_2 + (K_{k-1} \cup l K_{m-k-1})$$

*for some integer k between 2 and m/2.* □

Next we consider longest cycles through two given vertices.

**Theorem 7.4.7.** *Let $r \geqq 3$ be an integer and G a graph with property $E_r$. Then each pair of vertices of G is contained in a cycle of length at least $2r$.* □

Theorem 7.4.7 has been proved by R. Fuhrmann and myself (R. Fuhrmann, 1979; see also H.-J. Voss, 1977a). This somewhat lengthy proof was not published in a journal. An independent shorter proof is due to S. C. Locke, 1985.

We now turn our attention to bridges of longest cycles again. In the following let $r \geqq 3$ be an integer and $G$ a graph with property $E_r$.

If $G$ has a longest cycle $L$ of length $2r$ or $2r+1$, then by Theorem 7.2.2 this cycle can have only regular bridges of two or three different types when $r \neq 4$ or $r = 4$, respectively. If $G$ has a vertex $V$ so that each longest cycle $L_V$ through $V$ has length $2r$, then by Corollary 7.4.2 this cycle $L_V$ can also have regular bridges of only two different types. But if $L_V$ has length $2r+1$, then by Theorem 7.4.1 the cycle $L_V$ can have regular bridges of infinitely many different types.

All these bridges have length $r$ and except for finitely many of them all these bridges belong to two sequences, namely, $\{\mathfrak{B}(2; r-1, i) \mid i \geqq 2\}$ and $\{(\underline{I}_2 + K_1) + jK_1 \mid j \geqq 3\}$. The bridges of each of the sequences are "similar" in the following sense: $\mathfrak{B}(2; r-1, i_1)$ and $\mathfrak{B}(2; r-1, i_2)$, $i_1 \neq i_2$, differ only in the number of $K_{r-1}$'s with the "same neighbourhood", i.e. in the number of isomorphic bridges having the $K_{r-1}$'s as their kernels.

The same is true for $(\underline{I}_2 + K_1) + j_1 K_1$ and $(\underline{I}_2 + K_1) + j_2 K_1$, $j_1 \neq j_2$. They also differ only in the number of vertices (from the $K_1$'s) with the same neighbourhood, i.e. in the number of isomorphic bridges having the vertices of the $K_1$'s as their kernels.

If $G$ has two vertices $V_1$, $V_2$ so that each longest cycle $L_{V_1,V_2}$ through $V_1$ and $V_2$ has length $2r$ or $2r+1$, then such an assertion is not true. Moreover, we shall show that if $B$ is any graph with two non-adjacent vertices $A_1$, $A_2$ such that each vertex of $B - \{A_1, A_2\}$ has valency at least $r$ and is on an $A_1$,$A_2$-path of $B$, then $B$ can be a bridge of $L_{V_1,V_2}$ with the two vertices of attachment $A_1$, $A_2$. The graph $B$ is a bridge with respect to $H =_{\text{def}} (\{A_1, A_2\}, \emptyset)$.

Let $\Gamma$ denote the graph obtained from $\{A_1, A_2\} + ((B - \{A_1, A_2\}) \cup 2K_{r-1})$ by deleting all edges joining $\{A_1, A_2\}$ to $B - \{A_1, A_2\}$ which are not in $B$.

The graph $\Gamma$ has the following properties: if $V_i$ denotes an inner vertex of $B_i = \{A_1, A_2\} + K_{r-1}$ for $i = 1, 2$, then each longest cycle $L_{V_1,V_2}$ through $V_1$ and $V_2$ has length $2r$. But $B$ is a bridge of $L_{V_1,V_2}$, where $B$ is an arbitrary graph containing two non-adjacent vertices $A_1$, $A_2$ such that each vertex of $B - \{A_1, A_2\}$ has valency at least $r$ in $B$ and is on an $A_1$,$A_2$-path of $B$. These two vertices $A_1$, $A_2$ are the vertices of attachment of $B$.

If $V_3$ denotes an inner vertex of $B$, then it is obvious that $\Gamma$ has no cycle through $V_1$, $V_2$ containing $V_3$.

In order to have a cycle through any three given vertices it is useful to assume that $G$ is

3-connected, because G.A. Dirac, 1960, proved that any $k$ vertices of a $k$-connected graph are on a common cycle.

Really, there is the following nice result of Y. Egawa, R. Glas and S. C. Locke, 1987, generalizing Theorem 7.4.7.

**Theorem 7.4.8.** *Let $G$ be a k-connected graph, $k \geqq 2$, with minimum degree $\delta(G) \geqq r$ and of order $n(G) \geqq 2r$, and let $S$ be a set of k vertices of $G$. Then $G$ has a cycle $C$ of length at least $2r$ containing all vertices of $S$.* □

A version of Theorem 7.4.8 has appeared in R. Glas, 1987. A. Saito, 1987, has established a result relating the length of such a cycle to the length of a longest cycle of $G$. The class of graphs $I_r + (I_k \cup I_s)$, $s \geqq r \geqq k$, shows that the lower bound $2r$ cannot be improved because no cycle of length exceeding $2r$ can contain all vertices of $I_k$ (note that $I_r + (I_k \cup I_s)$ has circumference $2r$). In the graph $I_k + (k+1) K_{r+1-k}$ we form a set $S$ containing precisely one vertex from each copy of $K_{r+1-k}$. Hence no cycle contains all vertices of $S$ and therefore Theorem 7.4.8 cannot be improved in this direction. The only hope for a stronger theorem would lie in replacing the hypothesis that $G$ is $k$-connected by the hypothesis that $G$ is 2-connected and has circumference greater than $2r$, and the given vertices of $S$ are on a common cycle (compare with Theorem 7.5.3). An Ore-type version of Theorem 7.4.8 has been proved by R. Glas, 1987.

In this connection the following question arises:

*Which hypotheses on k-connected graphs, $k \geqq 2$, ensure that a longest cycle through certain given vertices has only regular bridges of finitely many different types?*

A partial answer will be given in sections 7.8 and 7.9.

## 7.5. Longest cycles through given edges and properties of their bridges

In this section we study bridges of cycles containing prescribed edges. First we shall prove the following simple lemma:

**Lemma 7.5.1.** *Let $e_1, \ldots, e_i$, $1 \leqq i \leqq 3$, be i edges in an $(i+1)$-connected graph $G$ forming no star $I_3 + K_1$. Then $G$ contains a cycle through $e_1, \ldots, e_i$.*

R. Häggkvist and C. Thomassen, 1982, proved this for arbitrary $i$.

*Proof.* Let $E_1^j$, $E_2^j$ be the ends of $e_j$. Some paths $p_h$ used in the proof can have length 0. We consider three cases:

Case 1. $i = 1$. Since $G$ is 2-connected, the connected graph $G - (E_1^1, E_2^1)$ contains an $E_1^1, E_2^1$-path $p$. Then the cycle $p \cup e_1$ contains $e_1$.

Case 2. $i = 2$. Since $G$ is 3-connected, there are two disjoint paths $p_1, p_2$ connecting $\{E_1^1, E_2^1\}$ and $\{E_1^2, E_2^2\}$. Then the cycle $p_1 \cup p_2 \cup e_1 \cup e_2$ contains $e_1$ and $e_2$.

Case 3. $i = 3$. We consider only the case where $e_1, e_2, e_3$ are independent. The other cases lead to cases 1 and 2. Since $G$ is 4-connected, the 3-connected graph $G - e_3$ contains three disjoint paths $p_1, p_2, p_3$ joining $\{E_1^1, E_2^1, E_1^3\}$ and $\{E_1^2, E_2^2, E_2^3\}$. If $E_1^3$, $E_2^3$ are not ends of the same path, then the cycle $p_1 \cup p_2 \cup p_3 \cup e_1 \cup e_2 \cup e_3$ contains $e_1, e_2, e_3$. If $E_1^3$, $E_2^3$ are ends of $p_3$, then $C_1 =_{\text{def}} p_1 \cup p_2 \cup e_1 \cup e_2$ and $C_2 =_{\text{def}} p_3 \cup e_3$ are two disjoint cycles. Since $G$

is 4-connected, there are three disjoint paths $v_1$, $v_2$, $v_3$ connecting $C_1$ and $C_2$. Now it can easily be proved that $C_1 \cup C_2 \cup v_1 \cup v_2 \cup v_3$ contains a cycle with edges $e_1$, $e_2$, $e_3$. □

For $i \geqq 4$ the problem dealt with in Lemma 7.5.1 and related questions are more difficult (see D. R. Woodall, 1977, and R. Häggkvist and C. Thomassen, 1982).

We shall now formulate the main theorem from which interesting results will be deduced later. For this purpose we need some definitions.

Let $\bar{G}$ denote the complement of the graph $G$, i.e. the graph with the same vertex set as $G$ and two vertices $X$ and $Y$ are adjacent in $\bar{G}$ if and only if they are not adjacent in $G$.

**Definition 7.5.2.** Let $G$ be a graph. If $e_1, \ldots, e_i$ $(1 \leqq i \leqq 3)$ are $i$ edges of $G \cup \bar{G}$ on a cycle of $G \cup \{e_1, \ldots, e_i\}$ of length at least $i+1$, then $L(e_1, \ldots, e_i)$ denotes a longest cycle through $e_1, \ldots, e_i$. Let $L_i(G)$ denote a smallest cycle $L(e_1, \ldots, e_i)$, where the minimum is taken over all edge sets $\{e_1, \ldots, e_i\}$ of $G \cup \bar{G}$ on a cycle of $G \cup \{e_1, \ldots, e_i\}$ of length at least $i+1$. A longest cycle of $G$ is denoted by $L_0(G)$, and $c_i(G) + i$ denotes the length of $L_i(G)$, $0 \leqq i \leqq 4$. Thus $c_i(G)$ counts the edges of $L_i(G)$ not in $\{e_1, \ldots, e_i\}$. □

An *s-vertices bridge* is a bridge with precisely $s$ inner vertices.

By Lemma 7.5.1 each set of $i$ edges $e_1, \ldots, e_i$ $(1 \leqq i \leqq 3)$ not forming a triangle or a star lies on a cycle. Then in this case $L(e_1, \ldots, e_i)$ exists.

I proved the following main theorem of this section (see H.-J. Voss, 1977a):

**Theorem 7.5.3.** *Let $k$, $i$, $r$ be integers with $2 \leqq k \leqq 4$, $0 \leqq i \leqq k-1$ and $r \geqq k$. Let $G$ be a $k$-connected graph with minimum degree $\delta(G) \geqq r$ and $c_i(G) \leqq (k-i)(r-k+2)$. Then the cycle $L_i(G)$ has only bridges of type $\underline{I}_k + K_{r+1-k}$ and s-vertices bridges for all $1 \leqq s \leqq k-i-1$, where the $(k-i-1)$-vertices bridges, $k-i-1>0$, have a unique structure, namely, they are of type $\underline{I}_{r-k+i+2} + K_{k-i-1}$.* □

With the aid of Theorem 7.5.3 all graphs $G$ which satisfy the conditions of this theorem can be constructed.

If $L_i(G)$ has a bridge $\underline{I}_k + K_{r+1-k}$ or $\underline{I}_{r-k+i+2} + K_{k-i-1}$, $k-i-1>0$, then obviously $c_i(G) = (k-i)(r-k+2)$. These results and their implications are formulated in Corollaries 7.5.4 and 7.5.5.

**Corollary 7.5.4.** *Let the graph $G$ satisfy the hypotheses of Theorem 7.5.3. If $L_i(G)$ has a bridge $\underline{I}_k + K_{r+1-k}$, then $c_i(G) = (k-i)(r-k+2)$ and $G = H + hK_{r+1-k}$, $h \geqq k-i+1$, where $H$ is an arbitrary graph of order $k$ and $H \cup \bar{H}$ contains the edges $e_1, \ldots, e_i$* (the case $k=4$ is depicted in Fig. 7.5.1). □

It can easily be checked that if $e_1, \ldots, e_i$ are $i$ edges, $1 \leqq i \leqq 3$, of $H \cup \bar{H}$ not forming a triangle $K_3$ or a star $K_1 + I_3$, then each cycle of $H + hK_{r+1-k}$ through $e_1, \ldots, e_i$ has length at most $c_i(G) + i = (k-i)(r-k+2) + i$.

**Corollary 7.5.5.** *Let the graph $G$ satisfy the hypotheses of Theorem 7.5.3. If $L_i(G)$ has a bridge $\underline{I}_{r-k+i+2} + K_{k-i-1}$, $k-i-1>0$, then $c_i(G) = (k-i)(r-k+2)$ and $G = H + hK_{k-i-1}$, $h \geqq r-k+3$, where $H$ is an arbitrary graph of order $r-k+i+2$ and $H \cup \bar{H}$ contains the edges $e_1, \ldots, e_i$* (the case $k=4$ is depicted in Fig. 7.5.2).

A bridge $\underline{I}_s + K_t$, $s$, $t \geqq 1$, of some subgraph $H$ is called a *stronghold* of $H$. So in Theorem 7.5.3 all $L_i(G)$-bridges are strongholds.

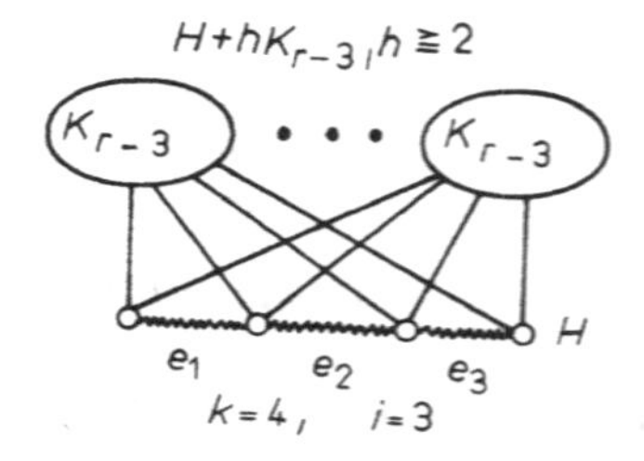


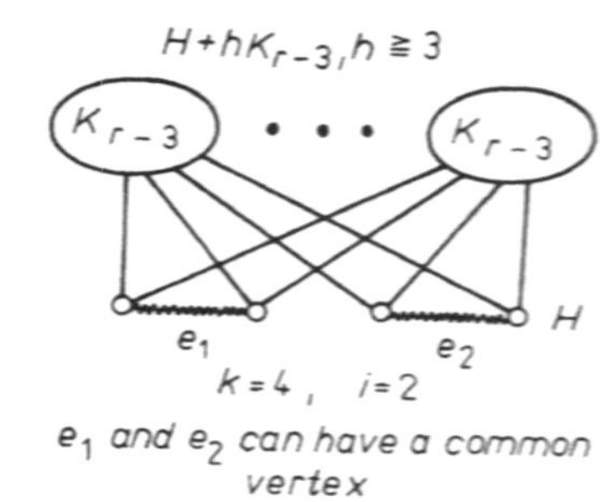


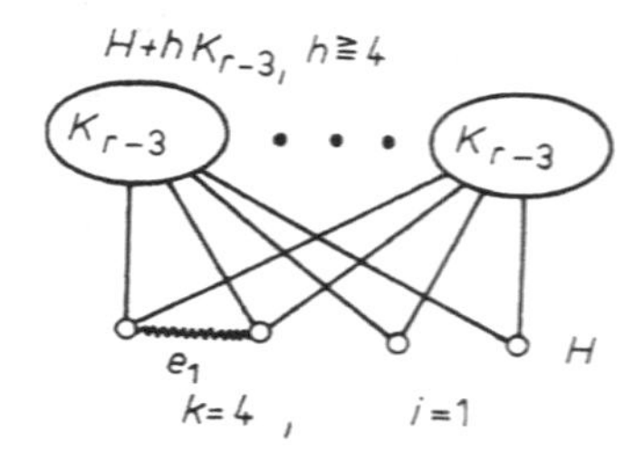


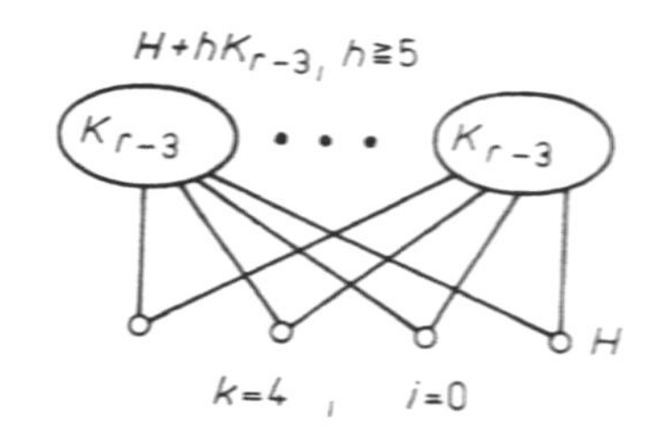


Fig. 7.5.1

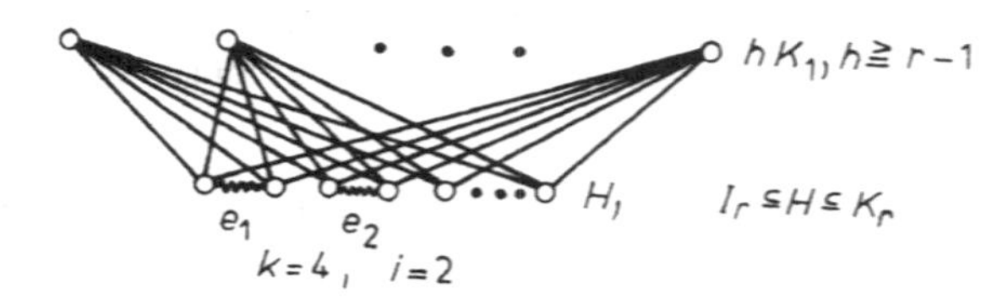


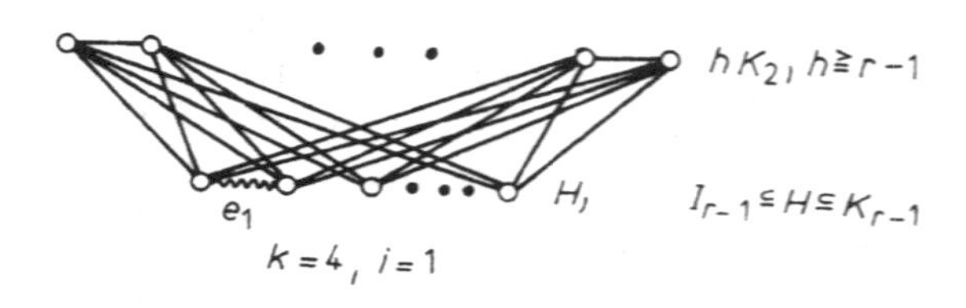


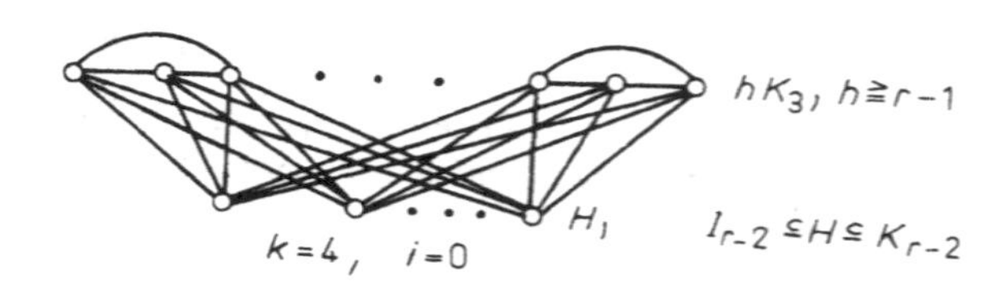


Fig. 7.5.2

Checkings of the proofs of Theorems 7.2.2, 7.2.5, 7.4.1, 7.5.3 and 7.6.2 (H.-J. Voss and C. Zuluaga, 1977) show that the hypothesis $\delta(G) \geqq r$ is only used for vertices in the kernel of the investigated bridge of a longest cycle. Taking this into consideration, H. A. Jung, 1986, formulated his result on cycles containing a certain edge of $G \cup \bar{G}$ as follows.

**Theorem 7.5.6.** *Let P and Q be distinct vertices and p a longest P,Q-path in the k-connected graph G, where $2 \leqq k \leqq 5$. Each p-bridge B with at least $k-2$ inner vertices contains an inner vertex V such that p contains a subpath p′ with end vertices in B and of length $\lambda(p') \geqq (k-1)(v(G:V)-k+2)$. Moreover, $\lambda(p') \geqq (k-1)(v(G:V)-k+2)+(k-1)$ unless B is a stronghold with at most $2k-4$ inner vertices or with precisely k vertices of attachment.* □

For $2 \leqq k \leqq 4$, $i = 1$ Theorem 7.5.6 is an improved version of Theorem 7.5.3.

The second assertion of this theorem includes a hypothesis on the structure of the bridge $B$ of a longest $P,Q$-path $p$, namely, that $B$ is not a stronghold. H. A. Jung, 1981, 1986, proved further results which strengthen known theorems by taking into consideration the structure of bridges of longest cycles or $P,Q$-paths.

Next we draw some conclusions from Theorems 7.5.3 and 7.5.6.

For each pair $(i, k)$ of integers with $0 \leqq i < k$, $2 \leqq k \leqq 4$, Theorem 7.5.3 and Corollaries 7.5.4-5 imply *nine* results. Corollaries 7.5.4-5 imply that if $c_i(G) < (k-i)(r-k+2)$, then $G$ has no $L_i(G)$-bridge of type $\underline{I}_k + K_{r+1-k}$ and no $L_i(G)$-bridge of type $\underline{I}_{r-k+i+2} + K_{k-i+1}$. By definition the length of $L_i(G)$ is equal to $c_i(G) + i$.

We consider the four cases $i = k-1$, $i = k-2$, $i = k-3$, and $i = k-4$. The two cases $i = k-1$ and $i = k-2$ are dealt with in this section (Theorems 7.5.7-8); the last two cases $i = k-3$ and $i = k-4$ are considered in Section 7.9 (Theorems 7.9.1, 7.9.3-4). Thus the *nine* results are presented in four theorems. The next theorem is a consequence of the case $i = k-1$ of Theorem 7.5.3.

**Theorem 7.5.7.** *Let k and r be integers, $2 \leqq k \leqq 4$, $r \geqq k$. Let $e_1, \ldots, e_{k-1}$ be $k-1$ edges of a k-connected graph G with $\delta(G) \geqq r$. Suppose that $e_1, \ldots, e_{k-1}$ do not form a triangle $K_3$ or a star $K_1 + I_3$. Then G has a Hamiltonian cycle through $e_1, \ldots, e_{k-1}$ or there is a cycle through $e_1, \ldots, e_{k-1}$ of length at least $r+1$.* □

For $k = 2$ this is a result of P. Erdős and T. Gallai, 1959, that each pair $P$, $Q$ of distinct vertices of a 2-connected graph with $\delta(G) \geqq r$ is joined by a Hamiltonian $P,Q$-path or by a $P,Q$-path of length at least $r$ (apply Theorem 7.5.7 with $k = 2$ to $G \cup (P, Q)$). Another proof is due to L. Lovász, 1979. We remark that by Corollaries 7.5.4-5 we can improve Theorem 7.5.7 when the cycle in question has length $r+1$. Then $G$ has a Hamiltonian cycle through $e_1, \ldots, e_{k-1}$ or $G$ is one of the graphs of Corollary 7.5.4.

Further, some of the $e_i$'s can be an edge of $\bar{G}$. Theorem 7.5.7 is also valid in this case, because $G \cup e_i$ satisfies the hypotheses of Theorem 7.5.7.

The consequence of the case $i = k-2$ of Theorem 7.5.3 is the following result:

**Theorem 7.5.8.** *Let k and r be integers, $2 \leqq k \leqq 4$, $r \geqq k$. Let $e_1, \ldots, e_{k-2}$ be $k-2$ edges of a k-connected graph G with $\delta(G) \geqq r$. Then G has a Hamiltonian cycle through $e_1, \ldots, e_{k-2}$ or there is a cycle through $e_1, \ldots, e_{k-2}$ of length at least $2r+2-k$.* □

For $k = 2$ this is again G. A. Dirac's Theorem 7.2.1. For $k = 3$ this is a result due to M. Grötschel, 1977. A stronger version will be derived from Theorem 7.5.3:

**Theorem 7.5.9.** *Let $r \geqq 3$ be an integer and $G$ a 3-connected graph with property* $E_r$. *Then*

(i) *each pair $P$, $Q$ of distinct vertices of $G$ is joined by a $P,Q$-path of length at least $2r-2$.*

(ii) *If $G$ contains a pair of distinct vertices which is only joined by $P,Q$-paths of length at most $2r-2$, then a longest such path has only bridges of types $\underline{I}_3 + K_{r-2}$ and $\underline{I}_r + K_1$.* □

H. A. Jung's Theorem 7.5.6 implies that if $G$ has a longest $P,Q$-path $p$ with a bridge which is not a stronghold, then $p$ has length at least $2r$. From Theorem 7.5.9 we can easily construct all graphs satisfying (ii).

**Theorem 7.5.10.** *Let $r \geqq 3$ be an integer and $G$ a 3-connected graph with property* $E_r$. *If $G$ contains a pair $P$, $Q$ of distinct vertices which is only joined by $P,Q$-paths of length at most $2r-2$, then either*

$$I_r + I_s \subseteqq G \subseteqq K_r + I_s, \qquad s \geqq r,$$

*or*

$$I_3 + hK_{r-2} \subseteqq G \subseteqq K_3 + hK_{r-2}, \qquad h \geqq 3.$$

S. C. Locke, 1985a, has proved a sharper version of Theorem 7.5.9(i).

**Theorem 7.5.11.** *Let $r \geqq 3$ be an integer and $G$ a 3-connected graph with property* $E_r$. *Then for any three distinct vertices $P$, $Q$, $R$ of $G$ there is a $P,Q$-path through $R$ of length at least $2r-2$.* □

Theorems 7.5.11 and 7.5.10 have been extended to Ore-type conditions by H. Enomoto, 1984.

**Theorem 7.5.12.** *Let $G$ be a 3-connected graph of order greater than $m$ with $\sigma(G) \geqq m$. Then*

(i) *each pair $P$, $Q$ of distinct vertices of $G$ is joined by a $P,Q$-path of length at least $m-2$;*

(ii) *if $G$ contains a pair $P$, $Q$ of distinct vertices which is only joined by $P,Q$-paths of length at most $m-2$, then either*

$$I_{m/2} + I_s \subseteqq G \subseteqq K_{m/2} + I_s, \qquad s > m/2,$$

*or*

$$I_3 + lK_{m/2-2} \subseteqq G \subseteqq K_3 + lK_{m/2-2}, \qquad l \geqq 3. \ \square$$

Note that Theorems 7.5.9-12 can be applied to $(k+2)$-connected graphs stating the existence of a cycle of certain length through any path of length $k$. The reduction can be made by deleting the $k-1$ inner vertices of such a path. The resulting graph is 3-connected.

A strengthening of Theorem 7.5.12(i) is due to J. A. Bondy and B. Jackson, 1985. They proved a theorem for 2-connected graphs.

**Theorem 7.5.13.** *Let $G$ be a 2-connected graph of order at least $m-2$ with $\sigma(G) \geqq m$. Let $P$, $Q$ be any pair of distinct vertices of $G$ such that neither $P$ nor $Q$ is contained in a 2-vertex cut set of $G$. Then $P$ and $Q$ are joined by a $P,Q$-path of length at least $m-2$.* □

H. Enomoto's Theorem 7.5.12(i) follows immediately from this result.

J. A. Bondy and B. Jackson, 1985, also proved the following result: let $G$ be a 2-connected graph of order at least $m-2$ with $\sigma(G) \geqq m$. Let $P_1$, $P_2$, $P_3$ be distinct vertices of $G$. Then $G$ contains a $P_i,P_j$-path of length at least $m-2$ for some $1 \leqq i < j \leqq 3$.

From Theorem 7.5.12 the following result of O. Ore, 1963, on Hamiltonian connected graphs can be derived.

**Theorem 7.5.14.** *Let $G$ be a graph of order $n$ with $\sigma(G) \geqq n$. If $G$ is not Hamiltonian connected* (i.e., there are two vertices that are not connected by a path of length $n-1$), *then either*

$$I_{n/2} + I_{n/2} \subseteqq G \subseteqq K_{n/2} + I_{n/2},$$

*or*

$$I_2 + (K_{k-1} \cup K_{n-k-1}) \subseteqq G \subseteqq K_2 + (K_{k-1} \cup K_{n-k-1})$$

*for some $k$ between $2$ and $n-2$.* □

We conclude this section with a generalization of Theorem 7.5.11 by Y. Egawa, R. Glas and S. C. Locke, 1987 (a version is due to R. Glas, 1987).

**Theorem 7.5.15.** *Let $G$ be a $k$-connected graph, $k \geqq 3$, with minimum degree $\delta(G) \geqq r$ and of order $n(G) \geqq 2r-1$. Let $S$ be a set of $k-1$ vertices of $G$. Then each pair $P$, $Q$ of distinct vertices of $G$ is joined by a $P,Q$-path of length at least $2r-2$ containing all vertices of $S$.* □

## 7.6. Longest odd and even cycles. Longest cycles in bipartite graphs

Theorem 7.2.1 (G. A. Dirac, 1952a) deals with valency conditions on a graph for the existence of cycles of given length. Here we shall generalize this result in the following sense: which valency conditions on a graph $G$ imply that $G$ contains a cycle of prescribed odd or even length? C. Zuluaga and I proved the following result (H.-J. Voss and C. Zuluaga, 1977; see also C. Zuluaga, 1977):

**Theorem 7.6.1.** *Let $r \geqq 3$ be an integer and $G$ a non-bipartite 2-connected graph of order greater than $2r$ with minimum degree $\delta(G) \geqq r$. Then $G$ contains an odd cycle and an even cycle, both of length at least $2r-1$.* □

We have succeeded in determining all bridges of a longest cycle in certain graphs (H.-J. Voss and C. Zuluaga, 1977). We formulate this result for bridges of longest cycles.

**Theorem 7.6.2.** *Let $G$ be a 2-connected graph with minimum degree at least $r \geqq 3$ which satisfies one of the following three conditions:*

(i) *$G$ is non-bipartite, $c(G) \geqq 2r+2$ and all odd cycles have length at most $2r-1$.*

(ii) *$G$ is non-bipartite, $c(G) \geqq 2r+3$ and all even cycles have length at most $2r$.*

(iii) *$c(G) \leqq 4r-4$ and $G$ has no triangle.*

*Then each longest cycle of $G$ has only so-called one-vertex bridges of type $\underline{I}_i + K_1$ with $r \leqq i \leqq \lfloor c(G)/2 \rfloor$, and bridges of type $(\underline{I}_2 \cup I_{r-2}) + I_s$, $s \geqq r$, and so-called star bridges of type $(\underline{I}_{r-1} \cup K_1) + I_j$, $j \geqq r$.* □

The graphs satisfying the hypotheses (i) or (ii) of Theorem 7.6.2 have circumference at least $2r+2$. Hence Theorem 7.2.2 is a good supplement of Theorem 7.6.2 because it deals with the graphs of circumferences $2r$ or $2r+1$.

Theorem 7.6.2 gives us the possibility of constructing all graphs satisfying the hypotheses (i) or (ii) or (iii). This construction has been accomplished in H.-J. Voss and C. Zuluaga, 1977, and proved in H.-J. Voss, 1988b, 1989a. The results are formulated in Theorems 7.6.3 and 7.6.4.

**Theorem 7.6.3.** *Let $r \geqq 3$ be an integer and $G$ a non-bipartite graph with property* E, *having circumference at least $2r+2$. Let all odd cycles of $G$ have length at most $2r-1$. Then $G$ consists of $H \cong K_2$ and at least two $H$-bridges of type $(\underline{I}_2 \cup I_{r-2}) + I_j$, $j \geqq r$* (for $r = 3$ an example is depicted in Fig. 7.6.1). □

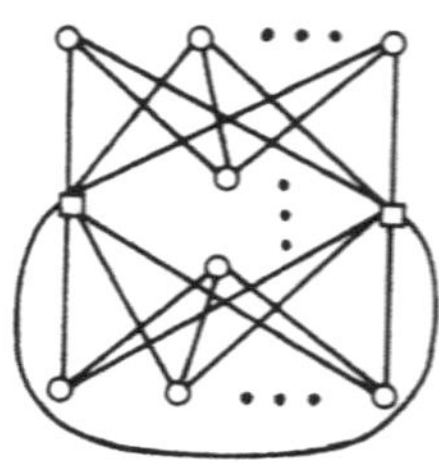

Fig. 7.6.1

The graphs $G$ satisfying all the hypotheses of Theorem 7.6.3 except that the circumference is $2r$ are the non-bipartite graphs of Theorem 7.2.3. Thus

$$I_r + I_s \subset G \subseteqq K_r + I_s, \qquad s \geqq r+1,$$

or

$$I_2 + hK_{r-1} \subseteqq G \subseteqq K_2 + hK_{r-1}, \qquad h \geqq 3.$$

Next we define two classes of graphs $\mathfrak{D}(r)$ and $\bar{\mathfrak{D}}(r)$:

A bipartite block $\mathfrak{B}$ is a $K_2$ or a 2-connected bipartite graph of order at least 4. A graph $R$ is called a *ring* of bipartite blocks if and only if there exists a sequence $V_1\mathfrak{B}_1V_2\mathfrak{B}_2\ldots V_s\mathfrak{B}_sV_1$, $s \geqq 3$, of vertices $V_i$ and blocks $\mathfrak{B}_i$ such that $R = \bigcup \mathfrak{B}_i$ and $\mathfrak{B}_i \cap \mathfrak{B}_{i+1} = \{V_{i+1}\}$ for all $1 \leqq i \leqq s-1$, $\mathfrak{B}_s \cap \mathfrak{B}_1 = \{V_1\}$, and $\mathfrak{B}_i \cap \mathfrak{B}_j = \emptyset$ for all $j \notin \{i-1, i, i+1\}$ or if and only if for $s = 2$ there exists a sequence $V_1\mathfrak{B}_1V_2\mathfrak{B}_2V_1$ such that $\mathfrak{B}_1 \cap \mathfrak{B}_2 = \{V_1, V_2\}$ and $V_1 \neq V_2$.

The cycles of $R$ which contain at least one edge from each block have the property that either all of them have odd length or all of them have even length. Then $R$ is called *odd* or *even*, respectively.

Let $\mathfrak{D}(r)$ denote the class of all odd rings $R$ of bipartite blocks of minimum degree $\delta(R) \geqq r$, the blocks of which are isomorphic to $K_2$ or $I_r + I_j$, $j \geqq r$ (for $r = 3$ examples with $s = 5$ are depicted in Fig. 7.6.2a).

Let $\bar{\mathfrak{D}}(r)$ denote the class of all graphs obtained from the odd rings of three bipartite blocks $V_1\mathfrak{B}_1V_2\mathfrak{B}_2V_3\mathfrak{B}_3V_1 \in \mathfrak{D}(r)$ by adding the path $V_1V_2V_3$, where $\mathfrak{B}_1$ is a copy of $I_r + I_{j_1}$, $j_1 \geqq r$, with $V_1, V_2 \in I_r$, $\mathfrak{B}_2$ is a copy of $I_r + I_{j_2}$, $j_2 \geqq r$, with $V_2, V_3 \in I_r$ and $\mathfrak{B}_3$ is a copy of $K_2$ with vertex set $\{V_3, V_1\}$ (for $r = 3$ this class is depicted in Fig. 7.6.2b).

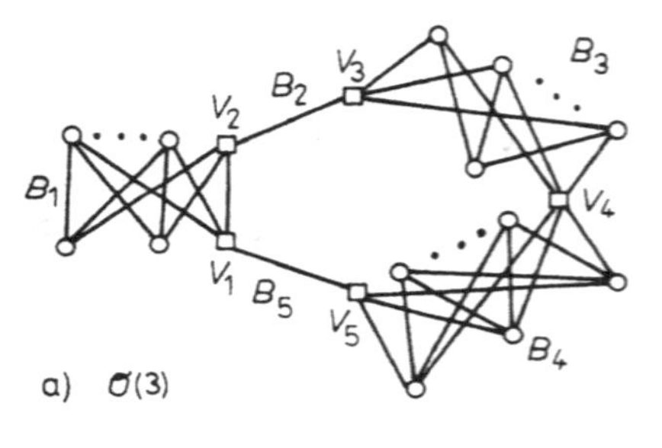


a) $\mathfrak{D}(3)$

B1 V2 B2 V1 B3 V3

b) $\bar{\mathfrak{D}}(3)$

Fig. 7.6.2

**Theorem 7.6.4.** *Let $r \geqq 3$ be an integer and G a non-bipartite graph with property* $E_r$ *having circumference at least* $2r+3$. *Let all even cycles of G have length at most* $2r$. *Then G is a 2-connected spanning subgraph of minimum degree at least r of a graph* $H \in \mathfrak{D}(r) \cup \bar{\mathfrak{D}}(r)$ (for $r = 3$ see the examples in Fig. 7.6.2).

The graphs $G$ satisfying all the hypotheses of Theorem 7.6.4 except that the circumference is $2r$ or $2r+1$ are the non-bipartite graphs of Theorem 7.2.5. These graphs are well described in section 7.2. Theorem 7.2.5 will not be repeated here.

The graphs $H$ of $\bar{\mathfrak{D}}(r)$ are edge-saturated, i.e., adding any edge to $H$ results in an even cycle of length greater than $2r$. The same is true for all graphs of $\mathfrak{D}(r)$ which are not subgraphs of a graph of $\bar{\mathfrak{D}}(r)$.

We conclude this section with some theorems on longest cycles in bipartite graphs. Let $B_{p,q}$ denote a bipartite graph with $p$ and $q$ vertices in the two classes, i.e., $B_{p,q} \subseteqq I_p + I_q$. From Theorem 7.6.2 the analogue of Dirac's Theorem 7.2.1 has been derived by H.-J. Voss and C. Zuluaga, 1977:

**Theorem 7.6.5.** *If* $B_{p,q}$ *is a 2-connected bipartite graph having minimum degree at least r, then* $c(B_{p,q}) \geqq 2\min\{p, q, 2r-2\}$. □

Dang Kai-Quian and Zhao Liang-Chang, 1989, improved Theorem 7.6.5:

$$c(B_{p,q}) \geqq 2\min\{p, q, 2(\sigma - r - 1)\}, \quad \text{where } \sigma = \sigma(B_{p,q}).$$

All graphs of Theorem 7.6.5 with circumference at most $4r-4$ can be characterized with the help of Theorem 7.6.2 (H.-J. Voss, 1988b).

**Theorem 7.6.6.** *Let $r \geqq 3$ be an integer and* $B_{p,q}$ *a 2-connected bipartite graph with* $p \geqq q \geqq r$, *minimum degree at least r, and circumference at most* $4r-4$. *Then either*

(i) *G is a 2-connected spanning subgraph of minimum degree at least r of* $I_p + I_q$, $p \geqq r$, $2r-2 \geqq q \geqq r$, *or*

(ii) *G consists of a graph* $H \cong I_{r-1}$ *and at least r H-bridges of types* $(\underline{I}_{r-1} \cup I_1) + I_j$, $j \geqq r$, *or*

(iii) *G consists of a graph* $H \cong I_2$ *and at least three bridges of types* $(\underline{I}_2 \cup I_{r-2}) + I_s$, $s \geqq r$. □

Theorem 7.6.6 shows that the lower bounds of Theorem 7.6.5 cannot be improved. B. Jackson, 1985, extended the result of Theorem 7.6.5.

**Theorem 7.6.7.** *Let* $B_{p,q}$ *be a 2-connected bipartite graph with bipartition (A, B), where* $p = |A| \geqq q = |B|$. *If each vertex of A has degree at least r, and each vertex of B has degree at least s, then G contains a cycle of length at least* $2\min\{p, q, 2r-2, r+s-1\}$. *Moreover, if* $r = s$ *and* $p = q$, *then* $B_{p,q}$ *contains a cycle of length at least* $2\min\{p, 2r-1\}$. □

The following graph $B_{p,q}$ shows that there exist 2-connected bipartite graphs which satisfy the hypotheses of Theorem 7.6.7 and whose longest cycle have length $2(r+s-1)$. Given integers $p, q, r, s \geqq 2$ such that $p \geqq q \geqq r+s-1$, $B_{p,q}$ is the graph obtained from two disjoint complete bipartite graphs $I_{p-s} + I_r$ and $I_s + I_{q-r}$ by joining every vertex in the $r$-set of $I_{p-s} + I_r$ to every vertex in the $s$-set of $I_s + I_{q-r}$.

## 7.7. Large cycles in regular graphs

Again we start with G. A. Dirac's well-known theorem:

**Theorem 7.7.1.** *Every 2-connected graph with minimum degree at least $r \geqq 3$ has a Hamiltonian cycle or contains a cycle of length at least $2r$.* □

By Theorem 7.2.5 there exist infinitely many graphs $G$ with $c(G) = 2r$. The class of all 2-connected $r$-regular graphs is a proper subset of the class of all 2-connected graphs with minimum degree at least $r \geqq 3$. Hence the question arises: is there an $r$-regular graph $G$ with $c(G) = 2r$?

The answer is "No"!

In Theorem 7.2.5 all graphs with property $E_r$ and $c(G) \leqq 2r + 1$ have been characterized. No one of these graphs is $r$-regular. This implies the following result (H.-J. Voss, 1980):

*Every 2-connected $r$-regular graph has a Hamiltonian cycle or contains a cycle of length at least $2r + 2$.*

G. Fan, 1985, R.-Y. Chen (see G. Fan, 1985) and H.-J. Voss, 1989b, independently proved:

**Theorem 7.7.2.** *Every 2-connected $r$-regular graph $G$ has a Hamiltonian cycle or contains a cycle of length at least $2r + 4$, or $r = 3$ and $G$ is the Petersen graph.* □

The proof of H.-J. Voss, 1989b, uses results on bridges of longest cycle and is similar to the proofs presented in section 7.3. P. Erdős and A. M. Hobbs, 1977, presented 2-connected $r$-regular graphs of order at least $3r + 5$ and circumference $2r + 4$. Thus the lower bound of Theorem 7.7.2 cannot be improved.

The graphs of P. Erdős and A. M. Hobbs, 1977, are the following. Let $h$ be an integer, $3 \leqq h \leqq r$. In $h$ $K_{r+1}$ choose $r$ independent edges $(A_i, B_i)$, $1 \leqq i \leqq r$, so that each copy of $K_{r+1}$ contains at least one of these edges. Add two new vertices $A$ and $B$ and for all $1 \leqq i \leqq r$ delete the edge $(A_i, B_i)$ and join $A_i$ to $A$ and $B_i$ to $B$ by an edge (an example with $r = 3$ is depicted in Fig. 7.7.1 a).

An open problem is to determine all 2-connected $r$-regular graphs $G$ with circumference $c(G) \in \{2r + 4, 2r + 5\}$.

If in the construction of P. Erdős and A. M. Hobbs, 1977, the copies of $K_{r+1}$ are replaced by copies of $K_{r,r}$, then 2-connected $r$-regular non-Hamiltonian bipartite graphs with circumference $4r + 2$ are obtained (an example with $r = 3$ is depicted in Fig. 7.7.1 b).

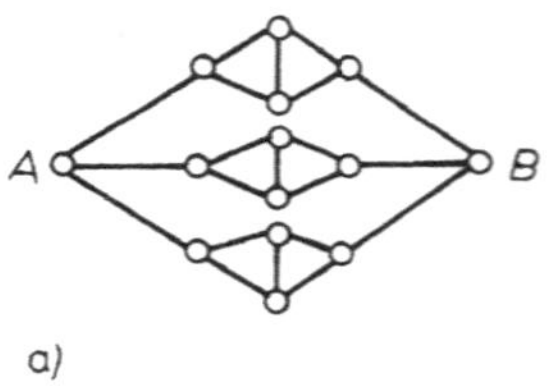

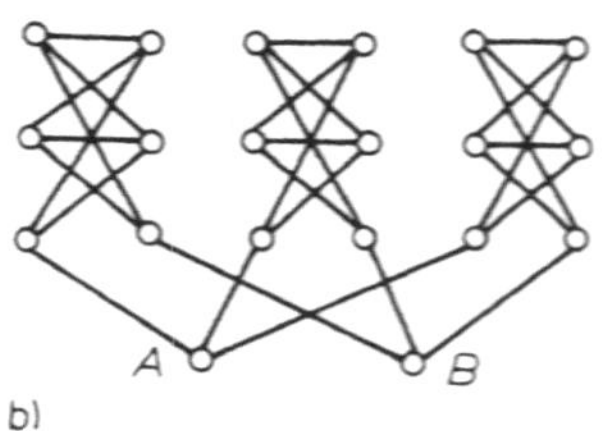


Fig. 7.7.1

R.-Y. Chen, 1985, has proved the following result:

**Theorem 7.7.3.** *Every 2-connected-r-regular bipartite graph has a Hamiltonian cycle or a cycle of length at least* $4r$. □

The bipartite graphs described above show that this bound cannot be greater than $4r+2$.

The well-known theorem of G. A. Dirac, 1952a, on Hamiltonian cycles is the following:

**Theorem 7.7.4.** *Every graph with minimum degree at least* $r \geqq 3$ *and order at most* $2r$ *has a Hamiltonian cycle.* □

Theorem 7.7.4 is not true for orders $n \geqq 2r+1$, as the graphs $K_{r,n-r} = I_r + I_{n-r}$ show.

We ask: what can be said if the graphs in question are *r*-regular? Can the condition that the order is at most $2r$ be weakened in this case? The answer is "Yes"!

Moreover, Szekeres asked for the maximum $s$ such that a 2-connected $(m-s)$-regular graph on at most $2m$ vertices is Hamiltonian.

C. St. J. A. Nash-Williams, 1970, has proved the following result:

**Theorem 7.7.5.** *Every r-regular graph of order at most* $2r+1$ *has a Hamiltonian cycle.*

Theorem 7.7.5 is not true for order $2r+2$, as the graph $2K_{r+1}$ shows. Considering connected graphs only we shall prove the following result:

**Theorem 7.7.6.** *Every connected r-regular graph of order at most* $2r+2$ *has a Hamiltonian cycle.*

Theorems 7.7.5 and 7.7.6 are not true for all orders $n$, $n \geqq 2r+3$. For each possible pair of integers $r$, $n$, $r \geqq 3$, $n \geqq 2r+3$, we shall construct a *connected* non-Hamiltonian *r*-regular graph of order $n$. (There is no *r*-regular graph of order $n$ if both $r$ and $n$ are odd integers!) We consider two cases.

Case 1. Let $r$ be an *even* number ($r \geqq 4$). Let $n$ be an integer, $n \geqq 2r+3$, and let $H_{n-r-2}$ be a Hamiltonian *r*-regular graph of order $n-r-2$, where $n-r-2 \geqq r+1$. In $K_{r+1} \cup H_{n-r-2}$ we choose $r/2$ independent edges $(A_i, B_i)$, $1 \leqq i \leqq r/2$, so that both $K_{r+1}$ and $H_{n-r-2}$ contain at least one of these edges. We add a new vertex $A$, delete the edge $(A_i, B_i)$ and join both $A_i$ and $B_i$ to $A$ by an edge for all $1 \leqq i \leqq r/2$. The resulting connected *r*-regular graph of order $n$, $n \geqq 2r+3$, has no Hamiltonian cycle (an example with $r=4$, $n=2r+3=11$ is depicted in Fig. 7.7.2a).

Case 2. Let $r$ be an *odd* number ($r \geqq 3$). Let $n$, $n \geqq 2r+4$, be an *even* integer, and let $H_{n-r-3}$ denote a Hamiltonian *r*-regular graph of order $n-r-3$, where $n-r-3 \geqq r+1$. In $K_{r+1} \cup H_{n-r-3}$ we choose $r-1$ independent edges $(A_i, B_i)$, $1 \leqq i \leqq r-1$, where $(A_i, B_i)$ is in $K_{r+1}$ for $1 \leqq i \leqq \frac{r-1}{2}$ and $(A_i, B_i)$ is in $H_{n-r-3}$ for $\frac{r+1}{2} \leqq i \leqq r-1$. We add two new vertices $A$, $B$ and the edge $(A, B)$. We join both $A_i$ and $B_i$ by an edge to $A$ when $1 \leqq i \leqq \frac{r-1}{2}$ and to $B$ when $\frac{r+1}{2} \leqq i \leqq r-1$. The resulting connected $r$-regular graph of order $n$, $n \geqq 2r+4$, $n$ even, has no Hamiltonian cycle (an example with $r=3$, $n=2r+4=10$ is depicted in Fig. 7.7.2b).

Next we give a common proof of Theorems 7.7.5 and 7.7.6.

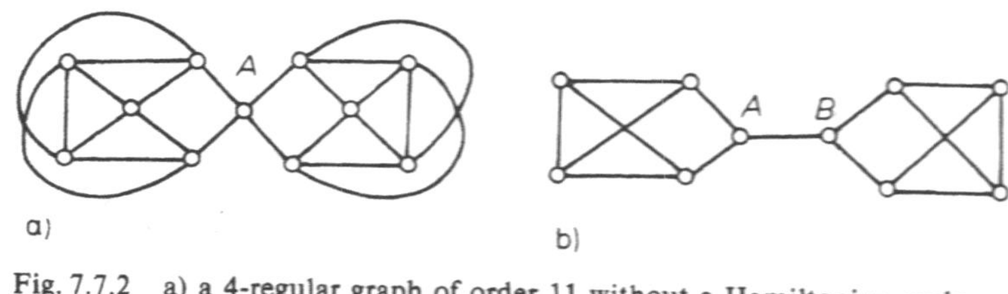


Fig. 7.7.2 a) a 4-regular graph of order 11 without a Hamiltonian cycle
b) a cubic graph of order 10 without a Hamiltonian cycle

***Proof of the Theorems 7.7.5 and 7.7.6.*** Let $G$ be an $r$-regular graph. We consider three cases.

Case 1. Let $G$ be disconnected. Then $G$ has two components $C_1$, $C_2$. Each vertex of $C_i$ ($i = 1, 2$) has $r$ neighbours in $C_i$. Hence $C_i$ has at least $r + 1$ vertices and G has order at least $2r + 2$.

Case 2. Let $G$ be connected and contain a cut vertex $A$. Then $G - A$ has two components $C_1$, $C_2$. Each vertex of $C_i$ ($i = 1, 2$) has $r$ neighbours in $C_i \cup \{A\}$. Hence $C_i$ has at least $r$ vertices. If $C_i$ has precisely $r$ vertices, then $A$ is joined by $r$ edges to vertices of $C_i$ and $v(G:A) > r$. Hence $C_i$ has at least $r + 1$ vertices and $G$ has order at least $2r + 3$.

Case 3. Let $G$ be 2-connected. By Theorem 7.2.5 all non-Hamiltonian 2-connected graphs with $\delta(G) \geqq r$, $n(G) \geqq 2r + 1$ and $c(G) \leqq 2r + 1$ are known. None of these graphs of orders $2r + 1$ and $2r + 2$ is regular. Hence there are no non-Hamiltonian 2-connected $r$-regular graphs of orders $2r + 1$ and $2r + 2$.

The Cases 1–3 imply the validity of Theorems 7.7.5 and 7.7.6. □

The transition from connected graphs to 2-connected graphs has a striking effect on the problem treated here. B. Jackson, 1980a, actually proved the following result:

**Theorem 7.7.7.** *If the 2-connected $r$-regular graph $G$ has order $n(G) \leqq 3r$, then $G$ has a Hamiltonian cycle.* □

Unlike the proofs of Theorems 7.7.5 and 7.7.6, the proof of B. Jackson's theorem is rather complicated. C. St. J. A. Nash-Williams, 1970, was the first to prove Theorem 7.7.7 for graphs of order $n(G) = 2r + 1$ (see Theorem 7.7.5). For orders $n(G) = 2r + 2$ and $n(G) \in \{2r + 3, 2r + 4\}$ this has been proved by L. Gordon (see P. Erdős and A. M. Hobbs, 1978) and P. Erdős and A. M. Hobbs, 1978, respectively. Proofs of Theorem 7.7.7 for orders $n(G) \leqq 2r + c\sqrt{r}$ ($c$ constant) and $n(G) \leqq 9r/4$ are due to P. Erdős and A. M. Hobbs, 1977, and B. Bollobás and A. M. Hobbs, 1978, respectively. J. Bondy and M. Kouider, 1988, gave a simpler proof of the result of B. Jackson, 1980a. Y. Zhu, Z. Liu and Z. Yu, 1985, showed that except for the Petersen graph every 2-connected $r$-regular graph on at most $3r + 1$ vertices is Hamiltonian. Y. Zhu, Z. Liu and Z. Yu, 1986, extended Theorem 7.7.7 to orders $n(G) \leqq 3r + 3$ if $r \geqq 6$. This extension is based on a result by G. Fan, 1985, which says that $c(G) \geqq \min\{n(G), 3r\}$ if either (a) $G$ is $r$-regular and 3-connected, or (b) $G$ is $r$-regular, 2-connected and has $n(G) \leqq 3r + 3$. Condition (a) has been independently proved by H. A. Jung (see note added in G. Fan, 1985). Condition (b) implies again the validity of Jackson's result. Independently, F. Hilbig, 1986, improved the result of Theorem 7.7.7 to orders $n(G) \leqq 3r + 3$ for all $r \geqq 3$, the sole exceptions being the Petersen graph and the graph $\mathfrak{P}$, where $\mathfrak{P}$ can be obtained by replacing one vertex of the Petersen graph by a triangle.

Next we investigate the bound $3r + 3$. This bound cannot be improved.

Above we presented the construction of non-Hamiltonian 2-connected $r$-regular graphs of order at least $3r + 5$ due to P. Erdős and A. M. Hobbs, 1977. If $r$ is odd, then there is no $r$-regular graph of order $3r + 4$. Hence we shall only construct non-Hamiltonian 2-connected $r$-regular graphs of order $3r + 4$ for even $r \geqq 4$.

We start with $2K_{r+1} \cup K_r$. In $K_r$ we partition the vertices into two classes $\{A_1, \ldots, A_{r/2}\}$ and $\{B_1, \ldots, B_{r/2}\}$ and in $2K_{r+1}$ we choose $r/2$ independent edges $(A_i, B_i)$, $r/2 + 1 \leqq i \leqq r$, so that each of the two copies of $K_{r+1}$ contains at least one of these edges. We add two new vertices $A$ and $B$ and for all $r/2 + 1 \leqq i \leqq r$ we delete the edges $(A_i, B_i)$ and for all $1 \leqq i \leqq r$ we join $A_i$ to $A$ and $B_i$ to $B$ by an edge. The graph thus obtained is 2-connected $r$-regular non-Hamiltonian and has order $3r + 4$. Now for $r \geqq 6$ the theorem of G. Fan, R.-Y. Chen, H.-J. Voss and the theorem of B. Jackson and Y. Zhu, Z. Liu, Z. Yu and F. Hilbig can be summarized as follows:

**Theorem 7.7.8.** *Let $G$ be a 2-connected $r$-regular graph, $r \geqq 6$. If $G$ has order at most $3r + 3$, then $G$ is Hamiltonian.*

*If $G$ has order at least $3r + 4$, then $c(G) \geqq 2r + 4$.*

*There are non-Hamiltonian graphs $G$ of order $3r + 5$ and of higher order.*

*There is no graph $G$ of order $3r + 4$ if $r$ is odd and there are non-Hamiltonian graphs $G$ of order $3r + 4$ if $r$ is even.* □

Some of these graphs are also Hamiltonian connected, i.e. each pair of distinct vertices is joined by a Hamiltonian path. Thus I. Tomescu, 1983, has proved that every $r$-regular graph $G$ of order $2r$, $r \geqq 3$, $G \neq I_r + I_r$, and every $r$-regular graph of order $2r + 1$, $r \geqq 4$, is Hamiltonian connected. From Theorem 7.7.7 of B. Jackson for $r$-regular graphs of higher connectivity the following conjecture of R. Häggkvist, 1976, can be derived (see also B. Jackson, 1979):

**Conjecture.** *If the $k$-connected $r$-regular graph $G$ has order at most $(k + 1)\, r$, $k \geqq 3$, then $G$ has a Hamiltonian cycle.*

By Theorem 7.7.7 the conjecture is true for $k = 2$. B. Jackson, 1980a, has mentioned that he has disproved it for large values of $k$ by constructing non-Hamiltonian $\frac{2}{3}r$-connected $r$-regular graphs on approximately $10r$ vertices.

Later B. Jackson and H. A. Jung independently constructed non-Hamiltonian $\left\lceil \frac{r-2}{2} \right\rceil$-connected $r$-regular graphs on $4r + 1$ or $4r + 2$ vertices (see Min Aung, 1988). Here this construction of Min Aung, 1988, is presented. In order to construct the examples of B. Jackson and H. A. Jung mentioned above one can start with the graph $(I_r + I_{r-1}) \cup K_{r+1} \cup K_s$, where $s = r + 1$ if $r$ is even, and $s = r + 2$ if $r$ is odd. One can introduce $r$ independent edges from $I_r$ to $K_{r+1} \cup K_s$, so that $2\left\lfloor \frac{r}{4} \right\rfloor$ of them end in $K_{r+1}$; then one can delete the edges corresponding to partial matchings in $K_{r+1}$ and $K_s$ (and the edges of some odd cycle in $K_s = K_{r+2}$ if $r$ is odd), so that the resulting graph is $r$-regular.

The graph so obtained is $2\left\lfloor \frac{r}{4} \right\rfloor$-connected, has $4r + 1$ or $4r + 2$ vertices, and circumference $4r$ and $4r + 1$ if $r$ is even or odd, respectively.

The conjecture of R. Häggkvist is still open for $k = 3$ (note that for 3-connected $r$-regular graphs the bound cannot exceed $4r - 3$ in general because of the Petersen graph). G. Fan, 1985, proved the bound $3r$. Recently, B. Jackson, H. Li and Y.-J. Zhu, 1989, showed that any longest cycle in a 3-connected $r$-regular graph on at most $4r$ vertices is dominating if $r \geqq 44$. Further Y. Zhu and H. Li, 1990, proved that every 3-connected $r$-regular graph on at most $3r + \frac{r}{7}$ vertices has a Hamiltonian cycle if $r \geqq 44$.

Moreover, P. Fraisse, 1986, made the following conjecture.

**Conjecture.** *Every 3-connected r-regular graph on at most $4r - 3$ vertices is Hamiltonian.*

For bipartite graphs R. Häggkvist, 1976, made the following conjecture.

**Conjecture.** *If the bipartite 2-connected r-regular graph G has order $n(G) \leqq 6r - 6$, then G has a Hamiltonian cycle.*

A. Chetwynd and R. Häggkvist, 1989, proved the conjecture for graphs with at most $\frac{21r}{5}$ vertices ($r \geqq 300$) and P. Ash, 1985b, provided a much better bound, namely $5r - 12$ without any constraint on $r$. A proof with constant $4r + 4$ is due to K.-Q. Dang, 1989b.

The constant of P. Ash's result was slightly improved by Min Aung, 1988, who proved the result of P. Ash with constant $5r - 8$, $r \geqq 6$.

Min Aung, 1988, mentioned that non-Hamiltonian r-regular bipartite graphs on $10r - 2$ vertices can be obtained by an adaption of the construction presented above for the general case.

In section 7.6 we stated (Theorem 7.6.1 due to H.-J. Voss and C. Zuluaga, 1977) that each 2-connected non-bipartite graph of order at least $2r$ with minimum degree at least $r \geqq 3$ contains an odd cycle and an even cycle, both of length at least $2r - 1$. Further, by Theorems 7.6.3 and 7.6.4 all these graphs with circumference at least $2r + 3$ and $c_{\text{even}}(G) = 2r$ or $c_{\text{odd}}(G) = 2r - 1$ are known. An exact investigation shows that the graphs of Theorems 7.6.3 and 7.6.4 are not regular. Therefore regular graphs satisfying the hypotheses of these theorems have longer odd and even cycles. By Theorem 7.7.2 we can formulate the following result:

**Theorem 7.7.9.** *Let G be a 2-connected r-regular graph, $r \geqq 3$, with $n(G) \geqq 2r + 4$. Then*
(i) *$c(G) \geqq 2r + 4$ or G is the Petersen graph;*
(ii) *if G is non-bipartite, then $c_{\text{even}}(G) \geqq 2r + 2$ and $c_{\text{odd}}(G) \geqq 2r + 1$.* □

I conjecture that the bound of Theorem 7.7.9 can be improved so that
(i) $c_{\text{even}}(G) \geqq 2r + 4$ or $G$ is the Petersen graph;
(ii) $c_{\text{odd}}(G) \geqq 2r + 3$.

These bounds, if true, would be best possible. This is shown by the graphs constructed by P. Erdős and A. M. Hobbs, 1977. We remark that G. A. Dirac, 1955, observed that the circumference in the class of all 2-connected $r$-regular graphs tends to infinity as the number of vertices does so.

We conclude this section by stating some results on 3-connected and 4-connected graphs. For these graphs larger lower bounds have been obtained. G. Fan, 1985, and H. A. Jung, 1989, independently proved:

**Theorem 7.7.10.** *Let G be a 3-connected r-regular graph of order n. Then* $c(G) \geqq \min(3r, n)$. □

Both problems will be discussed in the introduction of Chapter 11.

Recently Theorem 7.7.10 has been improved by Zh. Yu and Y. Zhu, 1990. They showed: *Let G be a 3-connected r-regular graph of order n. Then*

$$c(G) \geqq \min(3r + 5, n) \qquad \text{for any } r \geqq 71.$$

This result is essentially best possible, since there exist counterexamples as follows: a 3-connected $2m$-regular subgraph $G_1$ of $(2K_{2m+1} \cup 2K_{2m}) + 3K_1$ with $c(G_1) = 3(2m) + 5$ and a 3-connected $(2m-1)$-regular subgraph $G_2$ of $(3K_{2m} \cup K_{2m-1}) + 3K_1$ with $c(G_2) = 3(2m-1) + 6$, where $m$ is an integer.

If the graph $G$ of Theorem 7.7.10 is 4-connected or bipartite, then the lower bound can be improved. This was done by Min Aung, 1988.

**Theorem 7.7.11.** *Let G be a 4-connected r-regular graph on at least* $5r - 7$ *vertices. Then* $c(G) \geqq 4r - 4$. □

Min Aung, 1988, has conjectured that for *all* 4-connected $r$-regular graphs of order $n$ the circumference $c(G) \geqq \min(n, 4r - 4)$, and proved it for bipartite graphs.

**Theorem 7.7.12.** *Let G be a 3-connected r-regular bipartite graph of order n. Then* $c(G) \geqq \min(n, 6r - 14, 5r - 7)$. *Moreover, if* $n \geqq 8r - 24$, $r \geqq 5$, *then* $c(G) \geqq 6r - 14$. □

For 4-connected and 5-connected graphs H. A. Jung, 1989, improved Theorem 7.7.10 by stating a long cycle through each edge of $G$ or $G \cup \overline{G}$, respectively.

**Theorem 7.7.13.** *Let G be a 4-connected r-regular graph of order n. Then each edge of G is contained in a cycle C of length* $\lambda(C) \geqq \min(3r - 2, n)$. □

**Theorem 7.7.14.** *Let G be a 5-connected r-regular graph on at least* $6r - 13$ *vertices. Then each longest P,Q-path of G has length at least* $4r - 8$, *where P, Q is any pair of distinct vertices of G.* □

Similar results are independently proved by H. Li, 1990. A result analogous with Theorem 7.7.7 of B. Jackson has been proved by H. A. Jung, 1989, for Hamiltonian cycles through a given edge.

**Theorem 7.7.15.** *If the 3-connected r-regular graph G has order* $n(G) \leqq 3r - 1$, *then G has a Hamiltonian cycle through any edge of G.* □

## 7.8. Dominating cycles and paths in graphs with given order and minimum degree

In the preceding sections we have dealt with largest cycles in graphs with given minimum degree. The method of deriving results was a precise investigation of the bridges of a longest cycle. Here we shall deal with problems of finding a cycle $C$, and this can be a longest cycle, such that all regular $C$-bridges have precisely one inner vertex. Such a cycle is called a *dominating cycle (dominating cycle* by P. Ash and B. Jackson, 1984, and *D-cycle* by

B. N. Clark, Ch. J. Colbourn and P. Erdős, 1985). H. J. Veldman, 1983 b, calls a cycle $C$ a $D_\lambda$-*cycle* if $G-C$ contains only components with fewer than $\lambda$ vertices. P. Fraisse, 1986, calls such a cycle *$\lambda$-dominant.*

Obviously, a $D_1$-cycle is Hamiltonian and a $D_2$-cycle is dominating. We can also say that a cycle $C$ is *dominating* if and only if $G-C$ has no edge. Thus $C$ is dominating if and only if each edge of $G$ is adjacent to $C$. A cycle $C$ is said to be *vertex-dominating* if each vertex is either on $C$ or is adjacent to $C$.

A dominating cycle can be considered as a generalization of a Hamiltonian cycle, and a vertex-dominating cycle as a generalization of a dominating cycle. Therefore the smallest minimum degree that guarantees a vertex-dominating cycle should be smaller than that for a dominating cycle, which in turn should be smaller than the sufficiency condition with respect to $\delta(G)$ for Hamiltonian cycles. Similarly, dominating paths are defined.

Let $\varkappa(G)$ denote the connectivity number of the graph $G$, and $\alpha(G)$ the maximum number of pairwise non-adjacent (independent) vertices.

C. St. J. A. Nash-Williams, 1971 a, has proved the following result:

**Theorem 7.8.1.** *If $G$ is a 2-connected graph of minimum degree at least $r \geqq 3$ on at most $3r-2$ vertices, then any longest cycle of $G$ is dominating. Moreover, if $\delta(G) \geqq \alpha(G)$, then $G$ is Hamiltonian.* □

R. Häggkvist and G. G. Nicoghossian, 1981, proved more, namely, that every graph of Theorem 7.8.1 on at most $3r-\varkappa(G)$ vertices is Hamiltonian.

Theorem 7.8.1 has been strengthened by H. A. Jung, 1981, and by A. Bigalke and H. A. Jung, 1979. A. Bigalke and H. A. Jung, 1979, proved the following statement: every graph $G$ of Theorem 7.8.1 of order $3r-1$ or $3r$ having no component problem contains a dominating cycle, where a graph $G$ is said to have a component problem if there exists a non-empty set $X \subseteqq V(G)$ such that $G-X$ has at least $|X|+1$ components.

V. D. Hòa, 1986, determined all 2-connected graphs of order $3r-1$ and $3r$ with minimum degree $\delta(G) \geqq r$ which have a component problem. He improved Theorem 7.8.1 as follows: if the graph $G$ in Theorem 7.8.1 has order $3r-1$ or $3r$, then either any longest cycle of $G$ is dominating or any longest cycle of $G$ has precisely one bridge whose kernel is complete. A further improvement is due to B. Fassbender, 1989.

For 2-connected triangle-free graphs Min Aung, 1988 and 1989 b, obtained the following result:

**Theorem 7.8.2** *If $G$ is a 2-connected triangle-free graph of minimum degree at least $r \geqq 3$ with at most $6r-6$ vertices, then $G$ contains a longest cycle which is dominating.*

*Moreover, $G-C$ contains at most one edge for each longest cycle $C$.* □

The existence of a (not necessarily longest) dominating cycle in $G$ can easily be deduced from Theorem 3 of H. J. Veldman, 1983 a.

A similar theorem for $P,Q$-paths (for cycles containing a given edge) has been proved by H. A. Jung, 1986.

**Theorem 7.8.3.** *If $G$ is a 3-connected graph of minimum degree at least $r \geqq 3$ with at most $3r-4$ vertices, then any longest $P,Q$-path of $G$ is dominating, where $P,Q$ is any pair of distinct vertices of $G$.*

If $G$ has a longest $P,Q$-path $p$ with a bridge which is no stronghold, then $p$ is dominating even for graphs $G$ with at most $3r-1$ vertices. □

H. A. Jung, 1989, proved that if the graph $G$ is $r$-regular and 4-connected, then the first assertion remains true even for graphs $G$ with at most $4r-5$ vertices.

B. Jackson, 1980a (proof of Theorem 7.7.7), D. R. Woodall, 1978, and others applied the theorem of Nash-Williams in their papers. An improvement of Theorem 7.8.1 is due to J. A. Bondy, 1980b:

**Theorem 7.8.4.** *Let G be a 2-connected graph on at most $\sigma-2$ vertices in which the degree sum of any three independent vertices is at least $\sigma$. Then any longest cycle of G is dominating.* □

The bounds $3r-2$ and $\sigma-2$ of the orders of the graphs considered in the theorems of Nash-Williams and Bondy are best possible, as the graphs $I_2+3K_{r-1}$, $r \geqq 3$ (with $\sigma = 3r$), show. Each longest cycle avoids one $K_{r-1}$.

The bound $3r-4$ of Jung's theorem also cannot be improved. This can be seen by the graph $I_3+3K_{r-2}$. Each longest $P,Q$-path, $P,Q \in I_3$, avoids one $K_{r-2}$. A counterexample is $I_{r-1}+(r-1)K_2$. P. Ash and B. Jackson, 1984, proved a related result for bipartite graphs.

**Theorem 7.8.5.** *If G is a 2-connected bipartite graph of minimum degree at least $r \geqq 3$, and each set of the bipartition has at most $3r-3$ vertices, then G contains a longest cycle which is dominating.* □

This result is also a consequence of Aung's Theorem 7.8.2.

The bound $3r-3$ is best possible. This can be seen by the following graph: add two vertices $A$ and $B$ to the bipartite graph $3(I_{r-1}+I_{r-1})$ and join $A$ to each vertex of one set of the bipartition by an edge and $B$ to each vertex of the other set of the bipartition by an edge. Each longest cycle of this graph avoids one $I_{r-1}+I_{r-1}$.

The assertion that "there exists a largest dominating cycle" cannot be strengthened to the assertion that "any longest cycle is dominating". This is shown by the graph which is obtained from two copies $K$, $K'$ of $I_{r+2}+I_r$ in the following way: choose in $I_{r+2}$ of $K$ and $K'$ three vertices $A_1$, $A_2$, $A_3$ and $A'_1$, $A'_2$, $A'_3$, respectively, and add the edges $(A_1, A'_1)$, $(A_2, A'_2)$ and $(A_3, A'_3)$ to $K \cup K'$. This bipartite graph has a longest cycle avoiding both $A_3$ and $A'_3$. Thus it contains both a longest dominating and a longest non-dominating cycles.

P. Ash, 1985b, further proved that every longest cycle of the graph $G$ of Theorem 7.8.5 is a $D_3$-cycle.

For regular graphs Min Aung, 1988 and 1989b, obtained the stronger result that if $G$ is a 2-connected $r$-regular bipartite graph on at most $6r-6$ vertices, then each longest cycle of $G$ is dominating.

P. Fraisse, 1986, proved for $r$-regular (bipartite and non-bipartite) graphs that if $G$ is a 3-connected $r$-regular graph of order less than $4r-2$, then $G$ has a dominating cycle. Recently, this result has been strengthened by B. Jackson, H. Li and Y.-J. Zhu, 1989, who proved that each longest cycle of a 3-connected $r$-regular graph on at most $4r$ vertices is dominating if $r \geqq 44$.

Theorem 7.8.5 of P. Ash and B. Jackson has been extended by D. Bauer (see D. Bauer, 1985). The improved theorem follows easily from a theorem of H. J. Veldman, 1983a.

**Theorem 7.8.6.** *If G is a 2-connected bipartite graph of minimum degree at least $r \geqq 3$ and order at most $6r-5$, then G contains a dominating cycle.* □

P. Fraisse, 1986, presents more results on dominating cycles in balanced bipartite graphs, i.e., bipartite graphs in which both bipartition sets have the same cardinality.

For 4-connected graphs H. A. Jung, 1986, has proved the following result:

**Theorem 7.8.7.** *If $G$ is a 4-connected graph of minimum degree at least $r \geqq 3$ on at most $4r - 9$ vertices, then any longest $P,Q$-path of $G$ is a $D_3$-path, where $P$, $Q$ is any of pair of distinct vertices of $G$.*

*If $G$ has a longest $P,Q$-path $p$ with a bridge which is not a stronghold, then $p$ is a $D_3$-path even for graphs with at most $4r - 5$ vertices.* □

The bound $4r - 9$ of Jung's theorem cannot be improved. This can be seen by the graph $I_{r-2} + (r-2)K_3$. Each longest $P,Q$-path, $P,Q \in I_{r-2}$, avoids one $K_3$.

A graph is said to be *p-path connected* if and only if any two vertices are connected by a path of length at least $p$. J. A. Bondy, 1980 b, has proved a generalization of Theorem 7.8.2, presenting a theorem on bridges of a longest cycle.

**Theorem 7.8.8.** *Let $G$ be a $k$-connected graph on $n$ vertices. Let the degree sum of any $k+1$ independent vertices be at least $n + k(k-1)$. Then any longest cycle $C$ has the property that $G - C$ contains no $(k-1)$-path-connected subgraph (and, in particular, no $K_k$).* □

The graphs $I_{k+1} + hK_k$, $h \geqq k+2$, show that the bound cannot be relaxed. A generalization of theorem 7.8.8 was proved by B. Fassbender, 1989 a.

J. A. Bondy, however, conjectured that the conclusion can be strengthened substantially so that $G - C$ has no path of length $k - 1$ (see J. A. Bondy, 1980 b). For graphs containing no induced $K_1 + I_3$ subgraph C.-Q. Zhang, 1988, proved the validity of Bondy's conjecture. Moreover, these graphs have even a Hamiltonian cycle. A variant of this conjecture has been proved by P. Fraisse, 1986. (For $k = 1, 2$ see also H. I. Veldman, 1983 b.)

**Theorem 7.8.9.** *Let $G$ satisfy the hypotheses of Theorem 7.8.8. Then $G$ contains a cycle $C$ such that $G - C$ contains no path of length $k - 1$. Moreover, $G$ has a $D_k$-cycle.* □

We conclude this section with a result on vertex-dominating cycles.

Confirming a conjecture of B. N. Clark, Ch. J. Colbourn and P. Erdős, 1985, P. Fraisse, 1986, proved the following result:

**Theorem 7.8.10.** *If $G$ is a $k$-connected graph of minimum degree at least $r \geqq 3$ on at most $(k+1)r + 2k - 2$ vertices, then $G$ has a vertex-dominating cycle.* □

For $2 \leqq k \leqq 5$, B. N. Clark, Ch. J. Colbourn and P. Erdős, 1985, proved the theorem with a slightly weaker additive constant.

An independent proof of Theorem 7.8.10 is due to J. A. Bondy and G. Fan, 1987. They proved a stronger result:

**Theorem 7.8.11.** *If a $k$-connected graph $G$ on $n$ vertices has the property that every subset of $k+1$ vertices whose pairwise distances are at least 3 has degree sum at least $n - 2k$, then $G$ has a vertex-dominating cycle.* □

## 7.9. Dominating cycles and paths in graphs with given minimum degree

In this section we continue the investigations of section 7.8. Again we consider graphs with given minimum degree, but there is no restriction on the number of vertices.

G. A. Dirac, 1952a, proved that every 2-connected graph with $\delta(G) \geqq r \geqq 3$ has a Hamiltonian cycle or a cycle of length at least $2r$. The graphs $I_r + I_s$, $s \geqq r$, show that this theorem cannot be improved. But every longest cycle of $I_r + I_s$, $s \geqq r$, is dominating. The graphs $I_2 + h\,K_{r-1}$, $h \geqq 3$, are graphs that also satisfy the hypotheses of Dirac's theorem and have no dominating cycle. But these graphs have cut sets with two vertices. Hence we shall consider graphs with connectivity at least 3.

**Theorem 7.9.1.** *Let $G$ be a 3-connected graph with $\delta(G) \geqq r$. Then any longest cycle of $G$ is dominating or has length at least $3r-3$.* □

This theorem can be derived from Theorem 7.5.3 of H.-J. Voss, 1977a (case $i = k-3$, $k = 3$). The same theorem has been independently proved by H. A. Jung, 1981. The graphs $I_{r-1} + hK_2$, $h \geqq r \geqq 4$, show that for $r \geqq 4$ the bound of Theorem 7.9.1 cannot be improved. Each longest cycle avoids at least one $K_2$.

But if we consider graphs without triangles, we can state an improved bound of H.-J. Voss and C. Zuluaga, 1977, even for 2-connected graphs (see Theorem 7.6.2).

**Theorem 7.9.2.** *Let $G$ be a triangle-free 2-connected graph with $\delta(G) \geqq r \geqq 3$. Then any longest cycle of $G$ is dominating or has length at least $4r-4$.* □

The bound of Theorem 7.9.2 cannot be improved. This can be seen from the graphs which are obtained from

$$I_{r-1} + (I_{s_1} \cup I_{s_2} \cup \ldots \cup I_{s_h}), \qquad h \geqq r,\ s_i \geqq r,$$

for all $1 \leqq i \leqq h$, by adding the vertices $A_1, A_2, \ldots, A_h$ and joining $A_j$ to each vertex of $I_{s_j}$ by an edge for all $1 \leqq j \leqq h$.

Each longest cycle of this graph avoids at least one $I_{s_j} + K_1$. For 3-connected triangle-free graphs with $\delta(G) \geqq r \geqq 3$ Min Aung, 1988, has proved that any longest cycle has length at least $6r-12$ or all components of $G - C$ are trees of a diameter at most 3.

If the graphs of Theorem 7.9.1 are regular and bipartite, then a better lower bound for the circumference is obtained by Min Aung, 1988: let $G$ be a 3-connected $r$-regular bipartite graph, then any longest cycle of $G$ is dominating or has length at least $6r-14$.

If the graph $G$ of Theorem 7.9.1 is 4-connected, then there is a dominating cycle through any edge of $G \cup \bar{G}$. We formulate this as follows:

**Theorem 7.9.3.** *Let $G$ be a 4-connected graph with $\delta(G) \geqq r \geqq 4$. Then each longest $P,Q$-path of $G$ is dominating or has length at least $3r-6$, where $P$, $Q$ is any pair of distinct vertices of $G$.* □

This theorem is both a consequence of Theorem 7.5.3 of H.-J. Voss, 1977a, (case $i = k-3$, $k = 4$) and a consequence of Theorem 7.5.6 of H. A. Jung, 1986.

The graphs $I_{r-1} + hK_2$, $h \geqq r-1$, $r \geqq 5$, show that for $r \geqq 5$ the bound of Theorem 7.9.3 cannot be improved. Each longest $P,Q$-path, $P$, $Q \in I_{r-1}$, avoids one $K_2$.

For regular graphs Min Aung, 1989a, claims (see his theorem A') that if $G$ is a 4-connected $r$-regular graph, then each longest cycle of $G$ is dominating or has length at least $4r-4$.

Moreover, if $G$ has order $n$ and is non-Hamiltonian, then $n \geqq 3r+1$ and $c(G) \geqq \min\{4r-4, (n+3-2)/2\}$. H. A. Jung, 1989, proved: if $n \leqq 4r-5$, then each longest $P,Q$-path of $G$ is dominating, where $P,Q$ is any pair of distinct vertices of $G$.

The next two theorems state the existence of $D_3$-cycles and $D_3$-paths. By definition a cycle $C$ or a path $p$ is a $D_3$-cycle or a $D_3$-path if and only if $G - C$ or $G - p$, respectively, consists only of $K_1$'s and $K_2$'s.

**Theorem 7.9.4.** *Let $G$ be a 4-connected graph with $\delta(G) \geqq r \geqq 4$. Then any longest cycle of $G$ is a $D_3$-cycle or has length at least $4r - 8$.* □

This theorem can be derived from Theorem 7.5.3 (case $i = k - 4$, $k = 4$). It was also proved by H. A. Jung, 1981.

The graphs $I_{r-2} + h\,K_3$, $h \geqq r - 1$, $r \geqq 6$, show that for r $\geqq 6$ the bound of Theorem 7.9.4 cannot be improved. Each longest cycle misses some $K_3$.

For $k = 5$ the Theorem 7.5.6 of H. A. Jung, 1986, implies the following result:

**Theorem 7.9.5.** *Let $G$ be a 5-connected graph with $\delta(G) \geqq r \geqq 5$. Then each longest $P,Q$-path of $G$ is a $D_3$-path or has length at least $4r - 12$, where $P$, $Q$ is any pair of distinct vertices of $G$.* □

Again the bound $4r - 12$ cannot be improved, as the graphs $I_{r-2} + h\,K_3$, $h \geqq r - 2$, $r \geqq 7$, show. Each longest $P,Q$-path $P$, $Q \in I_{r-2}$, avoids one $K_3$.

We remark that by Theorem 7.5.6 of H. A. Jung Theorems 7.9.3 and 7.9.5 can be slightly improved, namely, when $G$ has a longest cycle or path, respectively, with a bridge that is not a stronghold.

For regular graphs H. A. Jung, 1989, proved a sharper result: if $G$ is a 5-connected $r$-regular graph, then each longest $P,Q$-path of $G$ is dominating or has length at least $4r - 8$.

I conjecture that for higher $k$ the following result holds:

**Conjecture.** *Let $G$ be a $k$-connected graph with $\delta(G) \geqq r \geqq 2k$. Then*

(i) *any longest cycle of $G$ is a $D_{k-1}$-cycle or has length at least $k(r - (k-2))$,*

(ii) *each longest $P,Q$-path of $G$ is a $D_{k-2}$-path or has length at least $(k-1)\,(r - (k-2))$.*

By Theorem 7.5.3 this conjecture is true for $k \leqq 4$, Theorem 7.5.6 implies the validity of (ii) for $k \leqq 5$. The validity of (i) for $k \leqq 4$ is also included in the following Theorem 7.9.6.

A tree $T$ is a *star (double star)* if all vertices of $T$ with at most *one (exactly two)* exceptions have degree 1. A *quasistar*, by definition, is a star or a double star. Let $\mathfrak{K}_5$ denote the class of all stars and $\mathfrak{K}_6$ the class of all double stars, further $\mathfrak{K}_k = \emptyset$ for $k < 5$.

Next we present a result recently obtained by H. A. Jung, 1990, which generalizes case $i = 0$ of Theorem 7.5.3 of H.-J. Voss, 1977a. As in Theorem 7.5.3, the new Theorem 7.9.6 combines the structure of the kernel of a bridge of a longest cycle $L$ with the length $\lambda(L)$.

**Theorem 7.9.6.** *Let $L$ be a longest cycle of the graph $G$, and let $K$ be a component of $G - L$ such that $|K| \geqslant k - 1$ ($k = 2, 3, 4, 5, 6$). Then there exists a vertex $V$ in $K$ such that*

(i) *$\lambda(L) \geqslant k\,v(G:V) - k(k-2)$ if $G$ is $k$-connected and $K \notin \mathfrak{K}_k$,*

(ii) *$|V(L \cup K)| \geqslant k\,v(G:V) - (k-1)\,(k-2)$ if $G$ is $(k-1)$-connected and $K \in \mathfrak{K}_k$.*

Parts of Theorem 7.9.6 are included in this book and have been proved by G. A. Dirac, 1952a ($k = 2$: Theorem 7.2.1), C. St. J. A. Nash-Williams, 1971a ($k = 3$, (ii): Theorem 7.8.1), H.-J. Voss, 1976a ($k = 2$: Theorem 7.2.2), H.-J. Voss, 1977a ($k \leqq 4$: Theorem 7.5.3), H. A. Jung, 1981 ($k \leqq 3$), and G. Fan, 1987 ($k \leqq 4$).

We define: for a 2-connected graph $H$ let $D(H)$ be the maximum integer $s$, such that for any two distinct vertices $P, Q$ in $H$ there is a $P,Q$-path $p$ of length at least $s$. For a complete graph $K_n$ $(n \leqq 2)$ we put $D(K_n) = n - 1$. If $H$ is connected and has cut vertices, we put

$$D(H) = \max\{D(H') \mid H' \text{ endblock of } H\}.$$

Extending the definition to arbitrary graphs we finally put

$$D(H) = \max\{D(H') \mid H' \text{ component of } H\}.$$

Next we present a further result, recently found by P. Fraisse and H. A. Jung, 1989.

**Theorem 7.9.7.** *Let $L$ be a longest cycle in a non-Hamiltonian graph $G$, which has connectivity $k \geqslant 2$. Further let $D = D(G - L)$. Then there exist independent vertices $V_1, V_2, \ldots, V_{k+1}$ with $v(G:V_1) \geqslant \ldots \geqslant v(G:V_{k+1})$ such that*

(i) $\lambda(L) \geqslant v(G:V_1) + \ldots + v(G:V_t) - tD + k(D + 2 - t)$ *for all $t$,* $0 \leqslant t \leqslant \min\{D + 1, k + 1\}$.

(ii) *Moreover, if $D + 1 \leqslant k$, then*

$$\lambda(L) \geqslant v(G:V_1) + v(G:V_2) + \ldots + v(G:V_{D+1}) + v(G:V_{k+1}) - D(D + 2).$$

This result is sharp, as can be seen from the graph $(k + 1)K_q + K_k$.

More on $D_\lambda$-cycles can be found in H. J. Veldman, 1983 a, b, P. Fraisse, 1986, and H. J. Broersma, 1988.

We conclude this section with an interesting conjecture due to H. Fleischner, 1982.

**Conjecture.** *Every 3-regular cyclically 4-edge-connected graph has a dominating cycle.*

The algorithmic aspects can be seen in Ch. J. Colbourn and L. K. Stewart, 1985. They proved that the vertex-dominating and the dominating cycle problems are NP-complete on many classes of graphs.

# Chapter 8

# Cycles with diagonals in graphs of minimum degree at least 3

## 8.0. Introduction

In Chapters 8-10 we deal with diagonals of cycles and paths.

A *diagonal* of a subgraph $H$ in a graph $G$—also called a *chord* or a *degenerate bridge* of $H$—is defined as an edge of $G$ not in $H$ joining two vertices of $H$. Thus diagonals are the simplest bridges a subgraph can have.

Investigations related to diagonals of cycles in graphs are devoted to the question: which properties imply the existence of cycles with certain diagonals and, vice versa, which properties are implied by the existence of cycles with certain diagonals? This question will be studied in Chapters 8-10.

Diagonals of certain cycles play a role in well-known graph-theoretic problems, for example in investigating perfect graphs and minimally 2-connected graphs, in the Whitney construction of all 2-connected graphs, in Hadwiger's conjecture and Kuratowski's theorem. This and the solution of two conjectures of P. Erdős will be discussed in 1°-5° of this section. The main purposes of Chapters 8-10 are the detailed proofs of the problems and results outlined in 3°. But also (simple) proofs of theorems related to 2° and 4° will be included.

### 1° Perfect graphs and the Strong Perfect Graph Conjecture

The perfect graphs form a subset of the set of all graphs whose odd cycles of length at least 5 have at least one diagonal. In introducing perfect graphs we need the following parameters of a graph:

$\chi(G)$, the *chromatic number*, and

$\mathrm{cl}(G)$, the *clique number* of $G$, that is, the order of the largest complete subgraphs of $G$.

Obviously, every graph $G$ satisfies $\mathrm{cl}(G) \leqq \chi(G)$. There are graphs $G$ with $\mathrm{cl}(G) < \chi(G)$ (see for instance the odd cycles of length at least 5). If no induced subgraph $H$ of $G$ satisfies the inequality $\mathrm{cl}(H) < \chi(H)$, then $G$ is called a *perfect graph*. These graphs have been introduced by C. Berge, 1961, and intensively investigated by many authors (see M. Ch. Golumbic, 1980). Thus the perfect graphs $G$ are defined by the property:

(P) $\mathrm{cl}(H) = \chi(H)$ for every induced subgraph $H$ of $G$.

C. Berge, 1961, conjectured and L. Lovász, 1972a and b, and D. R. Fulkerson, 1972, proved the following theorem.

**Perfect Graph Theorem.** *A graph is perfect if and only if its complement is perfect.*

This theorem had been known for a long time (from 1961 to 1972) as the weak perfect graph conjecture.

The non-perfect graphs are called *imperfect*. Evidently, the odd cycles $C_{2k+1}$ of length $2k+1$ are imperfect, where $k \geqq 2$ is an integer. Moreover, $\text{cl}(C_{2k+1}) < \chi(C_{2k+1})$ and for each of its proper induced subgraphs $H$ equality holds: $\text{cl}(H) = \chi(H)$. Thus the odd cycles $C_{2k+1}$, $k \geqq 2$, are critically imperfect (an imperfect graph is called *critically imperfect* if each proper induced subgraph is perfect). By the perfect graph theorem the complement $\bar{C}_{2k+1}$, $k \geqq 2$, of the odd cycle $C_{2k+1}$ is also critically imperfect. A conjecture of C. Berge, 1961, is the well-known strong perfect graph conjecture: the only critically imperfect graphs are the odd cycles $C_{2k+1}$, $k \geqq 2$, and their complements $\bar{C}_{2k+1}$. This can be reformulated as follows.

**Strong Perfect Graph Conjecture (SPGC).** *A graph $G$ is perfect if and only if in $G$ and $\bar{G}$ every odd cycle of length at least 5 has a diagonal.*

The non-trivial direction of this assertion is: if neither $G$ nor $\bar{G}$ contains odd cycles of length at least 5 without a diagonal, then $G$ is perfect. Until now this assertion could not be proved, but for some partial classes of graphs satisfying similar conditions (non-symmetric in $G$ and $\bar{G}$) the SPGC could be verified in the affirmative.

Trivially, the bipartite graphs are perfect.

A graph is said to be *triangulated* if and only if every cycle of length at least 4 has at least one diagonal (also called a *rigid circuit graph* by G.A. Dirac, 1961, and a *chordal graph* by F. Gavril, 1974). C. Berge, 1960, showed that triangulated graphs satisfy (P); A. Hajnal and J. Suranyi, 1958, proved that the complement of any tringulated graph satifies (P).

Since 1972 we know that the perfect graph theorem is true. It implies that the two results of C. Berge and A. Hajnal/J. Surányi are equivalent.

T. Gallai, 1962, proved that every graph in which each odd cycle of length at least 5 has two non-overlapping diagonals is perfect (a shorter proof was published by L. Suranyi, 1968). The corresponding case that each odd cycle of length at least 5 has two overlapping diagonals was proved by E. Olaru, 1969 (see H. Sachs, 1970b). H. Meyniel, 1976, proved the general case including the results on triangulated graphs:

*If $G$ is a graph such that every odd cycle of length at least 5 has two diagonals, then $G$ is perfect.*

The proof can be found in the book of M. Ch. Golumbic, 1980. Meyniel's theorem cannot be weakened to graphs with the following property:

(Q) Each odd cycle of length at least 5 has at least one diagonal.

The graph $\bar{C}_7$ (see Fig. 8.0.1) has the property (Q), but $\bar{C}_7$ is not perfect; it is critically imperfect.

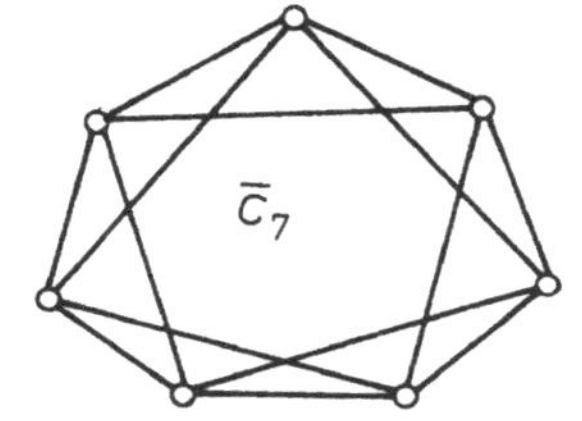


Fig. 8.0.1 $\chi(\bar{C}_7) = 4$, $\text{cl}(\bar{C}_7) = 3$

Property (Q) implies that for each odd cycle $C$ there is a triangle contained in the subgraph induced by $C$. This implies the validity of the SPGC for 3-chromatic graphs (A. C. Tucker, 1977).

We conclude 1° with the remark that there is an interesting characterization of the so-called comparability graphs by a diagonal condition: $G$ is a comparability graph if and only if each even walk $(V_1, V_2, \ldots, V_i, \ldots, V_{2k})$ of $G$ has no triangular chord, i.e., no chord $(V_i, V_{i+2})$, $i = 1, 2, \ldots, 2k$ (indices mod $2k$) (M. Ch. Golumbic, 1980). For further results see M. Ch. Golumbic, 1980, and "Topics on Perfect Graphs", 1984.

## 2° The structure of 2-connected and 3-connected graphs

By H. Meyniel, 1976, every graph whose odd cycles have at least two chords is perfect. The graphs $K_r$, $r \geqq 4$, and $K^+_{n,m}$, $n$, $m \geqq 3$ (obtained from $K_{n,m}$ by adjoining one edge) have the Meyniel property.

G. A. Dirac, 1960, proved that every 2-connected graph whose even cycles have at least two diagonals is complete (and perfect) or is an odd cycle (and critically imperfect). G. A. Dirac, 1967, and M. Plummer, 1968, considered 2-connected graphs whose cycles have no chord. They proved that these are even the so-called minimally 2-connected graphs, where a graph is said to be *minimally 2-connected* if and only if it is 2-connected and the deletion of any edge destroys its 2-connectivity In this chapter we shall show that these graphs have many vertices of degree 2.

One of the most important goals of the theory of $k$-connected graphs is to compile a list of all $k$-connected graphs. A natural way of achieving this would be to find some operations producing $k$-connected graphs from $k$-connected graphs such that every $k$-connected graph can be derived from a set of simple $k$-connected graphs by repeated applications of the operations (B. Bollobás, 1978a, and W. T. Tutte, 1984). This task has been accomplished by H. Whitney, 1932a, for 2-connected graphs (see also G. A. Dirac, 1967, and M. Plummer, 1968). For 3-connected graphs this was done by W. T. Tutte, 1961, D. W. Barnette and B. Grünbaum, 1969, and V. K. Titov, 1967; for 4-connected graphs—by P. J. Slater, 1966; for 3-connected graphs containing a subdivision of $K_{3,3}$—by A. K. Kelmans, 1981; for cyclically 4-edge-connected graphs—by A. K. Kelmans, 1981; and for cyclically 4-vertex-connected graphs—by N. Robertson, 1984a.

Such results serve two main purposes: (1) illustrating the structure of graphs in the class by isolating a type of extremal graph, and (2) by providing a set-up so that induction on the edge set $|E(G)|$ can be carried out effectively within the class.

It is obvious that if a path $p$ is added to a 2-connected graph $H$, having precisely its ends in $H$, the resulting graph $G$ is again 2-connected. $H$ contains a cycle through the two end vertices of $p$. Hence, $G$ is obtained from $H$ by adding a "topological diagonal" to one cycle of $H$.

H. Whitney, 1932a, was the first who observed that

*the class of all 2-connected graphs can be obtained from all cycles by repeatedly adding topological diagonals to some cycles.*

A 2-connected graph $G$ is said to be *minimally 2-connected* if and only if each proper spanning subgraph is not 2-connected. We shall prove that

*the class of all minimally 2-connected graphs can be obtained from all cycles by repeatedly adding topological diagonals to cycles so that no cycles with diagonals are formed.*

In section 8.6 the two results will be derived from a result of J. Czipszer who observed that every graph $G$ with minimum degree at least 3 contains a cycle with at least one diagonal (see L. Lovász, 1979, p. 376).

We write $G > H$ (or $G \vdash H$) if $G$ is obtained from $H$ by repeatedly adding topological diagonals to some cycles (to some cycles so that no cycle with a diagonal is formed). The relations $>$ and $\vdash$ are partial orders on the set of all 2-connected graphs. By the above results, in both cases the minimal elements are the cycles. In other words, the cycles constitute the basis of the class of all 2-connected graphs with respect to $>$ and $\vdash$ (for the concept of basis see also L. Lovász, 1976).

A more difficult question is the construction of all 3-connected graphs from a basis of simple 3-connected graphs. A simple class of 3-connected graphs consists of the wheels: a *wheel* is a cycle with an additional vertex joined to each vertex of the cycle by an edge.

The famous result of W. T. Tutte, 1961, is that all 3-connected graphs can be obtained from the set of all wheels by repeatedly applying the following two operations:

$O_1$: *the addition of a diagonal to a cycle,*

and the "dual" operation

$O_2$: *the splitting of a vertex of degree at least* 4, i.e. *the replacement of a vertex $X$ of degree at least* 4 *by two adjacent vertices $X'$, $X''$ and joining every neighbour of $X$ to exactly one of $X'$, $X''$ in such a way that both $X'$ and $X''$ will have degree at least* 3.

Thus the wheels form a basis for all 3-connected graphs with regard to the operations $O_1$ and $O_2$.

It is obvious that from a 3-connected graph $G$ a new 3-connected graph is obtained by the following procedure O: in an arbitrarily chosen cycle zero, one or two edges are subdivided by exactly one vertex and a diagonal is added to this cycle so that the resulting graph again has minimum degree at least 3.

Conversely, D. W. Barnette and B. Grünbaum, 1969, and V. K. Titov, 1967, showed that

*every* 3*-connected graph $G$ contains an edge $e$ (which can be interpreted as a diagonal of a cycle) so that $G - e$ is a subdivision of a* 3*-connected graph.*

These two statements imply that up to isomorphism all 3-connected graphs can be obtained from $K_4$ by repeated applications of procedure O. Consequently, up to isomorphism $K_4$ forms a basis for the class of all 3-connected graphs with respect to O. This result was used by A. K. Kelmans, 1981, in his new proof of the theorem of Kuratowski. A. K. Kelmans, 1981, also showed that all 3-connected graphs containing a subdivision of $K_{3,3}$ can be generated from $K_{3,3}$ by repeated applications of procedure O.

Finally we note that R. W. Dawes, 1983, defined three operations on 3-connected graphs and showed that by repeated applications of these three operations to $K_4$ the class of all minimally 3-connected graphs can be generated. This corresponds to the above result that the class of all minimally 2-connected graphs can be generated from the class of all cycles.

The class of all 2-connected graphs having no cycles with a diagonal coincides with the class of all minimally 2-connected graphs.

We conclude 2° with a diagram illustrating the containment relation between the graph classes described in 1° and 2°:

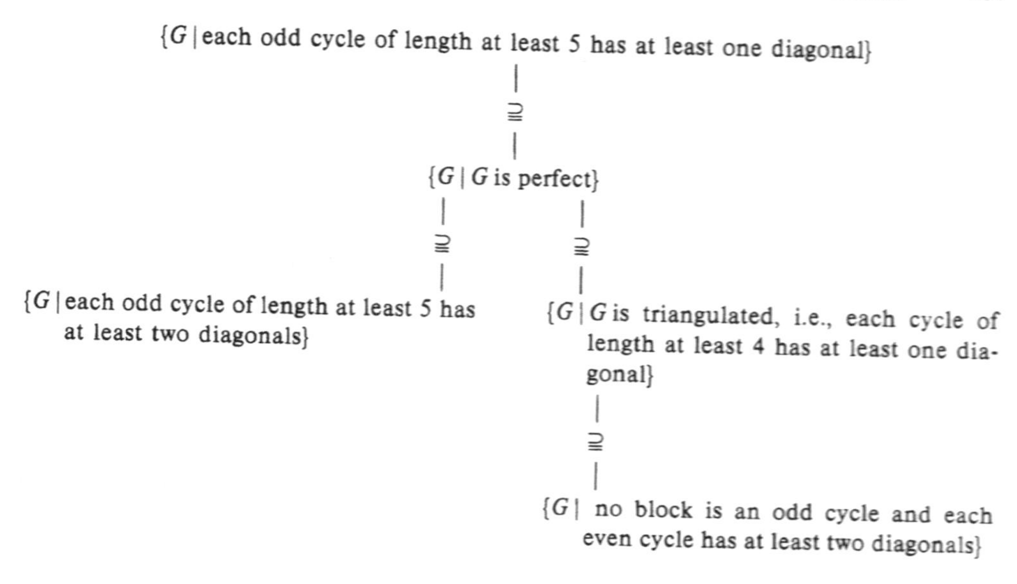


### 3° Cycles with diagonals

An imperfect graph always contains an induced subgraph $H$ with $\mathrm{cl}(H) < \chi(H)$. W. T. Tutte (also B. Descarte, 1948 and 1954), A. A. Zykov, 1949, J. Mycielski, 1955, and J. Nešetřil, 1966, constructed graphs of given chromatic number containing no triangles, i.e. $\mathrm{cl}(G) = 2$. Moreover, the graphs of W. T. Tutte do not contain 4- and 5-cycles and the graphs of J. Nešetřil include no cycles of lengths at most 7. For the general case we have the following statement:

*For each pair of integers $k \geqq 3$, $t \geqq 3$ there exists a k-chromatic graph of girth at least t.*

This was first proved by P. Erdős, 1959, using probabilistic arguments, and later by L. Lovász, 1968, with the aid of constructive methods. A short proof is due to J. Nešetřil and V. Rödl, 1979.

Though graphs of high chromatic number do not necessarily contain small cycles, one may ask whether all such graphs contain cycles with a certain number of diagonals. The answer is in the affirmative.

Above we stated that *each* odd cycle of length at least 5 of a perfect graph contains at least one diagonal. Here we deal with the existence of *at least one* odd cycle with a certain number of diagonals.

Moreover, answering two conjectures of P. Erdős, 1977, (see also P. Erdős, 1978) in the affirmative, we shall show that

(i) *every graph $G$, $G \ngeqq K_4$, with chromatic number at least 4 contains odd and even cycles with at least two diagonals;*

(ii) *every graph $G$ with chromatic number at least 4 and girth at least $t \geqq 4$ contains odd and even cycles with at least $c2^{t/12}$ diagonals (where $c$ is an appropriate constant).*

The results (i) and (ii) can be derived from results on graphs with given minimum degree by the following observations: it is well known that every $k$-chromatic graph contains a critically $k$-chromatic graph; if $k \geqq 4$, this critically $k$-chromatic graph is non-bipartite, 2-connected and has minimum degree at least $k-1 \geqq 3$.

The main result of this chapter is the proof of assertion (i) of the following theorem (assertion (ii) will be proved in Chapter 9):

(i) *every non-bipartite 2-connected graph $G$ of order at least 7 with minimum degree at least 3 contains odd and even cycles with at least two diagonals;*

(ii) *every non-bipartite 2-connected graph $G$ with minimum degree at least 3 and girth at least $t \geqq 4$ contains odd and even cycles with at least $c_1 2^{t/12}$ diagonals and odd and even cycles of length at least $c_2 2^{t/4}$*, where *$c_1$ and $c_2$ are appropriate constants.*

The existence of large odd and even cycles with many diagonals in graphs satisfying the hypothesis (ii) was proved independently by C. Thomassen, 1983a, and H.-J. Voss, 1980. The cycles investigated above have many diagonals. One may ask: when does a cycle exist with so many diagonals that some of them necessarily form a new cycle?

P. Erdős conjectured and B. Bollobás, 1978b, proved that there is a constant $c$ such that every graph on $n$ vertices and with at least $cn$ edges contains nested cycles, i.e. edge-disjoint cycles $C_1$ and $C_2$ with $V(C_2) \subseteqq V(C_1)$.

### 4° The Hadwiger Conjecture

Let $e$ be an edge of a graph $G$. If $e$ is deleted and its end vertices are identified, then we say that the resulting graph is *obtained from $G$ by contracting the edge $e$.*

A graph $H$ is a *contraction* of $G$ if $H$ can be obtained by contracting some edges of $G$. A graph $H$ is a *subcontraction* of $G$ if $H$ is a subgraph of a contraction of $G$.

In 1943 H. Hadwiger published his famous conjecture (H. Hadwiger, 1943 and 1958):

**Conjecture.** *Every $k$-chromatic graph has $K_k$ as a subcontraction.*

With respect to this conjecture see also section 6.0, 2°.

Hadwiger's conjecture can easily be verified for $k = 2$ and $k = 3$. G. A. Dirac, 1952a, proved its validity for $k = 4$.

For $k = 5$ K. Wagner, 1937, discovered the following theorem which closely relates the four-colour conjecture (4CC) to Hadwiger's conjecture:

*The 4CC is true if and only if $\chi(G) = 5$ implies that $K_5$ is a subcontraction of $G$.*

A consequence of the computer-aided proof of the 4CC by K. Appel and W. Haken, 1976, 1977 and 1989, is that the theorem of K. Wagner implies the validity of Hadwiger's conjecture for $k = 5$. The case $k > 5$ remains unsolved.

I agree with B. Bollobás, 1978a, who wrote: "Perhaps the most important problem concerning graph colourings is the connection between the chromatic number and the contractions onto complete subgraphs asserted by Hadwiger's conjecture."

In the case $k = 4$ G. A. Dirac, 1952a, answered Hadwiger's conjecture in the affirmative by showing that every 4-chromatic graph contains a subdivision of $K_4$.

A subdivision of $K_4$ can be considered as a cycle $C$ with two overlapping "topological" diagonals. Some questions arise:

1) *When are one or both topological diagonals chords,* i.e., *when do they have length one?*

2) *When can the cycle $C$ be chosen so that $C$ contains prescribed edges or has a prescribed parity?*

If $G$ contains a subdivision $S$ of $K_4$ in which one edge $e$ is left undivided, then $e$ is a chord of some cycle $C \subseteqq S$. C. Thomassen, 1974a, U. Krusenstjerna-Hafstrøm and B. Toft, 1980, and C. Thomassen and B. Toft, 1981, dealt with such problems.

They presented necessary conditions for the existence of subdivisions of $K_4$ in which certain edges are left undivided. Such subgraphs are called *semitopological* by B. Bollobás, 1978a. In sections 8.2-8.4 we deal with these results and, in particular, with those generalizations which guarantee the existence of a cycle with specified properties having some diagonals.

### 5° An extension of Kuratowski's theorem

In Chapter 1, K. Kuratowski's theorem together with a proof was presented: a graph is non-planar if and only if it contains a subdivision of $K_5$ or of $K_{3,3}$. Again it can be asked: are there semitopological $K_5$'s or $K_{3,3}$'s?

A. K. Kelmans, 1984, and C. Thomassen, 1984, independently proved a conjecture of A. K. Kelmans, 1981, that every non-planar 3-connected graph $G$, $G \not\cong K_5$, contains a special subdivision of $K_{3,3}$, namely, a cycle with three pairwise overlapping diagonals (which are called *pairwise interwoven diagonals* by A. K. Kelmans, 1984).

## 8.1. Cycles with given diagonals

It is well known that, for any edge $e$ of a graph $G$, either $e$ is a separating edge of $G$ or else there is a cycle of $G$ through $e$. Now we ask for conditions guaranteeing that a given edge $e$ is a diagonal of some cycle $C$.

We start with an elementary statement. Let $G$ be a 2-connected graph and $e$ an edge of $G$ with ends $P$ and $Q$. If $G - e$ is also 2-connected, then $G - e$ has a cycle $C$ containing both $P$ and $Q$. Hence $e$ is a chord of $C$.

In the second case, if $G - e$ is not 2-connected, there is a cut set of $G$ consisting of $e$ and a cut vertex $V$ of $G - e$. Evidently, if $G$ is 3-regular, then $V$ is incident with a separating edge $e'$ of $G - e$. Thus $e$, $e'$ constitute an edge cut of size 2 in $G$.

**Proposition 8.1.1.** *For any edge e in a 2-connected 3-regular graph G, either e belongs to an edge cut of size 2 or e is the diagonal of some cycle C.* □

H. Fleischner, 1983, improved this result for planar graphs.

**Theorem 8.1.2.** *If the graph G of Proposition 8.1.1 is planar, then C can be chosen to be an even cycle.* □

We conclude this section with a simple observation. If in a $3s$-connected graph $G$, $s \geqq 1$, $s$ given independent edges $(P_1, Q_1)$, $(P_2, Q_2)$, ..., $(P_s, Q_s)$ are deleted, the remaining graph is $2s$-connected. It can easily be proved that in the reduced graph a cycle $C$ through $P_1, \ldots, P_s, Q_1, \ldots, Q_s$ exists (first observed by G. A. Dirac, 1960). Hence $C$ is a cycle of $G$ with diagonals $(P_i, Q_i)$, $1 \leqq i \leqq s$.

**Proposition 8.1.3.** *In a $3s$-connected graph $G$ for any set of $s$ independent edges $e_1, \ldots, e_s$, $s \geqq 1$, there is a cycle of $G$ with diagonals $e_1, \ldots, e_s$.* □

With the same arguments it can be proved that in each $(2s + 1)$-connected graph $G$ for any set of $s$ edges $e_1, \ldots, e_s$, $s \geqq 1$, all incident with the same vertex of $G$, there is a cycle of $G$ with diagonals $e_1, \ldots, e_s$.

A consequence of Proposition 8.1.3 is that if C is any cycle of length at most $s$, then there is an edge-disjoint cycle containing all vertices of $C$. Such cycles are called *nested cycles* (see B. Bollobás, 1978a).

## 8.2. Cycles with few diagonals

In Chapters 8 and 9 we shall deal with the following question: what is the largest number $f(k, r)$, $k \geqq 0$, $r \geqq 3$, such that every graph $G$ with connectivity number at least $k$ and minimum degree $r$ contains a cycle with at least $f(k, r)$ diagonals?

Let $f'(k, r)$, $k \geqq 2$, $r \geqq 3$, be the corresponding largest number such that through each edge of $G$ a cycle with at least $f'(k, r)$ diagonals exists. For $0 \leqq k \leqq 2$, $r = 3$, these and similar problems will be investigated in this chapter. Our attention will be concentrated on the result $f'(2, 3) = 2$ which will be proved in detail in sections 8.7-8.9 (see Theorem 8.3.7). Only those results on semitopological $K_4$'s will be included here which are closely related to the problems above.

In this section we now start with the construction of infinite classes of 2-connected graphs of minimum degree at least 3 having only cycles with a small number of diagonals. So we shall obtain upper bounds for $f(k, 3)$.

Two such classes are $K_{3,s}$ and $K^+_{3,s}$, where $K^+_{3,s}$, $s \geqq 3$, is obtained from $K_{3,s}$ by joining two vertices of the bipartition class of cardinality 3 by an edge. $K_{3,s}$ and $K^+_{3,s}$ are 3-connected bipartite and non-bipartite graphs, respectively, having only degrees from $\{3, s, s + 1\}$. The $K_{3,s}$ and the $K^+_{3,s}$ have only cycles or odd cycles with at most two diagonals, respectively.

Next infinite classes of 2-connected graphs $G$ will be constructed satisfying the following properties: 1) $G$ has only cycles with few diagonals; 2) $G$ has only degree 3 and 4; 3) $G$ has girth 3, 4 or 5.

Let $T$ be a tree with root $W$ and end vertices $Y_1, \ldots, Y_s$, where $W$ is an inner vertex and each inner vertex of $T$ has valency at least 3. Let $T'$ be a tree such that $T$ and $T'$ are disjoint and there is an isomorphism $\varphi$ from $T$ onto $T'$.

We introduce the graphs $H_3(T)$, $H_4(T)$, $H_5(T)$ and $\bar{H}(T)$:

$H_3(T)$: for all $1 \leqq i \leqq s$ we add to $T \cup T'$ a copy $C_i$ of $W_3^-$ and identify $Y_i$ with a vertex of valency 2 of $C_i$ and $\varphi(Y_i)$ with the other vertex of valency 2 of $C_i$, where $W_3^-$ results from $W_3 \cong K_4$ by omitting one edge (see Fig. 8.2.1a).

$H_4(T)$: for all $1 \leqq i \leqq s$ we add to $T \cup T'$ a copy $C'_i$ of $K_{3,3}$ and identify $Y_i$ with a vertex of $C'_i$ and $\varphi(Y_i)$ with another vertex of the same bipartition class of $C'_i$ (see Fig. 8.2.1b).

$H_5(T)$ and $\bar{H}(T)$: if in the construction of $H_3(T)$ the graph $W_3^-$ is replaced by $\mathfrak{P}^-$ or by $K_2$, then we obtain $H_5(T)$ or $\bar{H}(T)$, respectively, where $\mathfrak{P}^-$ results from the Petersen graph by omitting one edge (see Fig. 8.2.1c, d).

**Lemma 8.2.1.** *Each cycle of $H_i(T)$ or $\bar{H}(T)$ touches at least one $C_j$ and at most two $C_j$'s. If $A$ is any vertex of $T$, then each $A,\varphi(A)$-path touches precisely one $C_j$.*

*Proof.* Let $C$ be a cycle of $H_i(T)$ or $\bar{H}(T)$. We consider the maximal arcs $a_1, a_2, \ldots, a_q$ of $C$ completely contained in $T$ and the maximal arcs $a'_1, a'_2, \ldots, a'_q$ of $C$ completely contained

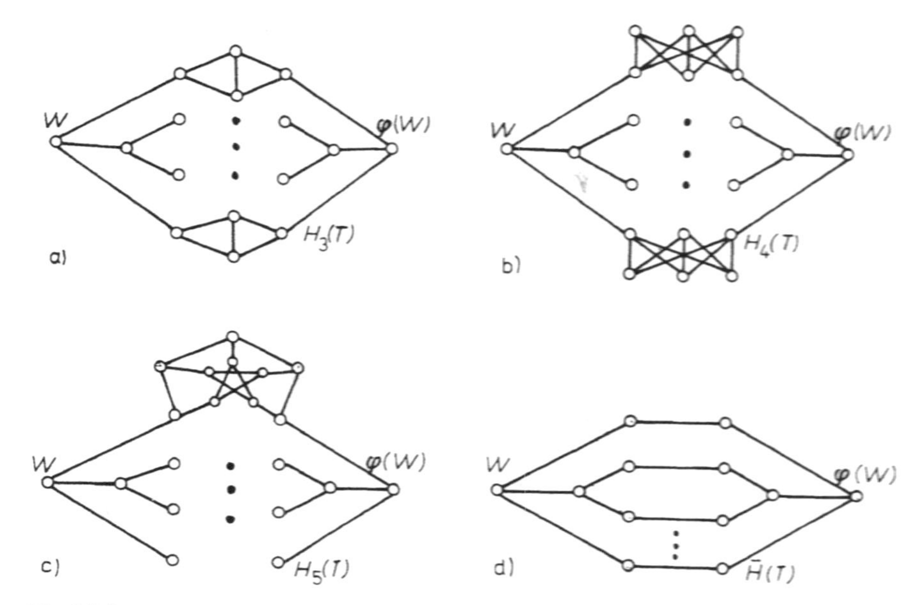


Fig. 8.2.1

in $T'$. The arcs $a_j$ and $a'_k$ join two vertices of $\{Y_1, \ldots, Y_s\}$ or $\{Y'_1, \ldots, Y'_s\}$, respectively. Thus $a_1, \ldots, a_q, \varphi^{-1}(a'_1), \ldots, \varphi^{-1}(a'_q)$ twice cover each edge of a certain subtree $S$ of $T$ with ends in $Y_1, \ldots, Y_s$. This subtree $S$ has no vertex of valency at least 3, because otherwise two of the arcs $a_1, \ldots, a_q$ or two of the arcs $a'_1, \ldots, a'_q$ have a common vertex, contradicting the definition of a cycle. Thus $S$ is a path and $q = 1$.

For an $A,\varphi(A)$-path, where $A$ is a vertex of $T$, the proof can be accomplished by the same arguments. □

By choosing different $T$'s, infinite classes of graphs $H_i(T)$, $\bar{H}(T)$ are obtained having the following properties:

**Proposition 8.2.2.** *For each tree $T$ the graphs $H_3(T)$, $H_4(T) \cup \{(W, \varphi(W))\}$, $H_5(T)$ are 2-connected non-bipartite graphs with minimum degree at least 3 having girth 3, 4 and 5, respectively. $H_4(T)$ is a 2-connected bipartite graph of minimum degree at least 3 having girth 4. These graphs have the following properties:*

(i) *If $A$ is any vertex of $T$, each $A,\varphi(A)$-path has at most two diagonals in $H_3(T)$ and $H_4(T)$ and at most three diagonals in $H_5(T)$;*

(ii) *every even cycle has at most four diagonals in $H_3(T)$, $H_4(T)$ and at most six diagonals in $H_5(T)$;*

(iii) *every odd cycle has at most two diagonals in $H_3(T)$, $H_4(T) \cup \{(W, \varphi(W))\}$ and at most five diagonals in $H_5(T)$.*

*The graph $\bar{H}(T)$ is 2-connected and has no cycle with a diagonal.*

## 8.3. Cycles with a given number of diagonals and semitopological $K_4$'s

Next we present lower bounds for $f(k, r)$. The simplest result, namely, $f(0, 3) \geqq 1$, is that by L. Pósa, a short proof is due to J. Czipszer (see L. Lovász, 1979):

**Theorem 8.3.1** (J. Czipszer, 1963). *Every graph G with minimum degree at least 3 contains a cycle with at least one diagonal.*

*Proof.* Let $p = [X_0, X_1, \ldots, X_d]$ be a longest path in $G$. $X_d$ is adjacent to two vertices other than $X_{d-1}$; since $p$ is maximal, these two vertices lie on $p$; let, say, $X_d$ be adjacent to $X_i$ and $X_j$, $0 \leqq i < j \leqq d-2$. Then $[X_i, X_{i+1}, \ldots, X_d, X_i]$ is a cycle in $G$ which has the diagonal $(X_j, X_d)$. □

R. P. Gupta, J. Kahn and N. Robertson, 1980, improved the assertion of J. Czipszer by showing that the graphs of Theorem 8.3.1 contain a cycle with at least two diagonals. We shall prove a stronger result due to C. Thomassen and B. Toft, 1981:

**Theorem 8.3.2.** *Let G be a graph of order at least 2 in which all vertices have valency at least 3, except perhaps one vertex. Then G contains a cycle having two overlapping diagonals such that two end vertices of these diagonals are joined by an edge of the cycle.*

The disjoint union of $K_4$'s shows that the assertion of Theorem 8.3.2 is sharp.

We present a proof of this result which is simpler than the proof of C. Thomassen and B. Toft, 1981.

*Proof.* Let $X_0$ denote the vertex of valency 1 or 2 if $G$ has such a vertex, or else an arbitrary vertex of valency at least 3. Let $p = [X_0, X_1, \ldots, X_d]$ be a longest $X_0$-path in $G$. The vertex $X_d$ is adjacent to two vertices other than $X_{d-1}$; since $p$ is maximal, these two vertices lie on $p$. Let, say, $X_d$ be adjacent to $X_i$ and $X_j$, $0 \leqq i < j \leqq d-2$. Let $p$ be chosen so that in the set of all longest $X_0$-paths the path $p$ is one with smallest cycle $[X_j, X_{j+1}, \ldots, X_d, X_j]$. The vertex $X_{j+1}$ is an end vertex of the longest $X_0$-path $v = [X_0, X_1, \ldots, X_j, X_d, X_{d-1}, \ldots, X_{j+1}]$. Since $v$ is maximal, $X_{j+1}$ has a neighbour $X_k$, $0 \leqq k \leqq d$, $k \notin \{j, j+1, j+2\}$. By the minimality of the cycle $[X_j, X_{j+1}, \ldots, X_d, X_j]$ the index $k$ is smaller than $j$.

If $k \geqq i$, then $[X_i, X_{i+1}, \ldots, X_d, X_i]$ is a cycle with two overlapping diagonals $(X_k, X_{j+1})$, $(X_j, X_d)$ joined by the edge $(X_j, X_{j+1})$ of this cycle.

If $k < i$, then $[X_k, X_{k+1}, \ldots, X_j, X_d, X_{d-1}, \ldots, X_{j+1}, X_k]$ is a cycle with two overlapping diagonals $(X_i, X_d)$, $(X_j, X_{j+1})$ joined by the edge $(X_d, X_j)$ of this cycle. □

The assertion of Theorem 8.3.2 is stronger than the assertion of Theorem 8.3.1 of J. Czipszer. Next we shall improve Czipszer's theorem by weakening the hypothesis of Theorem 8.3.1. For these purposes we prove the following result.

**Lemma 8.3.3.** *Let G be a 2-connected graph which is not a cycle. Let L be a cycle of G which is the union $L = a \cup b$ of two edge-disjoint arcs such that all vertices of a have valency at least 3 and all inner vertices of b have valency 2. Then two neighbours on a are joined by an L,L-path of G or L is contained in a subdivision of $K_4$ with one undivided edge which belongs to a.*

*Proof.* We follow the proof of Theorem 10.6.3 of O. Ore, 1967a. Since $G$ is not a cycle, there is at least one $L,L$-path in $G$. Assume that no $L,L$-path of $G$ joins two neighbours of $a$. Since $G$ is 2-connected, $a$ has at least three vertices. Let $V_0$ denote an inner vertex of $a$. From $V_0$ we can find an $L,L$-path with ends $V_0$ and $V_1$, where $V_1$ is not a neighbour of $V_0$ on the arc $a$.

Each $L,L$-path divides $L$ into two arcs, one including $b$ and the other not. We select an $L,L$-path $p_1$ so that the arc $a_1$ not including $b$ is as short as possible. Let $V_0$, $V_1$ again denote the end vertices of $a_1$. We note that $a_1$ must have at least one inner vertex. Let $S_0$ be an inner vertex of $a_1$ such that $(V_0, S_0)$ is an edge of $a_1$. At $S_0$ there is an $L,L$-path $p_2$ with ends $S_0$, $S_1$. Under the given conditions $S_1$ is not a vertex on $a_1$, nor can $p_1$ and $p_2$ have a vertex in common. Then $L \cup p_1 \cup p_2$ has the desired property.

Lemma 8.3.3 immediately implies Theorem 8.3.4; from Theorem 8.3.4 the result of J. Czipszer (Theorem 8.3.1) can again be derived.

A *topological edge* of $G$ is a path of $G$ such that the end vertices have valency at least 3 and all inner vertices have valency 2.

**Theorem 8.3.4.** *Let $G$ be a 2-connected graph which is not a cycle. Let $F$ denote the subgraph of $G$ induced by all vertices of valency at least 3. If no cycle of $G$ has a diagonal, then $F$ is a forest with at least two components and the topological edges of $G$ of length at least 2 join only different components of $F$.* □

The graphs $\bar{H}(T)$ show that the assertion of Theorem 8.3.4 is sharp. Moreover, any given forest $F$ with at least two components can be embedded in a 2-connected graph $G$ having no cycle with a diagonal so that the set of all vertices of valency at least 3 induces $F$. This can be proved by the following construction: let $C_1$ denote a component of $F$ and $C_2$ the rest of $F$; choose a vertex $X_i$ of $C_i$, $i = 1, 2$, and join $X_i$ to the vertices of $C_{i+1}$ by topological edges of length 2 until all vertices of $F$ have valency at least 3 (indices modulo 2). The graph $G$ so obtained has the desired properties (note that $G$ has only cycles with precisely two topological edges of $G$). An example is presented in Fig. 8.3.1, where $F$ consists of the two components $C_1$ and $C_2$.

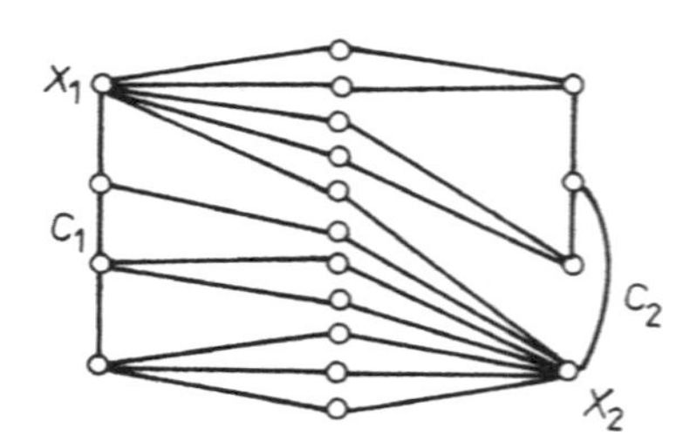


Fig. 8.3.1

For 2-connected graphs Pat Ash, 1985a, has proved the following result.

**Theorem 8.3.5.** *Every 2-connected graph with minimum degree at least 3 having at least five vertices contains a cycle with at least three diagonals.* □

Closely related to Theorem 8.3.2 is the following result independently proved by A. K. Kelmans, 1984, and C. Thomassen, 1984:

**Theorem 8.3.6.** *Every non-planar 3-connected graph contains a cycle with three pairwise overlapping diagonals.* □

Next we deal with the maximum number of diagonals of cycles through a given edge in 2-connected graphs having minimum degree at least 3. The main result is stated in the following theorem (H.-J. Voss, 1982a).

**Theorem 8.3.7.** *Let $G$ be a 2-connected graph with minimum degree at least 3. If $e$ is an arbitrary edge of $G$, then $G$ contains a cycle through $e$ with at least two diagonals.*

This theorem and Theorem 8.4.2 have a common proof which is presented in detail in sections 8.7–8.9.

By Proposition 8.2.2 the infinitely many graphs $H_3(T) \cup \{(W, \varphi(W))\}$ and $H_4(T) \cup \{(W, \varphi(W))\}$ have the property that each cycle through the edge $(W, \varphi(W))$ has at most two diagonals. Thus the bound 2 in Theorem 8.3.7 is best possible. All graphs of Theorem 8.3.7 having no 3-cycles or 4-cycles have a cycle with at least three diagonals through each edge (H.-J. Voss, 1982d).

**Theorem 8.3.8.** *Let $G$ be a 2-connected graph with minimum degree at least 3 having no 3-cycles or 4-cycles. If $e$ is any edge of $G$, then $G$ contains a cycle through $e$ with at least three diagonals.* □

The infinitely many graphs $H_5(T) \cup \{(W, \varphi(W))\}$ show that the bound 3 in Theorem 8.3.8 cannot be improved.

The search for cycles with diagonals in a graph $G$ is closely related to the investigation of semitopological $K_4$'s in $G$. If a 3-connected graph contains a triangle $C_3$, then it trivially contains a subdivision of $K_4$ in which only two adjacent edges are subdivided. But this is not true if $G$ only has connectivity 2. In general a graph without triangles and quadrangles can only contain subdivisions of $K_4$ in which the edges of a spanning tree are left undivided.

There are only two non-isomorphic spanning trees of $K_4$: a (Hamiltonian) path and a star.

A cycle with two overlapping diagonals from neighbouring vertices of the cycle is a subdivision of $K_4$ in which three edges of a Hamiltonian path of the $K_4$ are left undivided. We shall follow the notation of U. Krusenstjerna-Hafstrøm and B. Toft, 1980, and denote such a special subdivision by $K_4H$.

G. A. Dirac, 1960, stated that every graph with minimum degree at least 3 contains a subdivision of $K_4$. We now present an extension of this result due to C. Thomassen and B. Toft, 1981, which is a reformulation of Theorem 8.3.2 in terms of $K_4H$.

**Theorem 8.3.9.** *Let $G$ be a graph of order at least 2 in which all vertices have valency at least 3, except perhaps one vertex. Then $G$ contains a $K_4H$.*

A $K_4T$ is a subdivision of $K_4$, three adjacent edges of which are left undivided. If $G$ contains a $K_4T$, then $G$ contains a cycle $C$ and a vertex $X \notin C$ joined to $C$ by at least three edges. A related result is the following.

**Lemma 8.3.10.** *Let $G$ be a graph with minimum degree at least 3. Then $G$ contains a cycle $C$ having a regular bridge $\mathfrak{B}$ with at most two inner vertices. This bridge $\mathfrak{B}$ is either a one-vertex bridge* (see Fig. 8.3.2a) *or $\mathfrak{B}$ has two inner vertices $X$ and $Y$, vertices of attachment $X_1, Y_1, X_2, Y_2, \ldots, X_k, Y_k$ (in this cyclic order on $C$) and edges $(X, Y)$, $(X, X_i)$, $(Y, Y_i)$, $1 \leqq i \leqq k$, $k \geqq 2$* (see Fig. 8.3.2b).

*Proof.* Let $H$ be a terminal block of $G$. If $H$ has a cut vertex, denote it by $V$. Otherwise, let $V$ be an arbitrary vertex of $H$. Now let $L$ denote a shortest cycle through $V$. Hence $L$ has no diagonal. Since $L$ has at least two vertices of degree at least 3, the cycle $L$ has a regular bridge $\mathfrak{B}$. Let $\mathfrak{B}$ have more than one inner vertex.

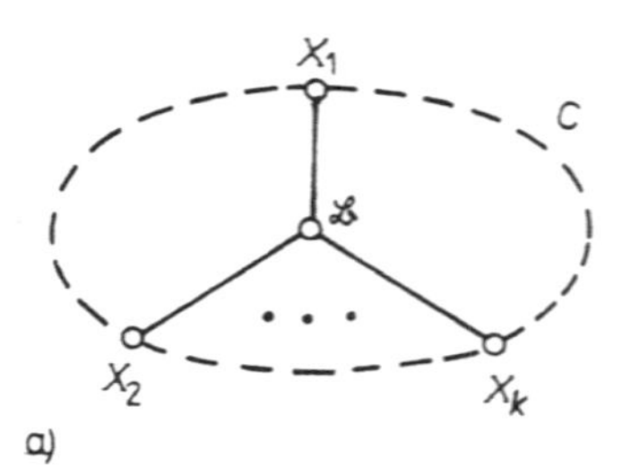


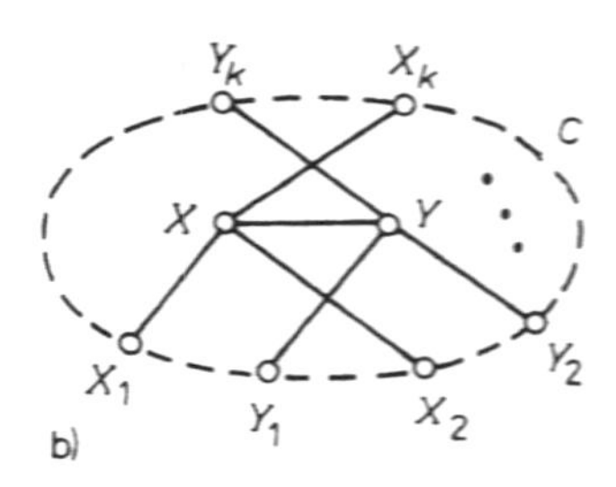


Fig. 8.3.2

If possible, choose an $L,L$-path $p$ of $\mathfrak{B}$ with ends $P$ and $Q$, say, having the following properties: 1) the path $p$ does not contain all inner vertices of $\mathfrak{B}$, 2) one $P,Q$-arc $a$ of $L$ is a segment of $\mathfrak{B}$, i.e. no inner vertex of $a$ is a vertex of attachment of $\mathfrak{B}$. Then $L - (a - \{P, Q\}) \cup p$ is a new cycle having a bridge $\mathfrak{B}'$ with $\ker(\mathfrak{B}') \subsetneqq \ker(\mathfrak{B}) - p$, where $\ker(\mathfrak{B})$ denotes the set of all inner vertices of $\mathfrak{B}$. Thus $\mathfrak{B}'$ has fewer inner vertices than $\mathfrak{B}$. Repeat this method until it cannot be applied any more.

This method fails if and only if for each $L,L$-path $p$ of $\mathfrak{B}$ we have $\ker(\mathfrak{B}) - p = \emptyset$, i.e. $p$ contains all inner vertices of $\mathfrak{B}$ or $\ker(\mathfrak{B}) - p \neq \emptyset$ and the two $P,Q$-arcs of $L$ are not segments. This implies that $\mathfrak{B}$ is a tree with two inner vertices of equal degrees $k$, $k \geqq 2$, joined to $2k$ vertices of $C$ such that neighbours of $X$ alternate with neighbours of $Y$ (see Fig. 8.3.2b). □

$K_{3,3}T$ denotes a subdivision of $K_{3,3}$ such that a spanning tree $T$ having two vertices of degree 3 in $T$ is left undivided.

**Theorem 8.3.11.** *Every graph with minimum degree at least 3 contains a $K_4T$ or a $K_{3,3}T$ such that each vertex incident with at least three undivided edges has all its neighbours in $K_4T$ or $K_{3,3}T$, respectively.* □

A related result is due to C. Thomassen and B. Toft, 1981.

**Theorem 8.3.12.** *Let $G$ be a graph with at least two vertices one of which is denoted by $X_0$. Suppose that all vertices have valency at least 3, except perhaps $X_0$. Then either $G$ contains a $K_4T$, or else $G - X_0$ contains an induced $K_{2,3}$ of which four vertices of a $K_{2,2}$ have valency 3 in $G$.*

*If $G$ is 3-connected, then in the second alternative all five vertices of the $K_{2,3}$ have valency 3 in $G$.* □

There is an infinite set of 3-connected cubic graphs containing no $K_4T$ (see C. Thomassen and B. Toft, 1981): take any 3-connected cubic graph $G$ and replace each vertex $X$ by a $K_{2,3}$ in such a way that each of the three vertices of the $K_{2,3}$ of degree 2 becomes incident with precisely one of the three edges incident with $X$ in $G$. The resulting graph is also a 3-connected cubic graph, and it contains no $K_4T$.

Theorem 8.3.2 of C. Thomassen and B. Toft states that every graph with minimum degree at least 3 contains a cycle with at least two overlapping diagonals. Now we ask: how can we characterize all graphs in which each cycle of length at least 4 has at least two chords?

G. A. Dirac, 1960, investigated graphs with the weaker condition that each even cycle has at least two diagonals (see also C. Berge, 1970, p. 172).

**Theorem 8.3.13.** *If each even cycle of a graph $G$ has at least two chords, then each block of $G$ is a complete graph or an odd cycle without diagonals.* □

The graphs with the condition that each cycle has at least one diagonal constitute a larger class of graphs. These graphs have been investigated as rigid circuit graphs (G. A. Dirac, 1961), chordal graphs (F. Gavril, 1974), and triangulated graphs (R. Halin, 1984). Triangulated graphs are not dealt with in this book. For a further reference see M. Ch. Golumbic, 1980.

We conclude this section with two very interesting conjectures. The first is due to C. Thomassen, 1986:

**Conjecture 8.3.14.** *In a 3-connected graph with minimum degree at least 3 each longest cycle has at least one diagonal.* □

A partial solution is due to Cun-Quan Zhang, 1987, who proved that the conjecture is true for cubic planar graphs and for planar graphs of minimum degree at least 4.

The 2-connected cubic graph $D(2j)$, $j \geqq 5$, of section 11.3 has the following property: each longest cycle has four diagonals, each cycle of length $c(G) - 1$ has two diagonals, but there are cycles of length $c(G) - 2$, which have no diagonal. Therefore it seems to be difficult to prove C. Thomassen's conjecture if true.

The proof of Theorem 8.3.1 of J. Czipszer can be used to show that every strongly 2-connected digraph contains a directed cycle with one diagonal. J. Bermond and C. Thomassen, 1981, made the following conjecture:

**Conjecture 8.3.15.** *Every strongly 2-connected digraph contains a directed cycle with at least two diagonals.*

## 8.4. Odd and even cycles with a given number of diagonals

In this section the parity of the cycles under consideration will be included in our investigations: a cycle is called *odd* or *even* if it has odd or even length, respectively.

Chapters 8 and 9 shall deal with the question: what are the largest numbers $f_o(k, r)$ and $f_e(k, r)$, $k \geqq 0$, $r \geqq 3$, such that every $k$-connected non-bipartite graph $G$ of order at least 7 and with minimum degree at least $r$ contains an odd cycle with at least $f_o(k, r)$ diagonals or an even cycle with at least $f_e(k, r)$ diagonals, respectively?

For $0 \leqq k \leqq 2$, $r = 3$ these and similar problems will be investigated here. Our attention will be concentrated on the results $f_o(2, 3) = 2$ and $2 \leqq f_e(2, 3) \leqq 4$; the common proof of the results $f'(2, 3) \geqq 2$ (see Theorem 8.3.7), $f_o(2, 3) \geqq 2$, $f_e(2, 3) \geqq 2$ will be given in sections 8.7–8.9. We include only one result on semitopological $K_4$'s which is closely related to the problem $f_o(2, 3) = 2$.

Next some definitions. An *r-wheel* $W_r$ denotes the graph obtained from an $r$-cycle by adding a new vertex $X$ and completely joining $X$ to $C$ by edges. The graph $W_r^-$ is obtained from $W_r$ by omitting one $X,C$-edge. A *prism graph* consists of the vertices and the edges of a prism, i.e., the only cubic planar graph on six vertices.

J. Larson, 1979, and C. Thomassen and B. Toft, 1981, proved the existence of an odd cycle with at least one diagonal in 2-connected non-bipartite graphs with minimum degree at least 3. Moreover, C. Thomassen and B. Toft, 1981, proved the following result:

**Theorem 8.4.1.** *Every 2-connected non-bipartite graph with at least 5 vertices and minimum degree at least 3 contains a subdivision of $K_4$ in which a 4-cycle becomes an odd cycle and one of the two remaining edges is left undivided.* □

For even cycles C. Thomassen and B. Toft, 1981, proved the following result:

*Let $G$ be a graph in which all vertices have degree at least 3, except perhaps one vertex. Then $G$ contains an even cycle with at least one diagonal.* □

Apart from three exceptional graphs, all 2-connected non-bipartite graphs contain both an odd and an even cycle with at least two diagonals. Moreover, we have the following result (H.-J. Voss, 1982a):

**Theorem 8.4.2.** *Let $G$ be a 2-connected graph with minimum degree at least 3. Then*

(i) *each pair of vertices $A$, $B$ of $G$ is joined by a path with at least two diagonals different from $(A, B)$,*

(ii) *if $G \neq W_4$, then $G$ contains an even cycle with at least two diagonals,*

(iii) *if $G$ is non-bipartite, $G \neq K_4$ and $G$ is not a prism graph, then $G$ contains an odd circuit with at least two diagonals.*

The second part of this chapter (sections 8.7-8.9) is devoted to the proof of this theorem. Note that assertion (i) is a new formulation of Theorem 8.3.7.

The classes $K_{3,s}$, $K^+_{3,s}$, $H_3(T)$ and $H_4(T)$, introduced in section 8.2, provide upper bounds for the number of diagonals in Theorem 8.4.2: the bounds of the first assertion and the third assertion are attained by an infinite set of graphs, but perhaps the maximum number of diagonals with respect to even cycles can be enlarged for almost all graphs of the second assertion.

By Theorem 8.3.5 of Pat Ash we know that every 2-connected bipartite graph with minimum degree at least 3 contains an even cycle with at least three diagonals, and this bound is sharp. The question remains an open problem for 2-connected *non-bipartite* graphs of minimum degree at least 3. The graphs $H_3(T)$ and $H_4(T) \cup \{(W, \varphi(W))\}$ provide only the upper bound 4.

Let us compare the assertions of Theorems 8.4.1 and 8.4.2. For 2-connected non-bipartite graphs $G$ of order at least 7 and minimum degree at least 3 the two theorems state that $G$ contains both an odd cycle with two diagonals and an odd cycle with one diagonal and an additional path forming a subdivision of $K_4$. There are such graphs, namely, the wheels $W_{2s+1}$, $s \geqq 2$, which contain the two configurations but do not contain an odd cycle with two diagonals constituting a subdivision of $K_4$. In this sense the two theorems can be regarded as independent results.

Here a topological diagonal $t$ of a cycle $C$ in $G$ is a path $t$ of $G$ such that only its ends belong to $C$. In a forthcoming paper St. Hübner, K. Ihle and H.-J. Voss will prove the following theorem, which generalizes both Theorems 8.4.1 and 8.4.2.

**Theorem 8.4.3.** *Let $G$ be a 2-connected non-bipartite graph with minimum degree at least 3 having at least five vertices. Then $G$ is a prism graph or $G$ contains an odd cycle $C$ with three topological diagonals, at least two of which have length 1, and two of them and $C$ form together a subdivision of $K_4$.* □

The proof of Theorem 8.4.3 is similar to the proof of Theorem 8.4.2 but is more complicated. Hence I prefer to present the proof of Theorem 8.4.2 in this chapter (see sections 8.7-8.9).

If the graphs of girths 3 and 4 are excluded, the cycles and paths in the graphs under consideration have at least three diagonals (H.-J. Voss, 1982d).

**Theorem 8.4.4.** *Let G be a 2-connected graph with minimum degree at least 3 having no 3-cycles or 4-cycles. Then*

(i) *each pair of vertices A, B of G is joined by a path with at least three diagonals different from (A, B);*

(ii) *if G is not a Petersen graph, then G contains an even cycle with at least three diagonals;*

(iii) *if G is non-bipartite, then G contains an odd cycle with at least three diagonals.* □

Note that assertion (i) is a reformulation of Theorem 8.3.8. The class $H_5(T)$ introduced in section 8.2 provides upper bounds for the number of diagonals in Theorem 8.4.4: the bound of the first assertion is attained by an infinite set of graphs; upper bounds for the maximum number of diagonals in assertions (ii) and (iii) are 6 and 5, respectively. Improving Theorem 8.4.2, I can prove the following result:

*Let G be a 2-connected graph in which apart from one exceptional vertex of valency 2 all vertices have degree at least 3. If $G \not\cong W_4^-$, then G also satisfies the assertions of Theorem 8.4.2.* □

This result implies the following:

**Theorem 8.4.5.** *Suppose that all the vertices of a graph G have valency at least 3. Then G contains an even cycle with at least two diagonals or one terminal block of G is a $W_4$, or two terminal blocks of G are $W_4^-$'s.* □

## 8.5. Cycles with a given number of diagonals in *k*-chromatic graphs

Assertions about *k*-chromatic graphs can be derived from the results of sections 8.3 and 8.4. We shall demonstrate this by deducing Theorem 8.5.3 from Theorems 8.4.1 and 8.4.2. Theorem 8.5.3 includes a solution of a conjecture of P. Erdős. Section 8.5 will be concluded by presenting some conjectures.

Graphs with a high chromatic number are closely related to graphs of high minimum degree. This observation is based on the following lemma.

**Lemma 8.5.1.** *Every graph G with chromatic number $\chi(G) \geqq 3$ contains a 2-connected non-bipartite subgraph with minimum degree at least $\chi(G) - 1$.*

*Proof.* An edge or a vertex of a graph $G$ is called *critical* if its deletion decreases the chromatic number of the graph. Let $k =_{\text{def}} \chi(G)$. By successively deleting non-critical edges and non-critical vertices of $G$ we arrive at a graph $H$ with the same chromatic number which has only critical edges and vertices. Thus $H$ is a critically *k*-chromatic graph.

It is well known that a critically *k*-chromatic graph is 2-connected and has minimum degree at least $k - 1$. Since $k \geqq 3$, the subgraph $H$ is non-bipartite. □

By Lemma 8.5.1 each result on graphs of high degree can be applied to obtain theorems on *k*-chromatic graphs. This can be done with the results of sections 8.3 and 8.4.

Only two applications will be discussed here. Concerning cycles with many diagonals in *k*-chromatic graphs P. Erdős, 1977, posed the following question at the International Colloquium on Graph Theory and Its Applications held at Oberhof (GDR) in April 1977 (see also P. Erdős, 1978).

**Conjecture 8.5.2.** *Every graph with chromatic number at least 4 not containing $K_4$ has an odd cycle with at least two diagonals.* □

Partial results with respect to Conjecture 8.5.2 are due to J. Larson, 1979, and C. Thomassen and B. Toft, 1981. J. Larson, 1979, proved the existence of an odd cycle with at least one chord and C. Thomassen and B. Toft, 1981, showed the existence of a subdivision of $K_4$ in which a 4-cycle becomes an odd cycle and one of the two remaining edges is left undivided. I was able to prove Conjecture 8.5.2 in the affirmative; the proof was announced in 1977 and published in 1982 (H.-J. Voss, 1977b and 1982a).

By Lemma 8.5.1 the result of C. Thomassen and B. Toft and my result are implied by Theorems 8.4.1 and 8.4.2, respectively.

**Theorem 8.5.3.** *Every graph G with a chromatic number at least 4 not having $K_4$ as a subgraph contains a subdivision of $K_4$ in which a 4-cycle becomes an odd cycle and one of the two remaining edges is left undivided and G includes both an odd and an even cycle with at least two chords.* □

(Note with respect to Theorem 8.4.2 that $W_4$ and the prism graph are 3-chromatic.)

The graph $W_5$ and the graph $\Gamma$, depicted in Fig. 8.5.1a, b, show that Theorem 8.5.3 cannot be improved: the two graphs have chromatic number 4 and each odd cycle of $W_5$ has zero or two diagonals and each even circuit of $\Gamma$ has one or two diagonals. If in Theorem 8.5.3 the 3-cycles and 4-cycles are suppressed, then by Theorem 8.4.4 the graph $G$ contains both an odd and an even cycle with at least three diagonals.

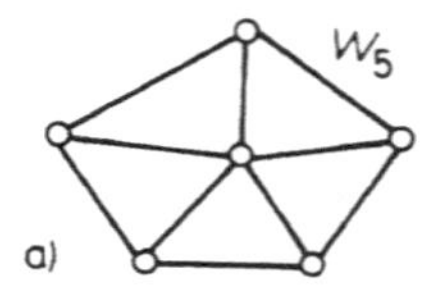


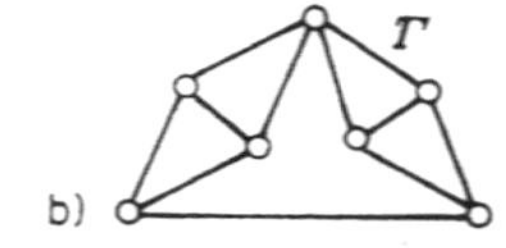


Fig. 8.5.1

Since the graphs $H_3(T)$, $H_4(T)$ and $H_5(T)$ are 3-colourable, they do not satisfy the hypotheses of Theorem 8.5.3. Adding to $W_5$ trees which have exactly one vertex in common with $W_5$ we obtain an infinite set of graphs satisfying the hypotheses of Theorem 8.5.3 and containing only odd cycles with zero or two diagonals, respectively.

I think that if we restrict ourselves to critical graphs, we cannot find such a set. Moreover, I make the following conjecture:

**Conjecture 8.5.4.** *Let $s_k(n)$ denote the largest integer $\alpha$ such that each k-critical graph, $k \geqq 4$, with n vertices has an odd cycle with at least $\alpha$ diagonals. Then $s_k(n) \to \infty$ as $n \to \infty$.* □

This conjecture corresponds to Corollary 6.3.8: if $f_k(n)$ denotes the largest integer $\beta$ such that each $k$-critical graph, $k \geqq 4$, with $n$ vertices has an odd cycle of length at least $\beta$, then $f_k(n) \to \infty$ as $n \to \infty$.

We conclude the discussion of Conjecture 8.5.4 with two problems related to the foregoing conjecture. I make the following conjectures (H.-J. Voss, 1985b):

**Conjecture 8.5.5.** *Every graph G with chromatic number at least 4 not having $K_4$ as a subgraph contains a cycle with at least four diagonals.* □

**Conjecture 8.5.6.** *If G is also 3-connected, then each pair of edges is contained in a circuit with at least two diagonals.* □

## 8.6. 2-connected graphs and the Whitney construction

In this section Theorem 8.3.1 of J. Czipszer will be applied to give a simple proof for the correctness of Whitney's construction of all 2-connected graphs from all cycles (see section 8.0, 2°). Properties of minimally 2-connected graphs will be investigated.

First some notation. A path $t$ of $G$ is called a *topological edge* of $G$ if and only if its inner vertices and its end vertices have valency 2 and valency at least 3, respectively. $G \ominus t$ denotes the graph obtained from $G$ by deleting all edges and all inner vertices of $t$. A *topological diagonal* of a cycle $C$ is a topological edge if only the ends of it belong to $C$. In order to show the correctness of Whitney's construction of all 2-connected graphs from all cycles it suffices to prove the following result.

**Theorem 8.6.1.** *Every 2-connected graph $G$ that is not a cycle contains a cycle with a topological diagonal. Consequently, $G \ominus t$ is also 2-connected.*

*Proof.* The graph $G$ is not a cycle, and $G$ is 2-connected. Hence $G$ is a subdivision of a 2-connected graph $H$ having minimum degree at least 3. Obviously, $H$ is not a loop. If $H$ has two (parallel) edges $e_1$ and $e_2$ with the same ends $A$ and $B$, then the corresponding topological edges $t_1$ and $t_2$ of $G$ form a cycle. Then $G$ contains a cycle $C$ such that $t_1$ is a topological diagonal of $C$.

If $H$ has no multiple edges, then by Theorem 8.3.1 the graph $H$ contains a cycle with a diagonal. This corresponds to a cycle of $G$ with a topological diagonal. □

We ask: what structure have the graphs containing no cycle with a diagonal?

We study this question for 2-connected graphs. By Theorem 8.3.1 we know that such a graph contains at least one vertex of degree 2.

A 2-connected graph $G$ is said to be *minimally 2-connected* if and only if $G$ has no proper 2-connected *spanning* subgraph. These graphs have been well investigated by G. A. Dirac, 1967, and M. D. Plummer, 1968; the latter author observed the following relation between minimally 2-connected graphs and graphs which have no cycles with diagonals:

**Theorem 8.6.2.** *A 2-connected graph is minimally 2-connected if and only if $G$ has no cycle with at least one chord.*

For minimally 2-connected graphs a variation of Whitney's construction can be introduced:

*The class of all minimally 2-connected graphs can be obtained from all cycles by repeatedly adding topological diagonals to cycles so that no cycle with a diagonal is formed.*

The next theorem implies the correctness of the variation of Whitney's construction.

**Theorem 8.6.3.** *Every minimally 2-connected graph $G$ that is not a cycle contains a cycle with a topological diagonal of length at least 2 (the end vertices of which are joined only by paths of length at least 2 without diagonals).*

*Proof.* Theorem 8.6.3 can be proved like Theorem 8.6.1 by taking into consideration the fact that $G$ has no cycles with diagonals. □

By Theorem 8.6.2 we can apply Theorem 8.3.4.

**Theorem 8.6.4.** *If $G$ is minimally 2-connected and $G$ is not a cycle, then the deletion of all vertices of valency 2 results in a forest $F$ with at least two components; the topological edges of $G$ join only different components of $F$.* □

An immediate consequence of Theorem 8.6.4 is the following (M. D. Plummer, 1968):

**Corollary 8.6.5.** *If $G$ is minimally 2-connected and $G$ is not a cycle, then the set of all vertices of valency 2 forms a cut set of $G$.* □

## 8.7. Some lemmas on bridges

The purpose of sections 8.7-8.9 is to prove Theorem 8.4.2. Its proof and the proofs of similar theorems are essentially based on lemmas concerning bridges of certain subgraphs. Hence four of them are presented here, where only Lemma 8.7.1 will be proved, namely in section 8.8. In the proof of Theorem 8.4.2 given in section 8.9 only this lemma will be used. First some notation. Let $H$ be a subgraph of a graph $G$. A path $p$ of $G$ is said to be an *$H$-path* if and only if one end vertex and no other vertices of $p$ belong to $H$. A path $v$ of $G$ is an *$H,H$-path* if and only if the two end vertices and no other vertices of $p$ belong to $H$.

In the following $\mathfrak{B}$ always denotes a regular $H$-bridge.

**Definition.** The length of the longest $H$-paths in $\mathfrak{B}$ is denoted by $d(\mathfrak{B})$.

If the end $E$ of an $H$-path $p$ that is not a vertex of $H$ is joined to $H - p$ by an edge, then an $H,H$-path of length $d(\mathfrak{B}) + 1$ is obtained. In general $E$ is not joined to $H - p$ by an edge, and therefore a longest $H,H$-path can have length less than $d(\mathfrak{B}) + 1$.

The following lemma, which is essential in the proof of Theorem 8.4.2, ensures the existence of $H,H$-paths of length $d(\mathfrak{B}) + 1$.

**Lemma 8.7.1.** *Let $G$ be a 2-connected graph. Let $\mathfrak{B}$ be a bridge of a subgraph $H$ in $G$ such that all inner vertices of $\mathfrak{B}$ have valency at least 3. Let $\mathfrak{B}$ have the property $\mathfrak{p}_1$: each $H,H$-path of $\mathfrak{B}$ has at most one diagonal in $\mathfrak{B}$. Then $\mathfrak{B}$ contains an $H,H$-path of length $d(\mathfrak{B}) + 1$.*

The example of Fig. 8.7.1 shows that the lemma is sharp. In this figure $H = \{A, B\}$, $d(\mathfrak{B}) = 10$, each $H,H$-path of $\mathfrak{B}$ has at most two diagonals, but its length is at most 7. An infinite set of such examples can be found in $H_3(T)$ and $H_4(T)$.

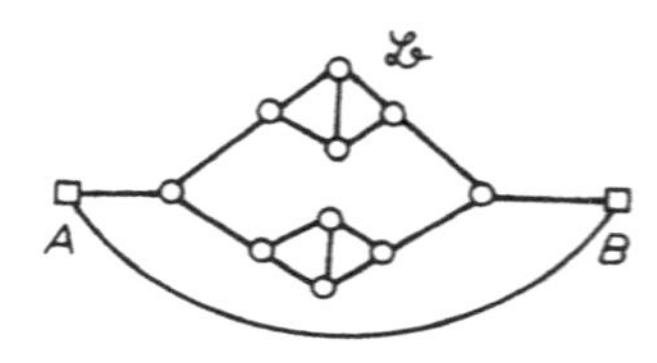


Fig. 8.7.1

If the girth of $G$ is at least 5, then a stronger lemma can be proved (H.-J. Voss, 1982d).

**Lemma 8.7.2.** *If graphs with triangles and quadrangles are excluded, then the assertion of Lemma 8.7.1 is also true for bridges $\mathfrak{B}$ with the weaker property $\mathfrak{p}_2$: each $H,H$-path of $\mathfrak{B}$ has at most two diagonals in $\mathfrak{B}$.* □

I have applied this lemma in the proof of Theorem 8.4.4 (H.-J. Voss, 1982d). With F. Eiselt, 1986, a student of mine, I investigated bridges $\mathfrak{B}$ with property $\mathfrak{p}_2$ in 3-connected graphs in which triangles and quadrangles are allowed.

**Lemma 8.7.3.** *Let G be a 3-connected graph. Let $\mathfrak{B}$ be a bridge of a subgraph H in G such that all inner vertices of $\mathfrak{B}$ have valency at least 3. Let $\mathfrak{B}$ have property $\mathfrak{p}_2$. Then by deleting appropriate inner vertices of $\mathfrak{B}$ a 3-connected graph G′ can be obtained such that the reduced bridge $\mathfrak{B}^*$ of H contains an H,H-path of length $d(\mathfrak{B}^*)+1$.* □

Together with St. Hübner and K. Ihle 1987, I was able to improve Lemma 8.7.1.

**Lemma 8.7.4.** *Let G be a 2-connected graph. Let $\mathfrak{B}$ be a bridge of a subgraph H in G such that all inner vertices of $\mathfrak{B}$ have valency at least 3. Let $\mathfrak{B}$ have property $\mathfrak{p}_1'$: no H,H-path of $\mathfrak{B}$ has two overlapping diagonals in $\mathfrak{B}$. Then $\mathfrak{B}$ contains an H,H-path of length $d(\mathfrak{B})+1$.* □

## 8.8. Proof of Lemma 8.7.1

First we repeat some notation. The valency of the vertex $V$ in a graph $G$ is denoted by $v(G:V)$. Let $V_1, V_2, \ldots, V_s$ be pairwise distinct vertices. Then $[V_1, V_2, \ldots, V_s, V_{s+1}]$ denotes the graph consisting of the vertices $V_1, V_2, \ldots, V_s, V_{s+1}$ and the edges $e_i = (V_i, V_{i+1})$, $1 \leqq i \leqq s$. If $V_{s+1} \notin \{V_1, V_2, \ldots, V_s\}$ or if $V_{s+1} = V_1$ or if $V_{s+1} \in \{V_1, \ldots, V_{s-2}\}$, then $[V_1, V_2, \ldots, V_s, V_{s+1}]$ is called a *$V_1,V_{s+1}$-path* or a *circuit* or a *$V_1$-lasso*, respectively.

The length of $[V_1, V_2, \ldots, V_s, V_{s+1}]$ is the number $s$ of its edges. If the graph $[V_1, V_2, \ldots, V_s, V_{s+1}]$ is a $V_1$-lasso, then $V_{s+1}$ is called its *branch vertex*.

If $P$ and $Q$ are vertices of a path $p$, then $p[P, Q]$ is defined to be the unique $P,Q$-path contained in $p$.

For brevity we write $V_1V_2\ldots V_{s+1} =_{\text{def}} [V_1, V_2, \ldots, V_{s+1}]$ and $PpQ =_{\text{def}} p[P, Q]$.

If $E$ is a subset of the edge set of $G$, then $G - E$ denotes the graph obtained from $G$ by deleting all edges of $E$. If $E = \{(P, Q)\}$, then we briefly write $G - (P, Q)$. If $H$ is a subgraph of $G$, then $G - H$ denotes the graph obtained from $G$ by deleting all vertices of $H$ and all edges incident with vertices of $H$. If $H$ is a vertex $V$, then we briefly write $G - V$. By $G \cup E$ and $G \cup H$ the unions of $G$ and $E$ and of $G$ and $H$, respectively, are denoted.

Let $C_1$ and $C_2$ be two cycles and let $E(C_1)$ and $E(C_2)$ be the set of edges of $C_1$ and $C_2$, respectively. Then $C_1 \triangle C_2$ denotes the subgraph of $G$ consisting of all edges of the symmetric difference $E(C_1) \triangle E(C_2)$ and all vertices of $G$ incident with an edge of $E(C_1) \triangle E(C_2)$.

Let $\mathfrak{B}$ denote a bridge of some subgraph $H$ of $G$. An $H$-path $p$ of $\mathfrak{B}$ has two end vertices, one in $H$ and the other in $\mathfrak{B} - H$, called the *origin* and the *terminal* of $p$, respectively.

We start with a general observation.

**Proposition 8.8.1.** *All neighbours of the terminal of any longest H-path p lie on $p \cup H$.*

*Proof.* Assume that there is an $H$-path $p$ of length $d$ in $\mathfrak{B}$ such that the terminal of $p$, say $Y$, has a neighbour $P$, $P \notin p \cup H$. Then $p \cup [Y, P]$ is an $H$-path of length $d+1$ in $\mathfrak{B}$: this contradicts the definition of $p$. □

For convenience we restate Lemma 8.7.1 as Lemma 8.8.2.

**Lemma 8.8.2.** *Let G be a 2-connected graph. Let $\mathfrak{B}$ be a bridge of a subgraph H of G such that all inner vertices of $\mathfrak{B}$ have valency at least 3. Let $\mathfrak{B}$ have property $\mathfrak{p}_1$: each H,H-path of $\mathfrak{B}$ has at most one diagonal in $\mathfrak{B}$. Then $\mathfrak{B}$ contains an H,H-path of length $d(\mathfrak{B})+1$.*

We remark that if the end vertices of an $H,H$-path $w$ of $\mathfrak{B}$ are joined by an edge $e$ of $G$, then $w$ can have two diagonals in $G$, namely, $e \notin \mathfrak{B}$ and a second chord which is also a chord in $\mathfrak{B}$.

*Proof of Lemma 8.8.2.* We consider a bridge $\mathfrak{B}$ with property $\mathfrak{p}_1$. Let $d =_{\mathrm{def}} d(\mathfrak{B})$. Assume that

(A) *the bridge $\mathfrak{B}$ contains no $H,H$-path of length $d+1$.*

From (A) it follows that

(1) *the terminal of each $H$-path of length $d$ has all its neighbours in $p$.*

*Proof of (1).* Assume that there exists an $H$-path $p$ of length $d$ whose terminal $E$ has a neighbour $E'$ not on $p$. Then, by Proposition 8.8.1, $E' \in H - p$. Therefore, $p \cup [E, E']$ is an $H,H$-path of length $d+1$, contradicting (A). □

In what follows a $V_1$-lasso $\Lambda$ is called an $H$-lasso if $V_1$ and no other vertices of $\Lambda$ belong to $H$.

By our hypotheses each inner vertex of $\mathfrak{B}$ has three neighbours. This is also true for the terminal of each $H$-path. Therefore, by (1) each $H$-path $p$ of length $d$ has a chord $e$ such that $p \cup e$ is an $H$-lasso. Consequently, each $H$-path of length $d$ is contained in an $H$-lasso of length $d+1$. If $\Lambda$ is a lasso, let $c(\Lambda)$ denote the length of the unique cycle of $\Lambda$. We define

$$c =_{\mathrm{def}} c(\mathfrak{B}) =_{\mathrm{def}} \max c(\Lambda),$$

the maximum being taken over all $H$-lassos of length $d+1$.

Let $\Lambda^*$ be an $H$-lasso satisfying $c(\Lambda^*) = c$. For $\Lambda^*$ we introduce some notation: $X$ is the unique common vertex of $\Lambda^*$ and $H$, $R$ is the branching vertex of $\Lambda^*$, $C$ is the circuit of $\Lambda^*$, and $Y$, $Z$ are the two neighbours of $R$ in $C$ such that $(Y, R)$, $(Z, R) \in C$ (see Fig. 8.8.1).

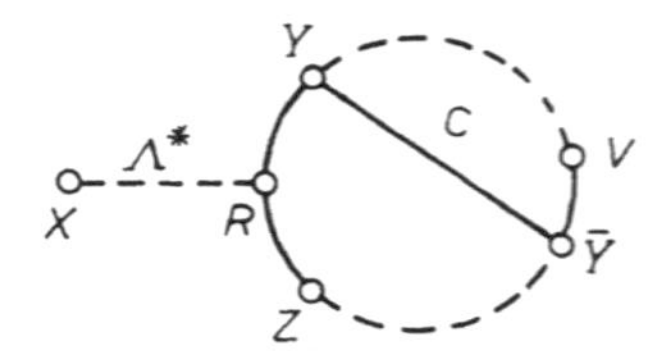


Fig. 8.8.1

Obviously, $\pi =_{\mathrm{def}} \Lambda^* - (Y, R)$ and $\pi' =_{\mathrm{def}} \Lambda^* - (Z, R)$ are two $H$-paths of length $d$ with terminals $Y$ and $Z$, respectively. Therefore, by (1) and the maximality of $C$, all neighbours of $Y$ and $Z$ belong to $C$.

Since $v(G:Y) \geqq 3$ and $v(G:Z) \geqq 3$, each of $Y$ and $Z$ has three neighbours on $C$. Hence there is a neighbour $\bar{Y}$ of $Y$ and there is a neighbour $\bar{Z}$ of $Z$ such that $(\bar{Y}, Y)$, $(\bar{Z}, Z)$ are chords of $C$.

In what follows we shall also use the neighbour $V$ of $\bar{Y}$ with $(\bar{Y}, V) \in \pi[\bar{Y}, Y)$. Since $v(G:V) \geqq 3$, there is a neighbour $\bar{V}$ of $V$ such that $(V, \bar{V}) \notin C$. The vertex $V$ is the terminal of the $H$-path $\pi^* =_{\mathrm{def}} X\pi\bar{Y}Y\pi V$.

Since $\pi^*$ contains all vertices of $\pi$, the path $\pi^*$ also has length $d$. Therefore, $V$ is the terminal of an $H$-path of length $d$ containing only vertices of $\pi$.

By (1), $\bar{V} \in \pi$.

If $\bar{V} \in \pi[X, R] - R$, then $\pi^* \cup (V, \bar{V})$ is an $H$-lasso of length $d+1$ with a cycle which is longer than $C$. This contradicts the maximality of $C$. Therefore, $\bar{V} \in C$. Consequently,

(2) *$(Y, \bar{Y})$, $(Z, \bar{Z})$ and $(V, \bar{V})$ are chords of $C$.*

Since $G$ is 2-connected, there are two (vertex-) disjoint paths $w$, $w'$ connecting $C$ and $H$. Since $R\pi X$ is a $C,H$-path, it can be assumed that

(3) *one of the two paths $w$, $w'$ starts in $R$.*

*Proof of (3).* Assume that neither $w$ nor $w'$ starts in $R$. If $R\pi X$ does not meet $w, w'$, then $R\pi X$ and $w$ are two disjoint $C,H$-paths with the desired property. Now let $R\pi X$ meet $w \cup w'$. Then we follow $R\pi X$ starting in $R$ until we meet a first vertex, say $S$, of $w \cup w'$. Assume that $S \in w$. Then $w'$ is disjoint from the $C,H$-path which is obtained by following $R\pi X$ from $R$ to $S$ and then $w$ from $S$ to $H$. □

Let $P$ denote the end vertex of $w$, $w'$ in $C$ different from $R$. Thus the two end vertices of $w$, $w'$ in $C$ are $P$ and $R$. By (2), $C \cup \{(Y, \bar{Y}), (Z, \bar{Z}), (V, \bar{V})\}$ is a subgraph of $G$.

Next we show that

(4) *$C \cup \{(Y, \bar{Y}), (Z, \bar{Z}), (V, \bar{V})\}$ contains a $P,R$-path having two chords.*

*Proof of (4). Remember that $\pi = \Lambda^* - (Y, R)$.* Since $\bar{Y} \neq R$, the edge $(\bar{Y}, V)$ splits the path $Z\pi Y$ into the two paths $Z\pi\bar{Y}$ and $V\pi Y$ (if $\bar{Y} = Z$, then $Z\pi\bar{Y} = \{Z\}$). The situation can be seen in Fig. 8.8.1. If $P \in Z\pi\bar{Y}$, then either $\bar{V} \in P\pi YR$ and $P\pi YR$ is a $P,R$-path with two distinct chords $(\bar{Y}, Y)$ and $(V, \bar{V})$ or $\bar{V} \in P\pi Z - \{P\}$ and $P\pi\bar{Y}Y\pi V\bar{V}\pi R$ is a $P,R$-path with two distinct chords $(\bar{Y}, V)$ and $(Y, R)$.

Assume that $P \in V\pi Y$. There are two cases:

Case 1. Assume that $\bar{Y} = Z$. It may happen that $(Y, \bar{Y}) = (\bar{Z}, Z)$. If $P \in V\pi Y - \{Y\}$, then either $\bar{V} \in P\pi R$ and $P\pi ZYR$ is a $P,R$-path with two distinct chords $(\bar{V}, V)$ and $(Z, R)$ or $\bar{V} \in P\pi Y - \{P\}$ and $P\pi V\bar{V}\pi YZR$ is a $P,R$-path with two distinct chords $(V, Z)$ and $(Y, R)$. If $P = Y$, then $\pi$ is a $P,R$-path with two distinct chords $(Y, R)$ and $(\bar{V}, V)$.

Case 2. Assume that $\bar{Y} \neq Z$. This implies that $(Y, \bar{Y}) \neq (\bar{Z}, Z)$. If $\bar{Z} \in P\pi YR\pi\bar{Y}$, then $P\pi Y\bar{Y}\pi R$ is a $P,R$-path with two distinct chords $(\bar{Z}, Z)$ and $(Y, R)$. If $\bar{V} \in P\pi Y - \{P\}$, then $P\pi V\bar{V}\pi Y\bar{Y}\pi R$ is a $P,R$-path with two distinct chords $(V, \bar{Y})$ and $(Y, R)$.

Now we assume that $P \in V\pi Y - \{Y\}$, $\bar{Z} \in P\pi V - \{P\}$ (see Fig. 8.8.2) and $\bar{V} \in P\pi R$. If $\bar{Z} \neq V$, then $(Z, \bar{Z}) \neq (V, \bar{V})$ and $P\pi R$ is a $P,R$-path with two distinct chords $(\bar{Z}, Z)$ and $(V, \bar{V})$. If $\bar{Z} = V$, then $P\pi\bar{Z}Z\pi\bar{Y}YR$ is a $P,R$-path with two distinct chords $(\bar{Z}, \bar{Y})$ and $(Z, R)$.

Thus (4) is proved. □

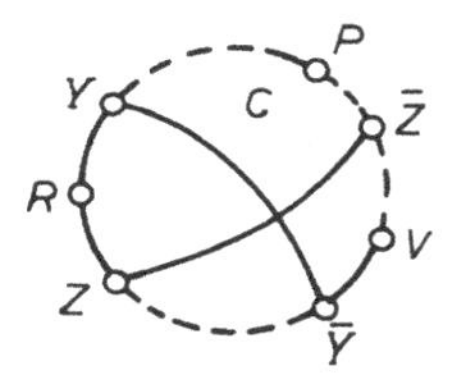

Fig. 8.8.2 It may be $\bar{Z} = V$

Let $u$ denote a $P,R$-path of $C \cup \{(Y, \bar{Y}), (Z, \bar{Z}), (V, \bar{V})\}$ having two chords. Then $w \cup u \cup w'$ is an $H,H$-path having two chords in $\mathfrak{B}$, contradicting (A). The proof of Lemma 8.8.2 is now complete. □

## 8.9. Proof of Theorem 8.4.2

The main purpose of Chapter 8 is to prove the validity of Theorem 8.4.2. This will be done in this section. We start with the proof of assertion (i) of Theorem 8.4.2.

**Theorem 8.9.1.** *Let $G$ be a 2-connected graph and $A$ and $B$ be two different vertices of $G$. Suppose that all vertices have valency at least 3, except perhaps $A$ and $B$. Then there is an $A,B$-path with at least two chords different from $(A, B)$.*

*Proof.* Let $G$ be a graph satisfying the hypotheses of Theorem 8.9.1. Since $G$ is 2-connected, $G$ has at least three vertices. Therefore there is a bridge $\mathfrak{B}$ of $\{A, B\}$ in $G$ containing at least one inner vertex $P$. Since $v(G:P) \geqq 3$, there is a vertex $Q \notin \{A, B\}$ which is joined to $P$ by an edge. Hence the kernel of $\mathfrak{B}$ contains at least two vertices. Thus $d = d(\mathfrak{B}) \geqq 2$.

(1) *The set of vertices of attachment of $\mathfrak{B}$ is $\{A, B\}$.*

*Proof of (1).* Since $G$ is 2-connected, each inner vertex $P$ of $\mathfrak{B}$ is joined to $A$ and $B$ by two internally disjoint paths. Hence $\{A, B\}$ is the set of vertices of attachment of $\mathfrak{B}$. Thus (1) is proved.

Let $H$ denote the subgraph of $G$ induced by $\{A, B\}$. By Lemma 8.8.2 we have only to investigate the case when $\mathfrak{B}$ contains an $A,B$-path $v$ of length $d+1$. In $v$ let $X$ and $Y$ be the neighbours of $A$ and $B$, respectively. Since $d \geqq 2$ we have $d+1 \geqq 3$, and $X \neq Y$. Since $v(G:X) \geqq 3$ and $v(G:Y) \geqq 3$, there is a neighbour $\bar{X}$ of $X$ and a neighbour $\bar{Y}$ of $Y$ such that $(X, \bar{X})$, $(Y, \bar{Y}) \notin v$. We know that $v - \{A\}$ and $v - \{B\}$ are two $H$-paths of length $d$ in $\mathfrak{B}$ with terminal $X$ and $Y$, respectively. Therefore, by Lemma 8.8.1 the vertices $\bar{X}$ and $\bar{Y}$ belong to $v$. If $\bar{X} \neq Y$, then the $A,B$-path $v$ has two distinct diagonals $(X, \bar{X})$ and $(Y, \bar{Y})$ in $\mathfrak{B}$ (different from $(A, B)$). Now assume that $\bar{X} = Y$, i.e., $(X, Y) \in \mathfrak{B}$. Let $V$ be the neighbour of $X$ with $(X, V) \in XvY$. Then $AXYvV$ is an $H$-path of length $d$. Thus $V$ is the terminal of an $H$-path of length $d$, and therefore by Proposition 8.8.1 all neighbours of $V$ are contained in $v$.

Since $v(G:V) \geqq 3$, the vertex $V$ has a neighbour $\bar{V} \in v$ such that $(V, \bar{V}) \notin v$. Consequently, $v$ is an $A,B$-path having two distinct chords $(X, Y)$ and $(V, \bar{V})$ (different from $(A, B)$). Thus the proof of Theorem 8.9.1 is complete. □

Now we shall prove the assertions (ii) and (iii) of Theorem 8.4.2.

**Theorem 8.9.2.** *Let $G$ be a 2-connected graph. Suppose that all vertices of $G$ have valency at least 3. Then $G$ contains an even cycle with two (or more) diagonals, or $G \cong W_4$.*

*Moreover, if $G$ is also non-bipartite, then $G$ contains an odd cycle with two (or more) diagonals, or $G$ is a $K_4$, or $G$ is a prism graph.*

*Proof.* Let $G$ be a graph satisfying the hypotheses of the theorem. An odd (even) cycle is said to have *parity* $p = 1$ $(p = 0)$. First we consider the case when $G$ is bipartite. Then $G$ contains only even cycles. Let $(A, B)$ be an edge of $G$. By Theorem 8.9.1 there is an $A,B$-path $p$ with two chords different from $(A, B)$. Hence $p \cup (A, B)$ is an even cycle with two chords. Therefore, Theorem 8.9.2 is true for bipartite graphs.

In the following discussion let $G$ be non-bipartite. Assume that $G$ has the property that each cycle of parity $p$ of $G$ has at most one diagonal. Then we have to prove the following result (if $H_1$ is isomorphic to $H_2$, we write $H_1 \cong H_2$):

(H) *Let $G$ be a 2-connected non-bipartite graph with minimum valency at least 3 containing no cycle of parity $p$ with two (or more) chords. Then for $p = 1$, $G \cong K_4$ or $G$ is a prism graph and for $p = 0$, $G \cong W_4$.*

Let $L$ denote a shortest odd cycle of $G$. Then

(1) *each $L,L$-path $w$ has at most one chord,*

because otherwise $w$ and one of the two arcs of $L$ between the two end vertices of $w$ would form a cycle of parity $p$ having two chords. □

Next we prove that

(2) *either $G \cong K_4$ or $L$ can be chosen so that there exists a bridge $\mathfrak{B}$ of $L$ containing at least two inner vertices.*

*Proof of (2).* The cycle $L$ has no chord, because otherwise $L$ could be split by such a chord into an even and an odd cycle both having a smaller length than $L$. This contradicts the definition of $L$.

Since $L$ has no chord and each vertex of $L$ has valency at least 3, there are edges with one vertex in $L$ and the other one in $G - L$. Let $A$ denote any vertex of $G - L$. If $A$ has at most two neighbours on $L$, then from $v(G:A) \geqq 3$ it follows that $A$ has a neighbour $B$ in $G - L$. Then $A$ and $B$ are contained in the same bridge $\mathfrak{B}$ of $L$.

Now let $A$ have (at least) three neighbours $A_1, A_2, A_3 \in L$. For all $1 \leqq i,j,k \leqq 3$ with $i \neq j \neq k \neq i$ the arc of $L$ between $A_i$, $A_j$, not containing $A_k$, is denoted by $L[A_i, A_j]$.

Let $C_i =_{\text{def}} L[A_i, A_{i+1}] \cup [A_{i+1}, A, A_i]$ for $1 \leqq i \leqq 3$ (subscripts taken modulo 3). Since $L = C_1 \triangle C_2 \triangle C_3$ has an odd length, at least one of $C_1$, $C_2$, $C_3$ is an *odd* circuit. Hence we may assume that $C_1$ has odd length. The two odd circuits $C_1$ and $L$ have $L[A_1, A_2]$ in common.

Since $C_1 = L[A_1, A_2] \cup [A_2, A, A_1]$ and since $L$ is a smallest odd circuit, the path $L[A_2, A_3] \cup L[A_3, A_1]$ has length 2; this means that both arcs of $L$ are edges and $C_2$ and $C_3$ are triangles. Consequently, $L$ is also a triangle and the subgraph of $G$ induced by $A, A_1, A_2, A_3$ is a $K_4$. Then either $G \cong K_4$ and (2) are true or there is a vertex $B \notin \{A, A_1, A_2, A_3\}$. Then from the 2-connectedness of $G$ it follows that $B$ is contained in a path $v$, where $v$ has the property that both its end vertices but none of its inner vertices are contained in $\{A, A_1, A_2, A_3\}$. We may assume that $A_2$ and $A_3$ are the end vertices of $v$. Then $[A, A_1, A_2, A]$ is a smallest odd cycle having a bridge $\mathfrak{B}$ containing $A_3$ and $B$. □

By Lemma 8.7.1 there is an $L,L$-path $v$ in $\mathfrak{B}$ of length $d + 1 = d(\mathfrak{B}) + 1$.

Let the end vertices of $v$ be denoted by $X'$ and $Y'$. Let $X$ and $Y$ denote the neighbours of $X'$ and $Y'$, respectively, so that $(X', X)$, $(Y, Y') \in v$.

Assertion (2) implies that $d = d(\mathfrak{B}) \geqq 2$. Therefore, $X \neq Y$. Since $v(G:X) \geqq 3$ and $v(G:Y) \geqq 3$, there is a neighbour $\bar{X}$ of $X$ and a neighbour $\bar{Y}$ of $Y$ such that $(X, \bar{X})$, $(Y, \bar{Y}) \notin v$. $v[X', Y]$ and $v[Y', X]$ are two longest $L$-paths with terminals $Y$ and $X$, respectively. Therefore, by Proposition 8.8.1, the vertices $\bar{X}$ and $\bar{Y}$ belong to $v \cup L$. There are four cases:

1) $\bar{X} \in v$, $\bar{Y} \in v$.
2) $\bar{X} \in L - v$, $\bar{Y} \in v$.
3) $\bar{X} \in L - v$, $\bar{Y} \in L - v$.
4) $\bar{X} \in v$, $\bar{Y} \in L - v$.

Since case 4 is symmetric to case 2, there remain only the first three cases to be considered.

Case 1. $\bar{X} \in v$, $\bar{Y} \in v$. If $\bar{X} \neq Y$, then the $L,L$-path $v$ has two distinct chords $(X, \bar{X})$ and $(Y, \bar{Y})$. This contradicts (1).

Now let $\bar{X} = Y$, i.e., $(X, Y) \in \mathfrak{B}$. Let $C$ denote the cycle $v[X, Y] \cup (X, Y)$. Then we have the following statement:

(3) *Label the vertices of $C$ with $Y_1, Y_2, \ldots, Y_s$, so that $Y_1 = X$, $Y_2 = Y$ and $C = [Y_1, Y_2, \ldots, Y_s, Y_1]$. Then for all integers $t$ satisfying $2 \leqq t \leqq s+1$ the following three conditions hold (indices modulo $s$):*

a) *there exist $t$ distinct vertices $Y_1', Y_2', \ldots, Y_t'$ lying on $L$ in this cyclic order;*

b) *$(Y_i, Y_i') \in G$ for all $1 \leqq i \leqq t$;*

c) *$C \triangle C_i$ has parity $p$ for all $1 \leqq i \leqq t-1$, where $C_i$ is the cycle formed by the path $[Y_i', Y_i, Y_{i+1}, Y_{i+1}']$ and the arc of $L$ between $Y_i'$, $Y_{i+1}'$ containing none of the $Y_j'$ $(j \neq i, i+1)$* (see Fig. 8.9.1).

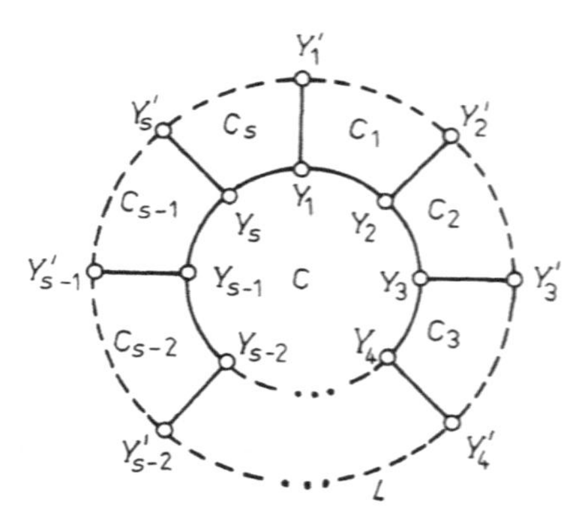


Fig. 8.9.1

*Proof of (3).* The proof is performed by induction on $t$. The path $v$ is an $L,L$-path dividing $L$ into an even and an odd arc. These two arcs together with the $L,L$-path $v$ generate precisely one even and one odd cycle. Thus exactly one of these two cycles has parity $p$. The cycle of parity $p$ is subdivided by the edge $(X, Y)$ into the cycle $C$ and a second cycle, say $C_1$. Let $Y_1 =_{\text{def}} X$, $Y_2 =_{\text{def}} Y$. With this notation, (3) is true for $t = 2$. Now assume that (3) is true for all $2 \leqq t \leqq \tau \leqq s$. It remains to prove (3) for $t = \tau + 1$.

Since $C$ has length $d$, the path $C - (Y_\tau, Y_{\tau+1})$ has length $d-1$. By induction, $(Y_\tau, Y_\tau') \in G$. Therefore, $[Y_\tau', Y_\tau] \cup C - (Y_\tau, Y_{\tau+1})$ is an $H$-path of length $d$ with terminal $Y_{\tau+1}$.

By Proposition 8.8.1 all neighbours of $Y_{\tau+1}$ are contained in $C \cup L$. It follows from $v(G: Y_{\tau+1}) \geqq 3$ that there is a vertex $Y_{\tau+1}'$ such that $(Y_{\tau+1}, Y_{\tau+1}') \in G$ and $(Y_{\tau+1}, Y_{\tau+1}') \notin C$. Since for $1 \leqq i \leqq \tau - 1$ the cycle $C \triangle C_i$ has parity $p$ and contains the chord $(Y_i, Y_{i+1})$, it follows that $(Y_{\tau+1}, Y_{\tau+1}')$, $Y_{\tau+1}' \notin C \triangle C_i$.

Consequently, $Y_{\tau+1}'$ is an inner vertex of the arc of $L$ between $Y_1'$ and $Y_\tau'$ not containing $Y_j'$, $2 \leqq j \leqq \tau - 1$. Thus (3a) and (3b) are also true for $t = \tau + 1$. Since the cycle $(C \triangle C_\tau) \triangle L$ has two distinct chords $(Y_\tau, Y_{\tau+1})$ and $(Y_1, Y_1')$ (note that $\tau \geqq 2$), this cycle has parity $p + 1$. Consequently, $C \triangle C_\tau$ has parity $p$. This completes the proof of (3). □

If $s = 3$, then $C$ and also $L$ are triangles. Consequently, $G$ contains a prism graph. If we add to a prism graph a path with the property that its end vertices but no inner vertices belong to the prism graph, then the graph so obtained contains a cycle of parity $p$ with two chords. Consequently, $G$ is a prism graph, and therefore $p = 1$. Now let $s \geqq 4$. If $p = 1$, either $C_1 \triangle C \triangle C_3$ or $C_1 \triangle C_2 \triangle C_3$ is an odd cycle with two distinct chords $(Y_1, Y_2)$, $(Y_3, Y_4)$ or $(Y_2, Y_2')$, $(Y_3, Y_3')$, respectively. Contradiction!

If $p = 0$, then $C \triangle C_i$ is an even cycle and all cycles $C, C_1, C_2, \ldots, C_s$ have the same parity.

Since $L = C \triangle C_1 \triangle \ldots \triangle C_s$, the odd cycle $L$ is composed of an odd number of cycles, i.e., $s$ is even, and all cycles $C, C_1, \ldots, C_s$ have odd lengths. Consequently, the odd cycle $C$ has the even length $s$. This contradiction establishes (H) in case 1.

Case 2. $\bar{X} \in L - v$, $\bar{Y} \in v$.

(a) Assume that $\bar{Y} \neq X'$. Let $C_1, C_2, C_3, C_4$ be defined in the same way as in the planar embedding of $\Gamma =_{\text{def}} L \cup v \cup (X, \bar{X}) \cup (Y, \bar{Y})$ in Fig. 8.9.2a.

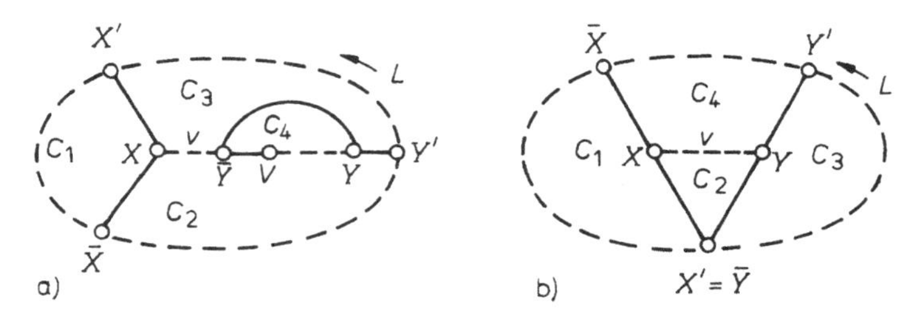


Fig. 8.9.2

In what follows the parity of a cycle $C_i$ is always denoted by the corresponding lowercase letter $c_i$. Since $C_1 \triangle C_2$ and $C_1 \triangle C_3 \triangle C_4$ are cycles having two chords each, we obtain the following system of equations over GF(2):

$$\begin{aligned} c_1 + c_2 \qquad\qquad &= 1 + p, \\ c_1 \qquad + c_3 + c_4 &= 1 + p, \\ c_1 + c_2 + c_3 + c_4 &= 1. \end{aligned}$$

The solution is $c_1 = 1$, $c_2 = c_3 + c_4 = p$. Since $C_1$ is an odd cycle, it cannot be longer than $L$. Therefore, the arc of $L$ between $X'$ and $\bar{X}$ containing $Y'$ has length 2, i.e., $(X', Y')$, $(\bar{X}, Y') \in L$.

Let $V$ denote the neighbour of $\bar{Y}$ such that $(\bar{Y}, V) \in v[\bar{Y}, Y]$. Then $X'v\bar{Y}Yv V$ is an $L$-path with terminal $V$ containing all vertices of $v[X', Y]$. Therefore, this $L$-path has length $d$, and by Proposition 8.8.1 all neighbours of $V$ belong to $v \cup L$. Because $v(G: V) \geqq 3$, the vertex $V$ is adjacent to a vertex $\bar{V} \in v \cup L$ such that $(V, \bar{V}) \notin v$. Since by $c_2 = c_3 + c_4 = p$ the cycles $C_2$ and $C_3 \triangle C_4$ have parity $p$ and contain the chord $(Y, \bar{Y})$, the vertex $\bar{V}$ is contained in $L - \{X', Y', \bar{X}\}$.

The same arguments applied to $\Gamma \cup (V, \bar{V}) - (Y, Y')$ yield $(X', \bar{V})$, $(\bar{X}, \bar{V}) \in L$.

Thus $L = [X', Y', \bar{X}, \bar{V}, X']$ is a cycle of length 4, contradicting the assumption that $L$ is odd.

b) Assume that $\bar{Y} = X'$. Let $C_1, C_2, C_3, C_4$ be defined in the same way as in the planar embedding of $\Gamma =_{\text{def}} L \cup v \cup (X, \bar{X}) \cup (Y, \bar{Y})$ in Fig. 8.9.2b.

Since each of the cycles $C_1 \triangle C_2 \triangle C_3$, $C_3 \triangle C_4$ and $C_1 \triangle C_4$ has two chords, the $c_i$'s satisfy the following system of equations over GF(2):

$$\begin{aligned} c_1 + c_2 + c_3 \phantom{{}+c_4} &= 1 + p, \\ c_3 + c_4 &= 1 + p, \\ c_1 \phantom{{}+c_2+c_3} + c_4 &= 1 + p, \\ c_1 + c_2 + c_3 + c_4 &= 1. \end{aligned}$$

The only solution is $c_1 = c_3 = 1$, $c_2 = 1 + p$, $c_4 = p$. Since $C_1$ and $C_3$ are odd cycles, their lengths are not smaller than the length of $L$. Thus $L$ must be the triangle $[X', \bar{X}, Y', X']$. But then the odd cycle $C_2 \triangle C_4$ has two distinct diagonals $(\bar{X}, X')$, $(Y', \bar{Y})$. For $p = 1$ this is a contradiction.

Now we consider the case $p = 0$. Let $V \in v$ such that $(V, Y) \in v[X, Y]$. If $V = X$, then $v[X, Y] = [X, Y]$ and $G$ contains a $W_4$. If we add to a $W_4$ a path with the property that its end vertices but no inner vertex belong to the $W_4$, then the graph so obtained contains an even cycle with at least two diagonals. Consequently, $G$ is a $W_4$.

Now let $V \neq X$. Since $v(G:V) \geqq 3$, there is a neighbour $\bar{V}$ of $V$ such that $(V, \bar{V}) \notin v$. If $\bar{V} \in v \cup L$, then $v$ or $\bar{Y}YvX\bar{X}$ is an $L,L$-path having two distinct chords $(X', Y)$, $(V, \bar{V})$ or $(\bar{Y}, X)$, $(V, \bar{V})$, respectively. This contradicts (1). Therefore, $\bar{V} \notin v \cup L$. Then $X'vV\bar{V}$ is an $L$-path of length $d$. By Proposition 8.8.1 all neighbours of $\bar{V}$ belong to $v \cup L - \{Y\}$. If $(\bar{V}, X') \in \mathfrak{B}$, then $[X', \bar{V}, V, Y, Y', \bar{X}, X']$ is a cycle of length 6 with two distinct chords $(X', Y)$ and $(X', Y')$. If $(\bar{V}, \bar{X}) \in \mathfrak{B}$, then $[X', Y, V, \bar{V}, \bar{X}, Y', X']$ is a cycle of length 6 with two distinct chords $(Y', Y)$ and $(X', \bar{X})$. If $(\bar{V}, Y') \in \mathfrak{B}$, then $[X', Y, V, \bar{V}, Y', \bar{X}, X']$ is a cycle of length 6 with two distinct chords $(Y', Y)$ and $(Y', X')$. Therefore, all neighbours of $\bar{V}$ belong to $v[X, Y]$. Since $v(G:\bar{V}) \geqq 3$ the vertex $\bar{V}$ has three neighbours $Y_1$, $Y_2$, $Y_3$ on $v[X, V]$ which occur in this order in a walk on $v$ from $X$ to $V$. Then $X'vY_2\bar{V}Y_3vY'$ is an $L,L$-path with two distinct diagonals $(X', Y)$ and $(\bar{V}, Y_1)$. This contradicts (1).

Case 3. $\bar{X} \in H - v$, $\bar{Y} \in H - v$. If $\bar{X} = \bar{Y}$, then exchanging the symbols $X'$ and $\bar{X}$ we arrive at Case 2b. Now assume that $\bar{X} \neq \bar{Y}$. We have to distinguish between the case when the paths $X'X\bar{X}$ and $Y'Y\bar{Y}$ overlap with respect to $L$ and the case when they do not overlap.

By possibly exchanging the symbols $X'$ and $\bar{X}$ we always arrive at the graph $\tilde{\Gamma}$ of Fig. 8.9.3b or the graph $\Gamma$ of Fig. 8.9.3a.

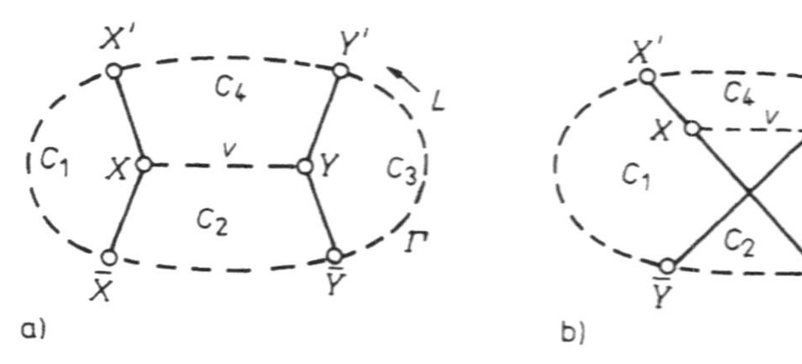


Fig. 8.9.3

Let $L[P, Q]$ be the arc of $L$ between $P$ and $Q$, its orientation coinciding with the orientation of $L$ (indicated by an arrow). In both cases, let $C_1 =_{\text{def}} [\bar{X}, X, X'] \cup L[X', \bar{X}]$, $C_3 =_{\text{def}} [Y', Y, \bar{Y}] \cup L[\bar{Y}, Y']$, $C_4 =_{\text{def}} v \cup L[Y', X']$, and let $C_2$ consist of $[\bar{Y}, Y] \cup v[Y, X] \cup [X, \bar{X}]$ and the arc of $L$ between $\bar{X}$ and $\bar{Y}$ not containing $X'$ and $Y'$.

Since the cycles $C_1 \triangle C_2 \triangle C_3$ and $C_1 \triangle C_3 \triangle C_4$ (in $\tilde{\Gamma}$ also the cycles $C_1 \triangle C_4$ and $C_3 \triangle C_4$) contain two chords, the $c_i$'s satisfy the following systems of equations over GF(2):

$$\Gamma: \begin{aligned} c_1 + c_2 + c_3 \phantom{{}+c_4} &= 1 + p, \\ c_1 \phantom{{}+c_2} + c_3 + c_4 &= 1 + p, \\ c_1 + c_2 + c_3 + c_4 &= 1, \end{aligned} \qquad \tilde{\Gamma}: \begin{aligned} c_1 + c_2 + c_3 \phantom{{}+c_4} &= 1 + p, \\ c_1 \phantom{{}+c_2} + c_3 + c_4 &= 1 + p, \\ c_1 + c_2 + c_3 + c_4 &= 1, \\ c_1 \phantom{{}+c_2+c_3} + c_4 &= 1 + p, \\ c_3 + c_4 &= 1 + p. \end{aligned}$$

In $\Gamma$ the solution is $c_1 + c_3 = 1$, $c_2 = c_4 = p$; in $\tilde{\Gamma}$ there is no solution. Then in $\Gamma$ either $C_1$ or $C_3$ is an odd cycle with a length smaller than the length of $L$: a contradiction! Thus the proof of Theorem 8.9.2 is complete. □

# Chapter 9

# Cycles with diagonals in graphs of minimum degree $r$ and girth $t$

## 9.0. Introduction

In Chapter 9 the investigations of Chapters 7 and 8 are continued by studying the circumference and the maximum number of diagonals of cycles in graphs of large girth and by also investigating the maximum number of diagonals of cycles in graphs of large minimum degree.

A well-known class of such graphs consists of the $r$-regular graphs of given girth with minimum vertex number. These graphs have been studied by W. T. Tutte, 1947, P. Erdős, and H. Sachs, 1963, H. Walther, 1965a, b, N. Sauer, 1967, and others (see H. Walther and H.-J. Voss, 1974).

### 1° Graphs of high chromatic number and large girth

Many authors have dealt with the construction of graphs of high girth with large chromatic number (see section 8.0, 3°). The general result is the theorem of P. Erdős, 1959:

*For each pair of integers $k \geqq 3$, $t \geqq 3$ there exists a $k$-chromatic graph of girth at least $t$.*

Though graphs of high chromatic number do not necessarily contain small cycles, one may ask whether all such graphs contain cycles with a certain number of diagonals. The answer is in the affirmative (see section 8.0, 3°). P. Erdős, 1977, conjectured that

*every graph $G$ with chromatic number at least 4 containing no triangle (or with girth at least $t \geqq 4$) has an odd cycle with many diagonals.*

We shall solve this conjecture in the affirmative by showing (assertion (ii) of section 8.0, 3°) that

*every graph $G$ with chromatic number at least 4 and girth at least $t \geqq 4$ contains both an odd and an even cycle with at least $2^{t/12-8}$ diagonals.*

### 2° Graphs of high minimum degree

In Chapter 7 cycles of maximum length in graphs with given minimum degree are well investigated. In this direction the best-known theorems are due to G. A. Dirac, 1952a:

*Every graph $G$ with minimum degree at least $r \geqq 3$ contains a cycle of length at least $r+1$. If $G$ is also 2-connected, then $G$ is Hamiltonian or contains a cycle of length at least $2r$.*

In such graphs some authors have studied the maximum number of diagonals in cycles. For arbitrary graphs $G$ with minimum degree $r \geqq 3$, G.P. Gupta, J. Kahn and N. Robertson, 1980, showed that

*G contains a cycle with at least* $\frac{1}{2}(r+1)(r-2)$ *diagonals.*

If $G$ is also 2-connected and of order greater than $2r$, then by P. Ash, 1985a,

*G has a cycle with at least* $r(r-2)$ *diagonals.*

Similar results are valid for cycles through a given edge and for cycles of a given parity.

## 3° Graphs of large girth

For graphs with girth at least $t \geqq 5$, O. Ore, 1967b, improved the above mentioned result of G. A. Dirac:

*If G is a 2-connected graph with minimum degree* $r \geqq 3$ *and girth* $t \geqq 5$, *then G is Hamiltonian or contains a cycle of length at least* $(t-2)(r-2)+5$.

The assertion of O. Ore will be strengthened by showing that

*G has even a cycle of length at least* $2^{t/4-7}$ *and a cycle with at least* $2^{t/12-8}$ *diagonals.*

We shall present an infinite family of 2-connected graphs with $\delta(G) \geqq r$ and $g(G) \geqq t$ having only cycles of length at most $2^{t+2}$ with at most $2^{t+2}$ diagonals. This implies that our result cannot be improved in the sense that the exponential function can be replaced by a function that increases more rapidly.

## 4° Graphs of high minimum degree and large girth

### a) *Large cycles with many diagonals through given edges*

Many authors have studied cycles through given vertices and edges. The first result is due to G.A. Dirac, 1960, who observed that in each k-connected graph any $k$ vertices lie on a common cycle. Strengthenings of this result are due to M. E. Watkins and D. M. Mesner, 1967, A. K. Kelmans and M. V. Lomonossow, 1982a, b, D. A. Holton et. al., 1982, R. Häggkvist and C. Thomassen, 1982, J. A. Bondy and L. Lovász, 1981 (presented in section 9.3). All these authors proved the existence of a cycle through a set of given vertices and edges.

M. Grötschel, 1977, H. Enomoto, 1984, Fan Geng-hua, 1984, and S. C. Locke, 1985, proved the existence of a *large cycle* through given vertices and edges in graphs with a given minimum degree (see Chapter 7).

Y. Egawa, G. Glas and S. C. Locke, 1987, proved: Let $G$ be a $k$-connected graph with minimum degree $r$ and at least $2r$ vertices. Then $G$ has a cycle of length at least $2r$ through any specified set of $k$ vertices. In this chapter we shall deal with similar problems in graphs of girth at least $t \geqq 4$. One of our results is:

*In every k-connected graph,* $k \geqq 2$, *with minimum degree at least* 3 *and girth* $t \geqq 3$ *any* $k-1$ *independent edges are contained in a cycle of length at least* $2^{t/4-7}$ *and in a cycle with at least* $2^{t/12-8}$ *diagonals.*

We shall also present a similar theorem on cyclic $k$-connected graphs.

b) *Large odd and even cycles with many diagonals*

B. Bollobás, 1978 a, C. Thomassen, 1983 a, and others dealt with the following question: when does a graph $G$ contain cycles of all lengths mod k, or all even lengths mod $k$ (if $G$ is bipartite). In Chapter 7 a theorem of C. Zuluaga and the author (H.-J. Voss and C. Zuluaga, 1977) was presented which states the existence of both an odd and an even cycle of length at least $2r - 1$ in non-bipartite 2-connected graphs of minimum degree at least $r$ and order at least $2r$.

Taking into consideration the girth $t$ of the graphs in question we can prove the following result (assertion (iv) of section 8.0, 3°):

*Each non-bipartite 2-connected graph $G$ with minimum degree at least 3 and girth at least $t \geqq 4$ contains both an odd and an even cycle of length at least $2^{t/4-7}$ and both an odd and an even cycle with at least $2^{t/12-8}$ diagonals.*

## 9.1. Cycles with diagonals in given classes of graphs

In Chapter 7 problems concerning largest circuits in graphs with a given minimum degree are well investigated. Hence in this section we deal only with cycles with many diagonals in those graphs.

In section 8.2, $f(k, r)$ has been introduced as the largest number $\alpha$ such that every $k$-connected graph of minimum degree at least $r$ contains a cycle with at least $\alpha$ diagonals. The case $r = 3$ and $0 \leqq k \leqq 2$ has been studied there. In Chapter 9 the investigations of $f(k, r)$ and of related problems will be continued for $r \geqq 4$ and $0 \leqq k \leqq 2$.

We start with the construction of infinite classes of 2-connected graphs of minimum degree at least 3 having only cycles with a relatively small number of diagonals. Thus we obtain upper bounds for $f(k, r)$.

Graphs with large girth will also be studied. Hence we first consider the well-known class of all $r$-regular graphs of given girth with minimum vertex number. This class has been well investigated (compare with section 9.0).

**Proposition 9.1.1.** *Let $G(r, t)$ be an $r$-regular graph, $r \geqq 3$, of given girth $t \geqq 3$ with minimum vertex number. Then $G(r, t)$ is 2-connected and each cycle and each path of $G(r, t)$ has at most $(r-1)^{t+1}$ diagonals and length at most $(r-1)^{t+1}$. If $t$ is odd, then $G(r, t)$ is non-bipartite.*

*Proof.* Let $G$ be a graph as defined in the proposition. H. Sachs proved that $G$ is 2-connected (see H. Walther and H.-J. Voss, 1974).

Let $K$ denote a cycle or path of $G$. H. Walther proved (see H. Walther and H.-J. Voss, 1974, p. 120) that the vertex number $n$ of $G$ is

$$n \leqq \left(2r - \frac{4}{r}\right) \frac{1}{r-2} \left\{(r-1)^{t-2} - 1\right\} + 2r$$

$$\leqq (r-1)^3 \{(r-1)^{t-2} - 1\} + 2r \leqq (r-1)^{t+1} - 1.$$

Thus each path and each cycle of $G$ has length at most $(r-1)^{t+1}$. Since $G$ is $r$-regular, the edge number $e$ of $G$ is $e = (r/2)\, n$. If $K$ is a cycle, then each vertex of $G$ is incident with two edges that are not diagonals of $K$. If $K$ is a path, then each vertex that is not an end vertex of $K$ is incident to two edges that are not diagonals of $K$ and each of the two end vertices of $K$ is incident with one edge that is not a diagonal of $K$.

Consequently, $G$ contains at least $n-1$ edges which are not diagonals of $K$. If $\varkappa$ denotes the number of diagonals of $K$, then

$$\begin{aligned}\varkappa &\leqq e-n+1 \leqq \frac{r-2}{2} n+1 \\ &\leqq \frac{r-2}{2}\left(2r-\frac{4}{r}\right)\frac{1}{r-2}\{(r-1)^{t-2}-1\}+r(r-2)+1 \\ &\leqq (r-1)^3\{(r-1)^{t-2}-1\}+(r-1)^2 \leqq (r-1)^{t+1}. \square\end{aligned}$$

Let $r$ and $t$ be fixed, where $r, t \geqq 3$. Then Proposition 9.1.1 states the existence of finitely many graphs $G(r, t)$ with cycles having at most $(r-1)^{t+1}$ diagonals and length at most $(r-1)^{t+1}$. Next, we shall construct infinitely many 2-connected non-bipartite graphs of minimum degree $r$ and girth $t$ each cycle of which has at most $2(r-1)^{t+1}$ diagonals and length at most $2(r-1)^{t+1}$.

Let $T_r$ be a tree with root $W$ and end vertices $Y_1, \ldots, Y_s$, where $W$ is an inner vertex of $T_r$ and all inner vertices of $T_r$ have valency at least $r$. Let $T'_r$ be a tree such that $T_r$ and $T'_r$ are vertex-disjoint and there is an isomorphism $\varphi$ from $T_r$ onto $T'_r$. Let $G^-(r, t)$ denote a graph obtained from an $r$-regular graph $G(r, t)$ (of girth $t$ with minimum vertex number) by deleting one edge, say $(P, Q)$.

We define $\underline{H}_t(T_r)$: for all $2 \leqq i \leqq s$ we add to $T_r \cup T'_r$ a copy $C_i$ of $G^-(r, t)$ and identify $Y_i$ with the vertex $P_i$ of $C_i$ and $\varphi(Y_i)$ with the vertex $Q_i$ of $C_i$. If $G^-(r, t)$ is non-bipartite, let $C_1$ also be a copy of $G^-(r, t)$. We identify $Y_1$ with the vertex $P_1$ of $C_1$ and $\varphi(Y_1)$ with the vertex $Q_1$ of $C_1$. If $G^-(r, t)$ is bipartite, let $C_1$ be a copy of $G(r, t)$ and let $P'$, $Q'$ be two vertices of the same bipartition class of $G(r, t)$. We identify $Y_1$ with the vertex $P'$ of $C_1$ and $\varphi(Y_1)$ with the vertex $Q'$ of $C_1$. By this procedure in both cases we arrive at a 2-connected non-bipartite graph of girth $t$ which has the following properties:

**Theorem 9.1.2.** *For each tree $T_r$ the graph $\underline{H}_t(T_r)$ is a 2-connected non-bipartite graph with minimum degree $r$ and girth $t$. If $G^-(r, t)$ is non-bipartite and $T_r$ has only inner vertices of degree $r$, then $\underline{H}_t(T_r)$ is $r$-regular. $\underline{H}_t(T_r)$ has the following properties (let $d(T_r)$ denote the diameter of $T_r$):*

(i) *if $A$ is any vertex of $T_r$, each $A, \varphi(A)$-path has at most $(r-1)^{t+1}$ diagonals and length at most $2d(T_r)+(r-1)^{t+1}$;*

(ii) *each cycle has at most $2(r-1)^{t+1}$ diagonals and length at most $2d(T_r)+2(r-1)^{t+1}$.*

For $t=3$ or $5$ and $r=3$ the construction here described is the same as in section 8.2, i.e. $\underline{H}_3(T_3)=H_3(T_3)$ and $\underline{H}_3(T_5)=H_3(T_5)$. For $t=4$ the graphs $\underline{H}_4(T_4)$ and $H_4(T_4)$ are non-isomorphic but very similar by construction.

Obviously, $G(r, 3) \cong K_{r+1}$. In this case Theorem 9.1.2 can be improved:

**Corollary 9.1.3.** *The graphs $\underline{H}_3(T_r)$ satisfy the following conditions:*

(i) *if $A$ is any vertex of $T_r$, then each $A, \varphi(A)$-path has at most $\frac{1}{2}(r+1)(r-2)$ diagonals and length at most $2d(T_r)+r$;*

(ii) *each cycle has at most $(r+1)(r-2)$ diagonals and length at most $2d(T_r)+2r$.*

It is a well-known result due to G. A. Dirac, 1952a, that the circumference of 2-connected graphs of order $n$ tends to infinity with $n$ provided that the degree of every vertex remains bounded for all $n$. This result can also be deduced from Theorem 4.2.3. Consequently, the lengths of the paths and cycles of $\underline{H}_t(T_r)$ are not bounded by a function depending only on $r$ and $t$.

*Sketch of the proof of Theorem 9.1.2.* We omit the easy proof that $\underline{H}_t(T_r)$ is a 2-connected non-bipartite graph with girth $t$ and minimum degree $r$ which is even $r$-regular if $G^-(r, t)$ is non-bipartite and $T_r$ has only inner vertices of degree $r$. Thus we have to prove the validity of assertions (i) and (ii). For $t = 3$, 5 and $r = 3$ the construction of $\underline{H}_t(T_r)$ is just the same as the construction of $H_t(T_3)$ in section 8.2. Hence Lemma 8.2.1 is also valid for the graphs $\underline{H}_t(T_r)$: each cycle $C$ of $\underline{H}_t(T_r)$ touches at most two $C_j$'s and if $A$ is any vertex of $T_r$, then each $A,\varphi(A)$-path $p$ of $\underline{H}_t(T_r)$ touches precisely one $C_j$.

By construction each diagonal of $C$ and of $p$ is an edge of $C_j$. Proposition 9.1.1 implies that any path of $C_j$ has at most $(r-1)^{t+1}$ diagonals and length at most $(r-1)^{t+1}$. Consequently, $C$ has at most $(2r-1)^{t+1}$ diagonals and length at most $2d(T_r) + 2(r-1)^{t+1}$ and $p$ has at most $(r-1)^{t+1}$ diagonals and length at most $2d(T_r) + (r-1)^{t+1}$. □

Since $G(r, 3) \cong K_{r+1}$, any path of $C_j$ has at most $\frac{1}{2}(r+1)(r-2)$ diagonals and length at most $r$. If $t = 3$, this leads to the better bounds of Corollary 9.1.3.

## 9.2. Cycles with many diagonals in graphs of small girth with given minimum degree

In this section we deal with lower bounds for $f(k, r)$ and similar problems. The theorems will be presented without proofs.

R. P. Gupta, J. Kahn and N. Robertson, 1980, proved that $f(0, r) = \frac{1}{2}(r+1)(r-2)$:

**Theorem 9.2.1.** *Every graph with minimum degree $r \geqq 3$ contains a cycle with at least $\frac{1}{2}(r+1)(r-2)$ diagonals.* □

The disjoint union of $K_{r+1}$'s shows that the assertion of Theorem 9.1.3 is sharp.

For 2-connected graphs P. Ash, 1985b, proved the following statement:

**Theorem 9.2.2.** *Every 2-connected graph with minimum degree $r \geqq 3$ having at least $2r$ vertices contains a cycle with at least $r(r-2)$ diagonals.* □

The complete bipartite graphs $K_{r,s}$ with $s \geqq r$ show that this bound is sharp. For large $s$ the $K_{r,s}$'s have vertices of large degree $s$. The bound $r(r-2)$ is also achieved by the Hamiltonian $r$-regular graphs of order $2r$. The number of these graphs is finite for fixed $r$.

By choosing different $T_r$'s, infinitely many graphs $\underline{H}_3(T_r)$ are obtained which have only degrees $r$ and $r+1$, satisfy the hypotheses of Theorem 9.2.2, and contain only cycles with at most $(r+1)(r-2)$ diagonals. An open question is the following:

**Problem 9.2.1.** *Can the bound in Theorem 9.2.2 be improved for r-regular graphs of order greater than $2r$?*

Next we deal with cycles with many diagonals through a given edge. By Theorem 9.1.2 the edge $(W, \varphi(W))$ of $\underline{H}_3(T_r) \cup \{(W, \varphi(W))\}$ is contained only in cycles with at most $\frac{1}{2}(r+1)(r-2)$ diagonals. The following theorem gives a lower bound for the maximum number of diagonals in cycles through a given edge in such graphs.

**Theorem 9.2.3.** *Let $G$ be a 2-connected graph with minimum degree at least $r \geqq 3$. If $e$ is any edge of $G$, then $G$ contains a cycle through $e$ with at least $\frac{1}{2}\binom{r-1}{2}$ diagonals.* □

For $r = 3$ or 4 Theorem 9.2.3 is a simple consequence of Theorem 8.3.7; the case $r \geqq 5$ has been proved in H.-J. Voss, 1985a.

A similar theorem holds with respect to odd and even cycles.

**Theorem 9.2.4.** *Let $G$ be a 2-connected graph with minimum degree $r \geqq 3$.* Then

(i) *$G$ contains an even cycle with at least $\frac{1}{2}\binom{r-1}{2}$ diagonals;*

(ii) *if $G$ is non-bipartite and $G \neq K_4$, then $G$ also contains an odd cycle with at least $\frac{1}{2}\binom{r-1}{2}$ diagonals.* □

For $r = 3$ or 4 Theorem 9.2.4 is a simple consequence of Theorem 8.4.2; the case $r \geqq 5$ has been proved in H.-J. Voss, 1985a.

The lower bounds for the number of diagonals in Theorems 9.2.3 and 9.2.4 are of order $r^2$. In examples we shall present upper bounds which are also of order $r^2$. Let $K^+_{r,s}$, $s > r$, be obtained from a $K_{r,s}$ by joining two vertices of the bipartition class of cardinality $r$ by an edge. $K^+_{r,s}$ and $K_{r,s}$ have degrees from $\{r, s, s+1\}$. Hence $K^+_{r,s}$ is an $r$-connected non-bipartite graph, each odd cycle of which has at most $(r-1)(r-2)$ diagonals, and $K_{r,s}$ is an $r$-connected bipartite graph, each (even) cycle of which has at most $r(r-2)$ diagonals. By Corollary 9.1.3 there are infinitely many 2-connected non-bipartite almost regular graphs $\underline{H}_3(T_r)$, each cycle of which has at most $(r+1)(r-2)$ diagonals and each cycle of $\underline{H}_3(T_r) \cup (W, \varphi(W))$ through $(W, \varphi(W))$ has at most $\frac{1}{2}(r+1)(r-2)$ diagonals.

## 9.3. Large cycles with many diagonals in graphs of large girth with given minimum degree

The starting point of the investigations of this section consists of the well-known theorems of G. A. Dirac, 1952a, already mentioned in this chapter. For convenience we repeat them here:

*Every graph with minimum degree at least $r \geqq 3$ contains a cycle of length at least $r+1$.*

*Every 2-connected graph with minimum degree at least $r \geqq 3$ contains a Hamiltonian cycle or a cycle of length at least $2r$.* □

O. Ore, 1967b, improved the second result for graphs with girth at least $t$:

*Every 2-connected graph with minimum degree at least $r \geqq 4$ and girth at least $t \geqq 5$ contains a Hamiltonian cycle or a cycle of length at least $(t-2)(r-2)+5$.* □

The assumption of 2-connectivity can be deleted (H.-J. Voss, 1976):

*Every graph with minimum degree at least $r \geqq 3$ and girth at least $t \geqq 3$ contains a circuit of length at least $(t-2)(r-1)+2$.* □

C.-Q. Zhang, 1989b, has proved that the length of a longest cycle of $G$ is at least $\max\{(t-2)(r-1)+2, (t-4)(2r-3)+4\}$. If $G$ is even 2-connected, then each pair of vertices of $G$ is joined by a path of length at least $\max\left\{\frac{1}{2}t(r-1), (t-4)\left(r-\frac{3}{2}\right)+2\right\}$.

In this book the existence of a cycle of length at least $\frac{1}{2}(r-1)^{t/4-6}$ and of a cycle having at least $2^{t/12-8}$ diagonals will be proved. Before stating these and other results we introduce two functions:

$$\gamma(r, t) =_{\text{def}} \begin{cases} 2^{t/12-8} & \text{for} \quad r = 3, \\ \frac{1}{2}(r-2)^{t/4-8} & \text{for} \quad r \geqq 4 \end{cases}$$

and

$$\varepsilon(r, t) =_{\text{def}} \frac{1}{2}(r-1)^{t/4-6}.$$

With these two functions we can state the following result:

**Theorem 9.3.1.** *Let $G$ be a 2-connected graph with minimum degree at least $r \geqq 3$ and girth at least $t \geqq 3$. Then each edge is contained in a cycle with at least $\gamma(r, t)$ diagonals and in a cycle of length at least $\varepsilon(r, t)$.*

I remark: if $G$ is 3-connected, then each edge is even contained in a cycle with at least $2^{t/8-10}$ diagonals.

The lower bounds for the maximum number of diagonals of cycles and the maximum length of cycles stated in Theorem 9.3.1 cannot be improved in the sense that the exponential function can be replaced by a function that increases more rapidly. This is shown by the graphs $\underline{H}_t(T_r)$ and $\underline{H}_t(K_{1,s})$, $s \geqq r$, because by Theorem 9.1.2 the graphs $\underline{H}_t(T_r)$ have only cycles with at most $2(r-1)^{t+1}$ diagonals and the graphs $\underline{H}_t(K_{1,s})$, $s \geqq r$, have length at most $4+2(r-1)^{t+1}$.

Theorem 9.3.1 is an immediate consequence of the more general Theorem 9.3.2.

**Theorem 9.3.2.** *Let $k$, $r$, $t$ be integers with $k \geqq 2$, $r \geqq \max\{3, k\}$ and $t \geqq 3$. Let $G$ be a $k$-connected graph with minimum degree at least $r$ and girth at least $t$. Then each $s$-tuple of vertices and edges, $s \leqq k-1$, lying on a cycle of length greater than $s$ is on a cycle with at least $2^{t/12-8}$ diagonals and on a cycle of length at least $2^{t/4-7}$.*

*If $r \geqq \max\{3, s+1\}+1$, then these edges are even on a cycle with at least $\gamma(r, t)$ diagonals and on a cycle of length at least $\varepsilon(r, t)$.*

For $k \geqq 3$ we have not derived upper bounds for the maximum number of diagonals of cycles of $k$-connected graphs with minimum degree at least $r$ and girth at least $t \geqq 5$.

Theorem 9.3.2 states results on cycles through specified vertices and edges. Many authors have dealt with conditions ensuring the existence of a cycle through a given set of vertices or edges, respectively.

The first result is due to G.A.Dirac, 1960, who observed that in each $k$-connected graph any $k$ vertices lie on a common cycle. M. E. Watkins and D. M. Mesner, 1967, and A.K.Kelmans and M.V.Lomonossow, 1982a, dealt with the additional conditions implying that $k+1$ and $k+2$ vertices are on a common cycle in a $k$-connected graph, respectively. Results concerning regular graphs are due to A.K.Kelmans and M.V.Lomonossow, 1982b, and D.A.Holton, 1982.

The result of G.A.Dirac and Theorem 9.3.2 imply that in every $k$-connected graph with minimum degree at least $r$ and girth at least $t$ any $k-1$ vertices are contained in a cycle with at least $\gamma(3, t)$ diagonals and on a cycle of length at least $\varepsilon(3, t)$. I conjecture that similar results can be derived with respect to $k$, $k+1$, ... vertices in graphs satisfying certain additional conditions.

R.Häggkvist and C.Thomassen, 1982, proved that in every $k$-connected graph any $k-1$ independent edges are in a common cycle (for a history of this problem see R.Häggkvist and C.Thomassen, 1982). A conjecture of L.Lovász is that in every $k$-connected graph any $k$ independent edges are in a common cycle. For $k=3$ L.Lovász and for $k=4$ Petér L.Erdős and E.Györi, 1985, proved this conjecture in the affirmative. A result concerning cycles through four independent edges in 3-connected cubic graphs is due to R.E.L.Aldred, D.R.Holton and C.Thomassen, 1985.

In Theorem 9.3.2 the cardinality $s$ of the set of edges contained in the cycles under consideration is bounded by $k-1$ and $k-1$ is bounded by $r-1$; hence $s \leqq r-1$. In order to overcome this situation we use the concept of cyclic $k$-vertex-connected graphs, as introduced by C.Thomassen, 1983a. A graph $G$ is said to be *cyclically k-vertex-connected* if and only if for any partition of $G = G_1 \cup G_2$ into two subgraphs $G_1$, $G_2$, each containing a cycle, the number of common vertices of $G_1$ and $G_2$ is at least $k$. It is obvious that a cyclically $k$-vertex-connected graph has girth at least $k$, it is cyclically $k$-edge-connected, the minimum degree can be smaller than $k$, and each pair of disjoint cycles is joined by $k$ independent paths (C. Thomassen, 1983a). As C. Thomassen, 1983a, pointed out, each graph-theoretic theorem with hypotheses on the girth and the cyclic connectivity number corresponds to a theorem with hypotheses on the minimum degree and the connectivity number.

Theorem 9.3.4. is a theorem of this type which corresponds to Theorem 9.3.2 (a sketch of the proof can be found in H.-J. Voss, 1986 and 1990).

**Theorem 9.3.4.** *Let $k$, $r$, $t$ be integers with $k \geqq 2$, $r \geqq 3$ and $t \geqq t_0(r, s)$,* where $t_0(r, s)$ *is an appropriate integer. Let $G$ be a cyclically $k$-vertex-connected graph with minimum degree at least $r$ and girth at least $t$. Then each $s$-tuple of edges, $s \leqq k-1$, lying on a cycle of length greater than $s$, is on a cycle of length at least $(r-1)^{t/4-5s-15}$. There is also a cycle with at least $(r-1)^{t/12-5s-15}$ diagonals.* □

## 9.4. Large odd and even cycles with many diagonals in graphs of large girth with given minimum degree

In this section the parity of the cycles under consideration will be included. We shall deal with the question: what are the largest numbers $f_o(k, r, t)$ and $f_e(k, r, t)$, $k \geqq 0$ and $r, t \geqq 3$, such that every k-connected *non-bipartite* graph $G$ with minimum degree at least $r$ and girth at least $t$ contains an odd cycle with at least $f_o(k, r, t)$ diagonals and an even cycle with at least $f_e(k, r, t)$ diagonals, respectively.

Theorem 9.2.4 implies that

$$f_o(2, r, 3) \geqq \frac{1}{2}\binom{r-1}{2} \quad \text{and} \quad f_e(2, r, 3) \geqq \frac{1}{2}\binom{r-1}{2}.$$

With respect to a conjecture of P. Erdős, 1977, relating to $(r+1)$-chromatic graphs, I showed that any 2-connected graph with minimum degree at least 3 and girth at least $17m$ contains a cycle with more than $m$ diagonals; moreover, if $G$ is non-bipartite, then $G$ contains both an odd and an even cycle with more than $m$ diagonals (H.-J. Voss, 1980; in the case when $G$ contains odd cycles, see also H.-J. Voss, 1979). Besides many other interesting properties of graphs with large girth C. Thomassen, 1983a, obtained the same result with lower bound $g(G) \geqq q(m)$ instead of $g(G) \geqq 17m$. Here we shall prove the validity of the lower bound $g(G) > 12(\log m + 8)$. This bound cannot be improved in the sense that the logarithmic function of $m$ can be replaced by a function of $m$ that decreases more slowly, because the graphs $\underline{H}_t(T_3)$ have girth at least $\log m - 2$ and each cycle of $\underline{H}_t(T_3)$ has at most $m$ diagonals (this result can easily be deduced from Theorem 9.1.2). If a 2-connected non-bipartite graph $G$ with minimum degree at least 3 and girth $t$ contains an odd (even) cycle with $m'$ diagonals and no odd (even) cycle with more than $m'$ diagonals, then the result mentioned above implies that the girth $t = g(G) \leqq 12(\log m' + 8)$, i.e. $m' \geqq 2^{t/12-8}$. Hence $G$ contains both an odd and an even cycle with at least $2^{t/12-8}$ diagonals. We shall prove the result mentioned above in this form.

By Theorem 7.6.1, every 2-connected non-bipartite graph of order greater than $2r$, $r \geqq 3$, with minimum degree at least $r$ contains both an odd and an even cycle of length at least $2r-1$. If in such graphs the girth is at least $t$, then they contain both an odd and an even cycle of length at least $\frac{1}{2}(r-1)^{t/4-6}$. Consequently, the lengths of these cycles grow exponentially with the girth $t$.

We remember that $\gamma(r, t) = 2^{t/12-8}$ for $r = 3$ and $\gamma(r, t) = \frac{1}{2}(r-2)^{t/4-8}$ for $r \geqq 4$.

**Theorem 9.4.1.** *Let $G$ be a 2-connected graph with girth at least $t \geqq 3$; besides one exceptional vertex, all vertices of $G$ have valency at least $r \geqq 3$. Then*

(i) *$G$ contains an even cycle with at least $\gamma(r, t)$ diagonals and an even cycle of length at least $\frac{1}{2}(r-1)^{t/4-6}$;*

(ii) *if $G$ is non-bipartite, then $G$ contains an odd cycle with at least $\gamma(r, t)$ diagonals and an odd cycle of length at least $\frac{1}{2}(r-1)^{t/4-6}$.*

This theorem is due to H.-J. Voss, 1980. It will be proved in sections 9.6–9.13.

The lower bounds stated in Theorem 9.4.1 cannot be improved in the sense that the exponential function of $t$ can be replaced by a function of $t$ that increases more rapidly. This is shown by the graphs $\underline{H}_t(T_r)$ and $\underline{H}_t(K_{1,s})$, $s \geqq r$, because by Theorem 9.1.2 the graphs $\underline{H}_t(T_r)$ have only cycles with at most $2(r-1)^{t+1}$ diagonals and the graphs $\underline{H}_t(K_{1,s})$, $s \geqq r$, have only cycles of length at most $4 + (r-1)^{t+1}$ (compare with the comment concerning Theorem 9.3.1 on page 187). Any graph $G$ contains a block $G'$ with at most one cut vertex $A$, where a block is either a maximal 2-connected subgraphs of $G$ or consists of two vertices and the edge joining them.

If $G$ has minimum degree at least $r$, then all vertices of $G' - A$ have degree at least $r$ in $G'$. Consequently, assertion (i) of Theorem 9.4.1 implies the following result:

**Theorem 9.4.2.** *Let G be a graph, not necessarily 2-connected, with minimum degree at least $r \geqq 3$ and girth at least $t \geqq 3$. Then G contains an even cycle with at least $\gamma(r, t)$ diagonals and an even cycle of length at least $\varepsilon(r, t)$.* □

A similar assertion with respect to $A,B$-paths and odd cycles is not true. In order to show this we consider $2k+1$ disjoint bipartite graphs $G_1, \ldots, G_{2k+1}$ with minimum degree at least $r$ and girth at least $2k+2$. We choose a vertex $X_i$ in each $G_i$ and add the circuit $C =_{\text{def}} [X_1, X_2, \ldots, X_{2k+1}, X_1]$. The graph thus obtained has minimum degree at least $r$ and girth $2k+1$. In this graph $X_i$, $X_j$ $(i \neq j)$ are only joined by paths having no diagonals and the only odd circuit $C$ also has no diagonal. If $G$ is 3-connected, then for $r = 3$ the exponent of the lower bounds of the maximum number of diagonals in even and odd cycles can be improved.

**Theorem 9.4.3.** *Let G be a 3-connected graph of girth at least $t \geqq 3$. Then*

(i) *G contains an even cycle with at least $2^{t/8-10}$ diagonals;*

(ii) *if G is non-bipartite, then G contains an odd cycle with at least $2^{t/8-10}$ diagonals.* □

G. A. Dirac, 1960, observed that in any 2-connected non-bipartite graph each vertex is included in some odd cycle. For even cycles this is not true, as can be seen by the graph which is obtained from a $K_{n,m}$ by adding a path $p$ of length 3 with ends $P$, $Q$ so that $K_{n,m} \cap p = \{P, Q\}$ and $P$, $Q$ are in the same bipartition class of $K_{n,m}$. A closely related result is the following (the proof will be published in a forthcoming paper):

**Theorem 9.4.4.** *Let G be a 2-connected non-bipartite graph with minimum degree at least r and girth at least t. Then*

(i) *each vertex is contained in an odd and an even cycle with at least $\gamma(r, t)$ diagonals and in an odd and an even cycle of length at least $\varepsilon(r, t)$;*

(ii) *if G is 3-connected, then with the exception of a subgraph of valency at most 1 each edge is included in both an odd and an even cycle with at least $\gamma(r, t)$ diagonals and in both an odd and an even cycle of length at least $\varepsilon(r, t)$.*

Assertion (ii) implies that with the exception of at most one edge all edges incident with a given vertex are in both an odd and an even cycle with at least $\gamma(r, t)$ diagonals. Let $\Gamma$ be a graph obtained from a 3-connected bipartite graph with minimum degree at least $r$ and girth at least $t$ by adding a new edge $e$ joining two vertices of distance at least $t$ contained in the same bipartition class. Hence $\Gamma$ is a 3-connected non-bipartite graph of minimum degree at least $r$. Since each cycle of $\Gamma$ through $e$ is odd, assertion (ii) of Theorem 9.4.4 cannot be improved.

It is an obvious consequence of Theorem 9.4.1 that in any 2-connected bipartite graph with minimum degree at least $r$ and girth at least $t$ each edge is contained in an even cycle with at least $\gamma(r, t)$ diagonals and in an even cycle of length at least $\varepsilon(r, t)$.

We conclude this section by mentioning that J. A. Bondy and L. Lovász, 1981, proved that any k-connected non-bipartite graph contains an odd cycle through any $k$ specified vertices, as conjectured by B. Toft.

## 9.5. Cycles with many diagonals in $k$-chromatic graphs

According Lemma 8.5.1 we know that

*every graph G with a chromatic number $\chi(G) \geqq 3$ contains a 2-connected non-bipartite subgraph with minimum degree at least $\chi(G) - 1$.*

Thus all results on 2-connected non-bipartite graphs with minimum degree at least $k - 1$, $k \geqq 4$, are also true for $k$-chromatic graphs.

We shall only formulate one result, namely, the assertion for $k$-chromatic graphs which can be deduced from Theorem 9.4.1.

Let us remember that $\gamma(r, t) = 2^{t/12-8}$ for $r = 3$ and $\gamma(r, t) = \frac{1}{2}(r-2)^{t/4-8}$ for $r \geqq 4$.

**Theorem 9.5.1.** *Every graph G with chromatic number at least $k \geqq 4$ and girth at least $t \geqq 3$ contains both an odd and an even cycle with at least $\gamma(k-1, t)$ diagonals and both an odd and an even cycle of length at least $\frac{1}{2}(k-2)^{t/4-6}$.* □

Theorem 9.5.1 answers P. Erdős's conjecture in the affirmative (P. Erdős, 1977; see also section 8.0, 3°):

*Every graph with chromatic number at least 4 containing no triangle (or with girth at least $t \geqq 4$) has an odd cycle with many diagonals.*

With respect to the cycle length in critically $k$-chromatic graphs we know more by Corollary 6.3.8: if $l_k(n)$ denotes the largest integer $\beta$ such that every critically $k$-chromatic graph, $k \geqq 4$, with $n$ vertices has an odd cycle of length at least $\beta$, then $l_k(n) \to \infty$ as $n \to \infty$.

## 9.6. The main lemma on bridges

The aim of the following sections is to prove the theorems of sections 9.3 and 9.4. In the proofs the most important role is played by the main lemma stated here. Its proof will be presented in sections 9.9–9.13.

Notation. Let $H$ be a subgraph of $G$. The subgraph $H$ has *diameter* $\delta$ if each pair of two distinct vertices is joined by a path of length at most $\delta$ and there is no pair of vertices joined by a path of length at most $\delta - 1$.

A path $p$ of $G$ is said to be an *H-path* if and only if one end vertex belongs to $H$ and no other vertices of $p$ belong to $H$.

A path $v$ of $G$ is said to be an *H,H-path* if and only if the two end vertices belong to $H$ and no other vertices of $p$ belong to $H$. The length of the longest $H$-paths in a bridge $\mathfrak{B}$ is denoted by $d(\mathfrak{B})$.

A *triple path* of a subgraph $H$ of a graph $G$ is defined as a path of length 3 in $G$, the middle edge of which is an edge of $H$ and the two end edges are diagonals of $H$.

Now we may formulate our main Lemma 9.6.1.

**Lemma 9.6.1.** *Let $m$, $r$, $t$ be integers with $m \geqq 0$, $r \geqq \max\{3, m+1\}$, and $t \geqq 3$. Let $G$ be a graph with girth $t$; apart from one exceptional vertex $P$, all vertices of $G$ have valency at least $r$.*

*Let $\mathfrak{B}$ be a non-degenerate bridge of a subgraph $H$ in $G$ such that no subgraph of $\mathfrak{B} - H$ of diameter less than $\lfloor t/2 \rfloor - 2$ is joined to $H$ by $m+1$ edges. Then $\mathfrak{B} - H$ consists only of the vertex $P$ of valency $v(G:P) \leqq m$ or $\mathfrak{B} - H$ contains a cycle $C$ with the following properties:*

(i) *the number of vertices of $C$ is at least $(r-1)^{\lfloor t/4 \rfloor - 4} - 1$;*

(ii) *the number of diagonals of $C$ is at least $(r-1)^{\lfloor (t-2)/8 \rfloor - 5} - 1$;*

(iii) *the number of internally disjoint triple paths of $C$ is at least $2^{\lfloor (t-6)/12 \rfloor - 5} - 1$ for $r = 3$ and at least $(r-2)^{(t-1)/4 - 6} - 1$ for $r \geqq 4$, $r \geqq m + 2$.*

The proof will be presented in sections 9.9-9.13. In section 9.8 from Lemma 9.6.1 we shall deduce that $\mathfrak{B}$ contains an $H,H$-path with many diagonals. With the aid of such $H,H$-paths the theorems of sections 9.3 and 9.4 will be proved.

Evidently, if $t$ is small, the assertions of Lemma 9.6.1 are not meaningful. Hence the $H$-bridges had to be studied more carefully. This was done in section 8.7, where for graphs with small girth similar lemmas, namely, Lemmas 8.7.1-8.7.4 were stated: Lemma 8.7.1 was proved there.

## 9.7. Two further lemmas

In the proofs of the theorems of sections 9.3 and 9.4 we also need the following two lemmas. They refer to the case when the cycle $C$ introduced in section 9.6 has a large number of diagonals. The problem consists in finding a path $v$ in $G\langle C \rangle$ connecting given vertices of $C$ so that $v$ has also many diagonals

If three distinct vertices $P$, $Q$, $R$ on $C$ are given, then there is such a path between two of them.

**Lemma 9.7.1.** *Let $P$, $Q$, $R$ be three vertices on a cycle $C$ which has $s$ diagonals. Then there is an arc of $C$ connecting two of these vertices having $\lceil s/3 \rceil$ diagonals.*

We leave to the reader the proof, which can easily be performed by pigeonhole arguments. □

The cycle $C$ of Fig. 9.7.1 has many diagonals and two given vertices $P$ and $Q$ such that each $P,Q$-path of $G\langle C \rangle$ has no diagonal.

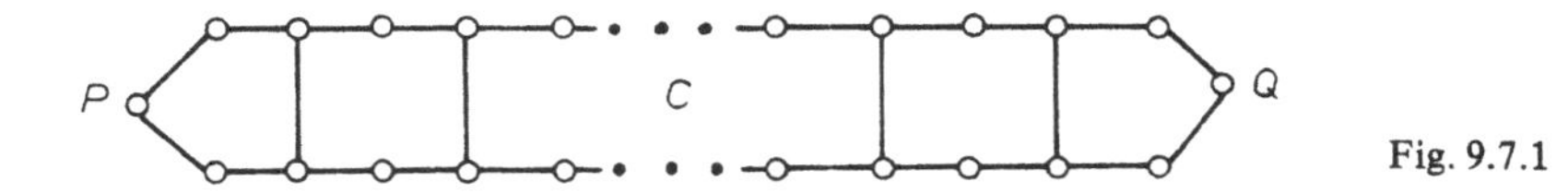

Fig. 9.7.1

Fortunately, in our proofs in section 9.8 a second condition concerning the diagonals of $C$ will occur: the diagonals can be arranged into disjoint pairs so that the diagonals of each pair are joined by an edge of $C$. The two diagonals of a pair together with the linking edge of $C$ form a so-called *triple path* of $C$.

**Lemma 9.7.2.** *Let $C$ be a cycle of a graph $G$. Let $G\langle C \rangle$ contain $\varrho$ internally disjoint triple paths of $C$. Then between any two different vertices $P$ and $Q$ of $C$ there is a $P,Q$-path of $G\langle C \rangle$ having at least $\lceil \varrho/2 \rceil$ diagonals.*

*Proof.* Let $C$ satisfy the hypotheses of the lemma. The cycle $C$ is split by $P$ and $Q$ into two arcs denoted by $v$ and $\bar{v}$. The notation $v$, $\bar{v}$ can be chosen so that $G\langle C\rangle$ contains $\lceil \varrho/2\rceil$ of the internally disjoint triple paths of $C$, the inner edges of them belonging to $v$. We denote these paths by $q_i = [P_1^i, P_2^i, P_3^i, P_4^i]$ $(1 \leqq i \leqq \lceil \varrho/2\rceil)$. Let the indices $i$ be chosen so that $P_1^i, P_4^i \in \bar{v} - \{P, Q\}$ *for all* $1 \leqq i \leqq \varkappa$ and $P_1^j \in v$ or $P_4^j \in v$ for all $\varkappa + 1 \leqq j \leqq \lceil \varrho/2\rceil$. Since for each $i$ with $1 \leqq i \leqq \varkappa$ the end vertices $P_1^i$, $P_4^i$ *of* $q_i$ lie on $\bar{v}$, they determine a unique partial path $\bar{v}[P_1^i, P_4^i]$ of $\bar{v} - \{P, Q\}$. Thus we have a set $\Sigma$ of $\varkappa$ partial paths of $\bar{v} - \{P, Q\}$. Among all subsets of pairwise vertex-disjoint paths of $\Sigma$ let $\Sigma^*$ denote one of maximum total length. We shall prove that

(1) *for each* $1 \leqq i \leqq \varkappa$ *at least one of the two end vertices* $P_1^i$, $P_4^i$ *of* $q_i$ *is contained in a path of* $\Sigma^*$.

*Proof of (1).* Suppose that (1) is not true. Let $j$ be an index with $1 \leqq j \leqq \varkappa$ such that neither $P_1^j$ nor $P_4^j$ is contained in a path of $\Sigma^*$. Then we add $\bar{v}[P_1^j, P_4^j]$ to $\sum^*$ and delete from $\Sigma^*$ all paths which are included in $\bar{v}[P_1^j, P_4^j]$. The set of vertex-disjoint paths of $\Sigma$ obtained in this way has a total length which is larger than the total length of $\Sigma^*$. This contradicts the maximality of $\Sigma^*$. □

Now let the paths $[P_1^i, \ldots, P_4^i]$ $(1 \leqq i \leqq \varkappa)$ be numbered so that

$$\Sigma^* = \{\bar{v}[P_1^i, P_4^i] | 1 \leqq i \leqq f\}.$$

We define: $C_i =_{\text{def}} [P_1^i, P_2^i, P_3^i, P_4^i] \cup \bar{v}[P_4^i, P_1^i]$ $(1 \leqq i \leqq f)$. The $C_i$'s are $f$ pairwise disjoint cycles. $u =_{\text{def}} v \triangle \left( \overset{f}{\underset{i=1}{\triangle}} C_i \right)$ is a $P,Q$-path of $G\langle C\rangle$. For all $1 \leqq i \leqq f$ the edge $(P_2^i, P_3^i)$ is a diagonal of $u$. For all $f + 1 \leqq i \leqq \lceil \varrho/2\rceil$ at least one of the two edges $(P_1^i, P_2^i)$, $(P_3^i, P_4^i)$ is a diagonal of $u$. Consequently, the $P,Q$-path $u$ has at least $\lceil \varrho/2\rceil$ diagonals. □

## 9.8. The proofs of theorems of sections 9.3-9.4

The aim of this section is to prove Theorems 9.3.2, 9.4.1 and 9.4.3. These theorems will be proved by using the main Lemma 9.6.1 and the two Lemmas 9.7.1 and 9.7.2. For these purposes we consider the class of all $k$-connected graphs of girth $t$ in which apart from perhaps one exceptional vertex $P$ all vertices have valency at least $r$, where $k \geqq 2$, $r \geqq \max\{3, k\}$ and $t \geqq 3$ if $s \geqq 2$, and $t \geqq 25$ if $s \geqq 3$. Let $e_1, \ldots, e_s$ be $s$ given edges, $0 \leqq s \leqq k - 1$, lying on a cycle of length at least $\max\{s + 1, 3\}$.

If $s \geqq 1$ or $s = 0$ and $G$ is bipartite, let $H$ denote a shortest cycle through $e_1, \ldots, e_s$ or an arbitrary shortest cycle, respectively. If $s = 0$ and $G$ is non-bipartite, let $H$ be a smallest odd cycle. Obviously, if $s \leqq 1$, then $H$ has no diagonal.

Case 1. Let $G - H$ consist of at most the vertex $P$. Let $M$ denote the set of all vertices of $H$ having distance at most $\lfloor t/2\rfloor - 2$ from a fixed vertex $W$. If $G$ contains $P$, we choose $W = P$. Then the induced subgraph $G\langle M\rangle$ of $G$ is a tree of height $\lfloor t/2\rfloor - 2$ and root $W$. Since $G$ is 2-connected, $v(G:P) \geqq 2$. By hypothesis all other vertices have valency at least $r$. Simple arguments show that the tree $G\langle M\rangle$ has at least

$$1 + 2 + \sum_{i=1}^{\lfloor t/2\rfloor - 3} (r-1)^{i-1} \geqq 3 + (r-1)^{\lfloor t/2\rfloor - 3}$$

vertices.

Since $G\langle M\rangle - W \subseteqq H$, the vertex number $n$ of $H$ (the length $n$ of $H$) is $n \geqq (r-1)^{t/2-3}+2$.

If $P \notin H$, then the number $e$ of edges of $G$ is $e \geqq \frac{r}{2}n + 1$ because all vertices $V$, $V \neq P$, of $G$ have valency at least $r$ and $v(G:P) \geqq 2$. Since $H$ contains at most $n$ edges and $v(G:P) \leqq r-1$, the cycle $H$ has at least

$$e - n - (r-1) \geqq \frac{r}{2}n - n - r + 2 = \frac{r-2}{2}n - r + 2$$

$$\geqq \frac{r-2}{2}(2 + (r-1)^{\lfloor t/2\rfloor - 3}) - r + 2 = \frac{r-2}{2}(r-1)^{t/2-3}$$

diagonals. If $P \in H$, then this inequality can be achieved in a similar way.

Case 2. Let $G - H$ contain a vertex different from $P$. Then $H$ has a non-degenerate bridge $\mathfrak{B}$ such that $\mathfrak{B} - H$ contains a vertex different from $P$.

(1) *No subgraph of $\mathfrak{B} - H$ of diameter less than $\lfloor t/2\rfloor - 2$ is joined to $H$ by $s+1$ edges if $s \geqq 1$, and by two edges if $s = 0$.*

To prove (1) suppose that there is a subgraph $S$ of $\mathfrak{B} - H$ of diameter less than $\lfloor t/2\rfloor - 2$ being joined to $H$ by at least $s+1$ edges if $s \geqq 1$, or by at least two edges if $s = 0$.

If $s \geqq 1$, then $H - \{e_1, \ldots, e_s\}$ consists of $s$ arcs (it may be that some arcs have length 0). Hence there is an arc joined by two edges $k_1$, $k_2$ to $S$. Consequently, for $s \geqq 0$, let $a$ and $a'$ denote the arcs of $H$ joining the end vertices of $k_1$, $k_2$ so that $a$ and $a'$ contain all edges $e_1, \ldots, e_s$ or none of these edges, respectively (we allow that $\lambda(a') = 0$).

Let $v$ denote a shortest path of $S$ connecting the two end vertices of $k_1$, $k_2$ in $S$. By the supposition we have $\lambda(v) < \lfloor t/2\rfloor - 2$. If $\lambda(a) \leqq \lambda(v) + 2$ or $\lambda(a') \leqq \lambda(v) + 2$, then $a \cup k_1 \cup v \cup k_2$ or $a' \cup k_1 \cup v \cup k_2$ is a cycle of length at most $2(\lambda(v) + 2) < 2\lfloor t/2\rfloor \leqq t$. This contradicts $g(G) \geqq t$!

Now let $\lambda(a)$, $\lambda(a') > \lambda(v) + 2$. If $s \geqq 1$ or $s = 0$ and $G$ is bipartite, then $a \cup k_1 \cup v \cup k_2$ is a cycle through $e_1, \ldots, e_s$ of length $\lambda(a) + \lambda(v) + 2 < \lambda(a) + \lambda(a') = \lambda(H)$. If $s = 0$ and $G$ is non-bipartite, then $a \cup k_1 \cup v \cup k_2$ or $a' \cup k_1 \cup v \cup k_2$ is an odd cycle of length less than $\lambda(a) + \lambda(a') = \lambda(H)$. In both cases the result obtained contradicts the minimality of $H$!

By Lemma 9.6.1 there is a cycle $C$ in $\mathfrak{B} - \mathfrak{H}$ with the following properties:

(2) *The length of $C$ is at least $(r-1)^{\lfloor t/4\rfloor - 4} - 1$.*

(3) *The number of diagonals of $C$ is at least $(r-1)^{\lfloor (t-2)/8\rfloor - 5} - 1$.*

(4) *The number of internally disjoint triple paths of $C$ is at least $2^{\lfloor (t-6)/12\rfloor - 5} - 1$ for $r = 3$ and at least $(r-2)^{(t-1)/4-6} - 1$ for $r \geqq 4$, $r \geqq s+2$.*

If $s \geqq 1$, then $H - \{e_1, \ldots, e_s\}$ consists of $s$ arcs. By hypothesis $\lambda(H) \geqq s+1$ and $H - \{e_1, \ldots, e_s\}$ has at least $s+1$ vertices. $t \geqq 25$ and assertion (2) imply that $\lambda(C) \geqq r \geqq s+1$. Consequently, in all cases when $s \geqq 0$ the lengths $\lambda(H)$ and $\lambda(C)$ are at least $\max\{s+1, 3\}$.

Since $G$ is $k$-connected, $k \geqq \max\{s+1, 2\}$, there are $\max\{s+1, 2\}$ vertex-disjoint $H,C$-paths. Since $H - \{e_1, \ldots, e_s\}$ consists of $s$ arcs, there is an arc joined by two of these paths to the cycle $C$, say $w_1$ and $w_2$.

Let $a$ and $a'$ denote the arcs of $H$ joining the end edges of $w_1$ and $w_2$ so that $a$ and $a'$ contain all edges $e_1, \ldots, e_s$ or none of these edges, respectively. Let $P$ and $Q$ denote the two vertices of $w_1$ and $w_2$ on $C$, respectively. Let $v$ denote the larger arc of $C$ with ends $P$ and $Q$. Then by (2) the cycles $a \cup w_1 \cup v \cup w_2$ and $a' \cup w_1 \cup v \cup w_2$ have length at least $\bar{\varepsilon}(r, t) = \frac{1}{2}(r-1)^{\lfloor t/4 \rfloor - 4} - 1$.

If $s \geqq 1$ or $s = 0$ and $G$ is bipartite, then $a \cup w_1 \cup v \cup w_2$ is a cycle through $e_1, \ldots, e_s$ of length at least $\bar{\varepsilon}(r, t)$. If $s = 0$ and $G$ is non-bipartite, then $a \cup w_1 \cup v \cup w_2$ and $a' \cup w_1 \cup v \cup w_2$ are an odd and an even cycle of length at least $\bar{\varepsilon}(r, t)$. By Lemma 9.7.2 the induced subgraph $G\langle C \rangle$ contains a $P,Q$-path $\bar{v}$, the number of diagonals of which is not less than half the number of triple paths of $C$. Then by (4) the cycles $a \cup w_1 \cup v \cup w_2$ and $a' \cup w_1 \cup v \cup w_2$ have at least $\bar{\gamma}(r, t)$ diagonals, where $\bar{\gamma}(3, t) = 2^{\lfloor (t-6)/12 \rfloor - 6} - 1$ if $r = 3$, and $\bar{\gamma}(r, t) = \frac{1}{2}(r-2)^{t/4-6} - 1$ if $r \geqq 4$, $r \geqq s + 2$.

If $s \geqq 1$ or $s = 0$ and $G$ is bipartite, then $a \cup w_1 \cup v \cup w_2$ is a cycle through $e_1, \ldots, e_s$ with at least $\bar{\gamma}(r, t)$ diagonals. If $s = 0$ and $G$ is non-bipartite, then $a \cup w_1 \cup v \cup w_2$ and $a' \cup w_1 \cup v \cup w_2$ are an odd and an even cycle with at least $\bar{\gamma}(r, t)$ diagonals.

Thus the proofs of Theorems 9.3.2 and 9.4.1 are complete.

In order to prove Theorem 9.4.3 we shall consider the following

Subcase $r = 3$, $k = 3$, $s = 0$. Since $\lambda(H), \lambda(C) \geqq 3$ and $G$ is 3-connected, there are three vertex-disjoint $H,C$-paths $w_1, w_2, w_3$. By Lemma 9.7.1 there are two end vertices $P$, $Q$ of $w_1, w_2, w_3$ on $C$ such that at least one third of the diagonals of $C$ are also diagonals of one of the two $P,Q$-arcs of $C$. Let this $P,Q$-arc be denoted by $v$. Without loss of generality, let $P$ and $Q$ be end vertices of $w_1$ and $w_2$. Let $a$ and $a'$ be the above defined arcs of $H$. Then (3) implies that $a \cup w_1 \cup v \cup w_2$ and $a' \cup w_1 \cup v \cup w_2$ are an odd and an even cycle, respectively, with at least $\frac{1}{3} 2^{\lfloor (t-2)/8 \rfloor - 5} - 1$ diagonals. Thus Theorem 9.4.3 is also proved. □

## 9.9. Begin of the proof of Lemma 9.6.1. Construction $C_S$ of the tree $D(p)$

In the remaining sections 9.9–9.13 of Chapter 9 we shall prove Lemma 9.6.1. The aim of the proof is to find a cycle $C$ with many internally disjoint triple paths of $C$ (note that if $C$ has many internally disjoint triple paths, it has also many vertices and diagonals).

We start with a longest $A$-path $p$ of $\mathfrak{B} - H$ of length at least 1, where $A$ is a fixed vertex of $\mathfrak{B} - H$.

In this section we prove that $G$ contains many internally disjoint triple paths of $p$. For this purpose we construct a tree $D(p)$ which contains many internally disjoint triple paths of $p$, each branch vertex $V$ of $D(p)$ has valency $v(D(p): V) = v((\mathfrak{B} - H): V) - 1$. First we present some notation.

The vertex $A$ of the $A$-path $p$ is called its *origin* and the other end vertex of $p$, the *terminus* of $p$. If $X$, $Y$, $Z$ are vertices of $p$ with $[X, Y, Z] \subseteqq p[A, Z]$, then $X$ and $Z$ are called the *predecessor* and the *successor* of $Y$ on $p$, respectively.

We begin with the following two assertions.

**Proposition 9.9.1.** *All neighbours of the terminus E of any longest A-path p lie on $p \cup H$.*

*Proof.* Suppose that $E$ has some neighbour $E'$ not on $p \cup H$. Then the $A$-path $p \cup [E, E'] \subseteqq B - H$ is longer than $p$, a contradiction! □

The following statement can also easily be proved:

**Proposition 9.9.2.** *Let $E$ be the terminus of a longest $A$-path $p$ which has a neighbour $E_i$ on $p$ with $[E_i, E] \nsubseteq p$. Let $\bar{E}_i$ denote the successor of $E_i$ on $p$. Then $\bar{E}_i$ is the terminus of the longest $A$-path $p \triangle [E, E_i, \bar{E}_i]$.* □

Later $E_i$ will be referred to as an intermediate vertex. Now let $G$ have girth $t \geqq 3$.

Proposition 9.9.2 shows how to replace a longest $A$-path $p$ by a new one, say $p_i$. Here the path $[E, E_i, \bar{E}_i]$ of length 2 is used, which consists of a diagonal and of an edge of $p$ (namely $(E, E_i)$ and $(E_i, \bar{E}_i)$, respectively). For convenience we assume that each vertex $X$ of $\mathfrak{B} - H$ has valency $v((\mathfrak{B} - H): X) \geqq 3$. Therefore the terminus $E$ of $p$ in Proposition 9.9.2 has at least one neighbour $E^1$ such that $[E^1, E] \nsubseteq p$. By Proposition 9.9.2 the path $p^1 =_{\text{def}} p \triangle [E, E^1, \bar{E}^1]$ is a longest $A$-path with terminus $\bar{E}^1$. Since $v(G - H: \bar{E}^1) \geqq 3$, there is a neighbour $E^2$ of $\bar{E}^1$ on $p^1$ with $E^2 \neq E^1$ and $[\bar{E}^1, E^2] \nsubseteq p^1$. Then $p =_{\text{def}} p^1 \triangle [\bar{E}^1, E^2, \bar{E}^2] = p \triangle [E, E^1, \bar{E}^1, E^2, \bar{E}^2]$ is a longest $A$-path with terminus $\bar{E}^2$. We recursively adapt this procedure until the path $v =_{\text{def}} [E, E^1, \bar{E}^1, \ldots, E^s, \bar{E}^s]$ has length $2s \in \{t-2, t-1\}$.

On the path $v$ diagonals and edges of $p$ alternate, and for all $1 \leqq j \leqq s$ the path $p \triangle [E, E^1, \bar{E}^1, \ldots, E^j, \bar{E}^j]$ is a longest $A$-path with terminus $\bar{E}^j$.

Let $p$ be again a longest $A$-path and $E$ its terminus. In Proposition 9.9.2 only one neighbour of $E$ on $p$ has been used. Now we take into consideration all neighbours of $E$ on $p$ besides the only one which precedes $E$ on $p$. Instead of the 2-path $[E, E_i, \bar{E}_i]$ a subgraph $D^0(p)$ of $G\langle p \rangle$ is obtained, where $G\langle p \rangle$ again denotes the subgraph of $G$ induced by the path $p$.

**Construction $C_S^0$ of $D^0(p)$.** *Let $E$ be the terminus of a longest $A$-path $p$ and $E_1, \ldots, E_{v-1}, E_v$ be its neighbours on $p$, where $E_v$ is the predecessor of $E$ on $p$. Let $\bar{E}_i$ denote the successor of $E_i$ on $p$. Then $D^0(p) =_{\text{def}} \bigcup_{i=1}^{v-1} [E, E_i, \bar{E}_i]$, and by Proposition 9.9.2 the vertex $\bar{E}_i$ is the terminus of the longest $A$-path $p \triangle [E, E_i, \bar{E}_i] = p \triangle D^0(p)[E, \bar{E}_i]$ for all $1 \leqq i \leqq v-1$.* □

Now let $g(G) = t \geqq 9$. Then $D^0(p)$ is a tree, and $E$ is called its *root.* The vertices $E_1, \ldots, E_{v-1}$ are called the *intermediate vertices* and $E, \bar{E}_1, \ldots, \bar{E}_{v-1}$ the *main vertices* of $D^0(p)$. Since the construction $C_S^0$ can be adapted to any longest $A$-path, we apply $C_S^0$ to $p_i =_{\text{def}} p \triangle D^0(p)[E, \bar{E}_i]$ for all $1 \leqq i \leqq v-1$. Going on in this way, we obtain for $g(G) = t \geqq 9$ a tree $D(p)$ with (even) height at most $2\lfloor t/4 \rfloor - 2$.

**Construction $C_S$ of $D(p)$.** *Let $p$ be a longest $A$-path and $W$ its terminus.*

Step 1. *Put $D_1 =_{\text{def}} D^0(p)$.*

Step 2. *Let $D_i$ be already constructed and have the property*

(P) *$p \triangle D_i[W, X]$ is a longest $A$-path with terminus $X$, for all end vertices $X$ of $D_i$.*

*Choose an end vertex $X_i$ of $D_i$ of rank at most $2\lfloor t/4 \rfloor - 4$ with $v((\mathfrak{B} - H): X_i) \geqq 3$ which is not already checked. Put $D_{i+1} =_{\text{def}} D_i \cup D^0(p_i)$, where $p_i =_{\text{def}} p \triangle D_i[W, X_i]$.*

Step 3. *Stop when step 2 cannot be implemented further, and put $D(p) =_{\text{def}} D_i$.* □

Note that if in step 2 the valency $v((\mathfrak{B} - H): X_i) = 2$, then $D_{i+1} = D_i \cup D^0(p_i) = D_i$.

Next we have to prove that the construction $C_S$ works. It suffices to show that $D_{i+1}$ has again the property (P). The tree $D_i$ has (P) by induction, and from Proposition 9.9.2 it follows that $D^0(p_i)$ has also the property (P), where $p_i = p \triangle D_i[W, X_i]$. The only end vertices of $D_{i+1}$ which are not in $D_i$ are the end vertices of $D^0(p_i)$.

Let $Y$ denote an end vertex of $D^0(p_i)$. Then

$$p \triangle D_{i+1}[W, Y] = p \triangle (D_i[W, X_i] \triangle D^0(p_i)[X_i, Y]) = p_i \triangle D^0(p_i)[X_i, Y].$$

By induction, $p_i$ is a longest $A$-path with terminus $X_i$, and from Proposition 9.9.2 it follows that $p_i \triangle D^0(p_i)[X_i, Y]$ is a longest $A$-path with terminus $Y$. Therefore $p \triangle D_{i+1}[W, Y]$ is also a longest $A$-path with terminus $Y$. This implies that $C_S$ works.

By $C_S$ a sequence $D_1, \ldots, D_i, D_{i+1}, \ldots, D(p)$ of trees in $G\langle p\rangle$ is recursively obtained, where $G\langle p\rangle$ is completely contained in the kernel of an $H$-bridge $\mathfrak{B}$ of $G$.

The vertex $W$ of $D_j$ and $D(p)$ is called their *root*. A vertex of $D_j$ and $D(p)$ is called an *intermediate* or a *main* one if it is an intermediate vertex or a main vertex of some tree $D^0(p_i)$, respectively. The root $W$ is also referred to as a main vertex. A vertex $V$ of $D_j$ or $D(p)$ has rank $q$ if the unique path $D(p)[W, V]$ has length $q$. An edge joining a vertex of rank $q$ to a vertex of rank $q+1$ in $D_j$ or $D(p)$ is called a *$q,q+1$-edge.*

By induction on $i$ and by Proposition 9.9.2, the following statement can easily be proved:

**Proposition 9.9.3.** *Let $G$ be a graph of girth $g(G) = t \geqq 9$ and $H$ one of its subgraphs. Suppose that $\mathfrak{B}$ is an $H$-bridge of $G$ and $A$ is a vertex of $\mathfrak{B} - H$. Let $p$ be a longest $A$-path in $\mathfrak{B} - H$ and $W$ its terminus. Then $G\langle p\rangle$ contains a tree $D =_{\text{def}} D(p)$ with root $W$ and height at most $2\lfloor t/4\rfloor - 2$ having the following properties:*

(i) *for each main vertex $X$ of $D$, i.e., for each vertex $X$ of even rank, $p \triangle D[W, X]$ is a longest $A$-path with terminus $X$;*

(ii) $$v(D:X) = \begin{cases} 2 & \text{if } X \text{ is an intermediate vertex,} \\ v(G\langle p\rangle : X) - 1 & \text{if } X \text{ is an main vertex and } X \text{ is no end vertex of } D, \\ 1 & \text{if } X \text{ is an end vertex of } D; \end{cases}$$

(iii) *every $2s-2, 2s-1$-edge and every $2s-1, 2s$-edge of $D$ is a chord or an edge of $p$, respectively.*

We remark that $D(p)$ possibly consists only of the vertex $W$.

## 9.10. Construction $\bar{C}_S$ of the tree $T(p)$

When we constructed $D^0(p)$ we have used all neighbours of $E$ on $p$ apart from the predecessor of $E$ on $p$. Hence for each main vertex $X$ of $D(p)$ which is not an end vertex of $D(p)$ the valency $v(D:X)$ is equal to $v((\mathfrak{B} - H):X) - 1 = v(G\langle p\rangle : X) - 1$. If $r \geqq 4$, then in general $v(D:X) \geqq r - 1 \geqq 3$. In the case when $r = 3$ the *branch vertices* of $D$ have only valency at least 2, and the tree $D(p)$ has not enough vertices. Therefore in section 9.10 we shall construct a second tree $T(p)$ of $\mathfrak{B} - H$, each branch vertex $V$ of $T(p)$ has valency $v(T(p):V) = v(\mathfrak{B} - H:V)$. Hence in the case when $r = 3$ the valency $v(T(p):V)$ is also equal to $v((\mathfrak{B} - H):V) \geqq r \geqq 3$.

The construction of $T(p)$ is more involved than that of $D(p)$. For instance, the tree $T = T(p)$ contains vertices of $\mathfrak{B} - H$ which are not in $p$. These vertices of $(\mathfrak{B} - H) - p$ do not belong to any triple path of $p$. Hence we shall present a condition when a path of $T$ contains a triple path of $p$. We start with the

**Construction $\bar{C}_S^0$ of $T^0(w)$.** *Let $E$ be the terminus of any A-path $w$ of length at least 1, and let $E_1, \ldots, E_{v-1}, E_v$ denote the neighbours of $E$ in $\mathfrak{B} - H$ so that $E_1, \ldots, E_q, E_v$ are on $w$, where $E_v$ is the only predecessor of $E$ on $w$, and $E_{q+1}, \ldots, E_{v-1}$ are not on $w$. For all $1 \leqq j \leqq q$ let $\bar{E}_j$ denote the successor of $E_j$ on $p$. Then*

$$T^0(w) =_{\text{def}} \bigcup_{j=1}^{q} [E, E_j, \bar{E}_j] \cup \bigcup_{k=q+1}^{v} [E, E_k].$$

Using Proposition 9.9.2 we can easily prove the following result:

**Proposition 9.10.1.** *For $1 \leqq j \leqq q$, the vertex $\bar{E}_j$ is the terminus of the A-path $w \,\Delta\, [E, E_j, \bar{E}_j]$ of length $\lambda(w)$. For $q+1 \leqq k \leqq v-1$, the vertex $E_k$ is the terminus of the A-path $w \,\Delta\, [E, E_k]$ of length $\lambda(w)+1$ and the vertex $E_v$ is the terminus of the A-path $w \,\Delta\, [E, E_v]$ of length $\lambda(w)-1$.*

In the further proof we shall use the following notation: $I^0(w) =_{\text{def}} \{E_j \mid 1 \leqq j \leqq q\}$ is called the set of intermediate vertices. The vertex $E$ is the root of $T^0(w)$. The remaining vertices are called main vertices, where $M_1^0(w) =_{\text{def}} \{\bar{E}_j \mid 1 \leqq j \leqq q\}$, $M_2^0(w) =_{\text{def}} \{E_k \mid q+1 \leqq k \leqq v-1\}$ and $M_3^0 =_{\text{def}} \{E_v\}$ are sets of main vertices of type 1, type 2 and type 3, respectively. Thus with Proposition 9.10.1 we can state that if $X$ is a main vertex of $T^0(w)$ or $X = E$, then $w' =_{\text{def}} w \,\Delta\, T^0(w)\,[E, X]$ has length

$$\lambda(w') = \begin{cases} \lambda(w) & \text{if } X \in M_1^0 \cup \{E\}, \\ \lambda(w)+1 & \text{if } X \in M_2^0, \\ \lambda(w)-1 & \text{if } X \in M_3^0. \end{cases}$$

If $p$ is a longest $A$-path, then $D^0(p)$ is the subgraph of $T^0(p)$ induced by $\{E\} \cup M_1^0(p) \cup I^0(p)$.

Now we shall describe the construction $\bar{C}_S$ of the tree $T(p)$. This method is similar to that of constructing $D(p)$: we have only to replace $D^0(p_i)$ by $T^0(p_i)$ in $C_S$ and to take into consideration that the $A$-path $p_i$ is not necessarily a longest $A$-path.

**Construction $\bar{C}_S$ of $T(p)$.** *Let $p$ be a longest A-path and $W$ its terminus, and let $h =_{\text{def}} \lfloor t/2 \rfloor - 1$.*

Step 1. *Put $T_1 =_{\text{def}} T^0(p)$.*

Step 2. *Let $T_i$ be already constructed and have the property*

($\bar{\text{P}}$) *$p \,\Delta\, T_i[W, X]$ is an A-path with terminus $X$, for all end vertices $X$ of $T_i$.*

*Choose an end vertex $X_i$ of $T_i$ of rank at most $h-2$ with $v((\mathfrak{B} - H) : X_i) \geqq 2$. Put $T_{i+1} =_{\text{def}} T_i \cup T^0(p_i)$, where $p_i =_{\text{def}} p \,\Delta\, T_i[W, X_i]$.*

Step 3. *Stop when step 2 cannot be implemented further, and put $T(p) =_{\text{def}} T_i$.*

Note that if in step 2 the valency $v((\mathfrak{B} - H) : X_i)$ is equal to 2, then

$$T_{i+1} = T_i \cup T^0(p_i) = T_i \,\Delta\, [X_i, X_i'],$$

where $X_i'$ is the predecessor of $X_i$ on $p_i$.

With the aid of Proposition 9.10.1 it can readily be shown that the construction $\bar{C}_S$ works. For convenience we write $T =_{\text{def}} T(p)$. Let $I =_{\text{def}} \bigcup_i I^0(p_i)$ be the set of all intermediate vertices of $T$, put $M_j =_{\text{def}} \bigcup_i M_j^0(p_i)$, and $M =_{\text{def}} \{W\} \cup M_1 \cup M_2 \cup M_3$ the set of all main vertices of $T$. If $X \in M$, then $\alpha_j(X)$ denotes the number of all main vertices of $T[W, X]$ belonging to $M_j$. Let $\alpha(X) = \alpha_1(X) + \alpha_2(X) + \alpha_3(X)$. Thus $\alpha(X) + 1$ is the number of main vertices of $T[W, X]$.

We remark that since $p$ is a longest $A$-path, $D(p)$ is the component of the subgraph of $T = T(p)$ induced by $\{W\} \cup M_1 \cup I$ which contains $W$.

By $\bar{C}_S$ a sequence $T_1, \ldots, T_i, T_{i+1}, \ldots, T(p)$ of trees of the kernel of an $H$-bridge $\mathfrak{B}$ is recursively obtained, where $\mathfrak{B}$ contains the path $p$. Hence, by induction on $i$ and by Proposition 9.10.1, the following statement can easily be shown:

**Proposition 9.10.2.** *Let $G$ be a graph of girth $g(G) = t \geqq 9$ and $H$ one of its subgraphs, and let $\mathfrak{B}$ be an $H$-bridge of $G$ and $A$ a vertex of $\mathfrak{B} - H$. Suppose that $p$ is a longest $A$-path in $\mathfrak{B} - H$ and $W$ is its terminus. Then $\mathfrak{B}$ contains a tree $T =_{\text{def}} T(p)$ with root $W$ and height at most $\lfloor t/2 \rfloor - 1$ having the following properties:*

(i) *for each $X \in M$, i.e., for each main vertex $X$, $p \Delta T[W, X]$ is an $A$-path with terminus $X$ of length $\lambda(p) - \alpha_3(X) + \alpha_2(X)$;*

(ii) $$v(T:X) = \begin{cases} 2 & \text{if } X \in I, \\ v((\mathfrak{B} - H):X) & \text{if } X \in M \text{ and } X \text{ is not an end vertex of } T, \\ 1 & \text{if } X \text{ is an end vertex of } T. \ \square \end{cases}$$

Since $p$ is maximum, the assertion (i) of Proposition 9.10.2 implies the following result:

**Corollary 9.10.3.** $\alpha_2(X) \leqq \alpha_3(X)$. □

The tree $D(p)$ only contains vertices of $p$; furthermore, chords and edges of $p$ alternate on $D(p)$. Unfortunately, the tree $T(p)$ has not such nice properties. But vertices, edges and chords of $p$ being in $T(p)$ play an important role in the proofs. Therefore we formulate a condition which states when a partial path of $T(p)$ contains a vertex or an edge or a chord of $p$ or a triple path of $p$, where a triple path of $p$ is defined as a path of length 3 in $G$ whose middle edge is an edge of $p$ and whose two end edges are diagonals of $p$. Two main vertices $Z_1, Z_2$ are said to be *consecutive* on $T[W, Z_2]$ if $T[Z_1, Z_2] \subseteqq T[W, Z_2]$ has length at most 2, and if $\lambda(T[Z_1, Z_2]) = 2$, then the only vertex between $Z_1$ and $Z_2$ is an intermediate vertex of $I$.

**Proposition 9.10.4.**

(i) *If $Z \in M_1 \cup M_3$, then $Z \in p$.*

(ii) *If $Z_1, Z_2 \in M_1$ are two consecutive main vertices of $T[W, Z_2]$, then $T[Z_1, Z_2]$ contains a chord of $p$.*

(iii) *If $Z_1, Z_2, Z_3 \in M_1$ are three consecutive main vertices of $T[W, Z_3]$, then $T[Z_1, Z_3]$ contains a triple path of $p$.*

*Proof.* Obviously, if $Z \in M_3$, then $Z \in p$. In order to complete the proof it suffices to show:

(1) *If $Z \in M_1$, then the unique path $[X, Y, Z] \subseteqq T[W, Z]$ has the property that $(X, Y)$ is a chord of $p \Delta T[W, X]$ and $(Y, Z)$ is an edge of $p$.*

The validity of (1) follows from the construction $\bar{C}_S$, especially from $\bar{C}_S^0$. □

## 9.11. Walks through $T$ in bridges with property $(Q_m)$

In the remaining sections of Chapter 9 we investigate trees $T = T(p)$ in graphs and bridges satisfying the hypotheses of Lemma 9.6.1.

Before formulating the content and the aim of section 9.11, we shall introduce the concept of a branch of the tree $T$. Let $V$ be a vertex of $T$ of rank greater than 0, and let $V'$ be the neighbour of $V$ with $(V', V) \in T[W, V]$. The component of $T - (V', V)$ containing $V$ is called the *branch* $T_V$ of $T$. This branch can again be considered to be a (rooted) tree with root $V$. Let $T_W =_{\text{def}} T$. Unfortunately, the tree $T = T(p)$ contains vertices of $\mathfrak{B} - H$ which are not in $p$. These vertices of $(\mathfrak{B} - H) - p$ do not belong to any triple path of $p$. Hence we have to look at subtrees of $T$ which contain a triple path of $p$. Indeed we shall prove that for each main vertex $V$ of $T$ the branch $T_V$ has an end vertex $E$ such that 1) $E$ has rank at least $h - 1$ in $T$, where $h$ denotes the height of the tree $T$; 2) the path $T_V[V, E] - \{V\}$ contains at most two main vertices of $M_3$; 3) if the distance of $W$ and $V$ in $T$ is sufficiently small, then $T_V[V, E]$ contains a triple path of $p$.

In the remaining sections of Chapter 9 we only consider graphs that satisfy the hypotheses of Lemma 9.6.1, which are repeated here: let $G$ be a graph with girth $t \geqq 3$; apart from one exceptional vertex $P$ all vertices of $G$ have valency at least $r \geqq 3$. Let $B$ be a non-degenerate bridge of a subgraph $H$ in $G$. The $H$-bridge $B$ is said to have the property.

$(Q_m)$ if no subgraph of $\mathfrak{B} - H$ of diameter less than $\lfloor t/2 \rfloor - 2$ is joined to $H$ by $m + 1$ edges $(m \leqq r - 1)$.

If $P$ is a vertex of $\mathfrak{B} - H$, then put $A =_{\text{def}} P$, otherwise let $A$ be any vertex of $\mathfrak{B} - H$. If $\mathfrak{B} - H$ contains precisely one vertex, then by $(Q_m)$ this only vertex is $P$ and the assertion of Lemma 9.6.1 is valid. Thus in the remaining part of Chapter 9 we may suppose that $\mathfrak{B} - H$ has at least two vertices and $\mathfrak{B} - H$ contains an $A$-path of length at least 1.

By the construction $\bar{C}_S$ of $T$ the vertex $P$ cannot be a main vertex of $T$.

For bridges with property $(Q_m)$ we prove the following statement:

**Proposition 9.11.1.** *Let $\mathfrak{B}$ have the property $(Q_m)$. Suppose that $Y \in M$ with $v(T: Y) \geqq 2$. Let $p_Y =_{\text{def}} p \triangle T[W, Y]$. Then $T^0(p_Y)$ contains an end vertex that is not in $T[W, Y]$ and has valency at least 2 in $\mathfrak{B} - H$.*

*Proof.* Suppose that Proposition 9.11.1 is not true. Then all end vertices of $T^0(p_Y)$ which are not in $T[W, Y]$ have valency 1 in $\mathfrak{B} - H$. Hence each of them has at least $r - 1$ neighbours in $H$.

Let $f$ denote the number of neighbours of $Y$ in $H$. By $(Q_m)$, the subtree $T^0(p_Y)$ has at most $m \leqq r - 1$ neighbours in $H$. Therefore, $0 \leqq f \leqq r - 1$.

By the hypothesis $v(T: Y) \geqq 2$, the tree $T^0(p_Y)$ has at least one end vertex $V$ which is not in $T[W, Y]$. Every end vertex of $T^0(p_Y)$ which is not in $T[W, Y]$ has at least $r - 1$ neighbours in $H$. If $f = r - 1$, then the vertices $Y$ and $V$ of $T^0(p_Y)$ are joined by at least $f + (r - 1) = (r - 1) + (r - 1) > r - 1$ edges to $H$. This contradicts $(Q_m)$!

Now let $0 \leqq f \leqq r - 2$. The subtree $T^0(p_Y)$ contains $v(T: Y) - 1 \geqq r - 1 - f$ end vertices which are not in $T[W, Y]$. Then $T^0(p_Y)$ is joined by at least $f + (r - 1 - f)(r - 1) > r - 1$ edges to $H$. This contradicts $(Q_m)$! □

In what follows let $\mathfrak{B}$ have the property $(Q_m)$.

Next we construct long paths in $T$ that have vertices of $M_3$ as few as possible. Such paths shall be called *walks*. Let $Y$ be an arbitrary main vertex of rank at most $h-2$ and valency at least 2 in $T$. A so-called *Y-walk* $w_Y \subseteqq T_Y$ will be introduced by the following

*Routine.* Assume that a path $w_i =_{\text{def}} T[Y, Y_i]$ has already been constructed so that $Y_i$ is a main vertex with $\text{rank}(Y_i) \leqq h-2$ and $v(T: Y_i) \geqq 2$.

For $i=0$ the vertex $Y_0 =_{\text{def}} Y$ and $w_0 =_{\text{def}} (Y_0) = T[Y, Y]$. Evidently, $w_0$ has the desired properties by definition.

In the general case, by Proposition 9.11.1 the subtree $T^0(p_{Y_i})$ contains an end vertex (of type 1, 2 or 3) which is not in $T[W, Y] \cup w_i$ and has valency at least 2 in $T$. If such a vertex of type 1 or 2 exists, let $Y_{i+1}$ be one of these vertices. If there is no such vertex, let $Y_{i+1}$ be the end vertex of type 3 with $v(T: Y_{i+1}) \geqq 2$. In both cases we define

$$w_{i+1} =_{\text{def}} w_i \cup T[Y_i, Y_{i+1}] = T[Y, Y_{i+1}].$$

The Routine stops with $w_Y =_{\text{def}} T[Y, Y_s]$ if and only if $\text{rank}(Y_s) > h-2$. □

We remark that contrary to $p_E$ where the suffix $E$ denotes the terminus of $p_E$ the suffix $Y$ of $w_Y$ means the initial vertex of $w_Y$.

Next we shall prove the following result:

**Proposition 9.11.2.** *Every main vertex $Y$ of $T$ with valency $v(T: Y) \geqq 2$ if $Y \neq W$ and valency $v(T: Y) \geqq 1$ if $Y = W$ is the initial vertex of a Y-walk $w_Y \subseteqq T_Y$ such that*

$$h-1 \leqq \lambda(T[W, Y] \cup w_Y) \leqq h.$$

*The path $w_Y - Y$ contains at most two main vertices of type* 3.

*Proof.* Since by the Routine $\text{rank}(Y_s) > h-2$, the inequalities hold. Next suppose that $Y_{i+1}$ is of type 3. By the Routine,

(1) *all end vertices $V$ of $T^0(Y_i)$ which are not in $T[W, Y_i]$, $V \neq V_{i+1}$, have valency* 1 *in* $\mathfrak{B} - H$. □

Let $f$ denote the number of neighbours of $Y_i$ in $H$. By $(Q_m)$ the subtree $T^0(p_{Y_i})$ has at most $r-1$ neighbours in $H$. Hence $0 \leqq f \leqq r-1$.

We shall prove that

(2) *the vertex $Y_i$ or an end vertex of $T^0(p_{Y_i})$ that is not in $T[W, Y_i]$ is joined to $H$ by at least $r-2$ edges.*

*Proof of (2).* If $f \geqq r-2$, then the assertion (2) is true. Now let $0 \leqq f \leqq r-3$. Since each vertex $V$ of $G$, $V \neq P$, has valency at least $r$, the subtree $T^0(p_{Y_i})$ contains $v(T: Y_i) - 2 = v(G: Y_i) - f - 2 \geqq r - f - 2$ end vertices which are not in $T[W, Y_i]$ and which are main vertices of types 1 and 2. By (1) all of them are linked by at least $r-1$ edges with $H$. Then $T^0(p_{Y_i})$ is joined by $\varphi \geqq f + (r-f-2)(r-1)$ edges to $H$.

Property $(Q_m)$ implies that $\varphi \leqq r-1$. Hence $f + (r-f-2)(r-1) \leqq r-1$. In all cases $(f, r) \neq (0, 3)$ we arrive at a contradiction! Consequently, in these cases we have $f \geqq r-2$.

Now let $f=0$ and $r=3$. In this case we have $v(T: Y_i) = 3$.

Since $Y_{i+1}$ is of type 3, the second end vertex $\bar{Y}$ of $T^0(p_{Y_i})$ which is not in $T[W, Y_i]$ is of type 1 or 2 and has valency 1 in $T$. Therefore, $\bar{Y}$ is joined by $r-1=2$ edges to $H$. □

The assertion (2) implies that the vertex $Y_i$ or an end vertex of $T^0(p_{Y_i})$ which is not in $T[W, Y_i]$ is joined to $H$ by at least $r-2$ edges.

If $r \geqq 4$, then by $2(r-2) > r-1$ and property ($Q_m$) no two of such vertices have distance less than $\lfloor t/2 \rfloor - 2$. Since $\bigcup_{j=0}^{s-1} T^0(p_{Y_j})$ can be split into two trees of diameter less than $\lfloor t/2 \rfloor - 2$, it does not contain three distinct vertices each of which is joined to $H$ by $r-2$ edges. Therefore, $w_Y - Y$ has at most two vertices of type 3.

If $r = 3$, then a more careful investigation leads to the same result. □

We remark that this result can be slightly improved for $r \geqq 4$.

**Proposition 9.11.3.** *Let $Y$ be a main vertex of $T$ with valency $v(T\colon Y) \geqq 2$. If $w_Y$ does not contain a triple path of $p$, then*

$$\lambda(T[W, Y] \cup w_Y) \leqq 6\alpha(Y) + 24.$$

*If $w_Y$ does not contain a chord of $p$, then*

$$\lambda(T[W, Y] \cup w_Y) \leqq 4\alpha(Y) + 14.$$

*If $w_Y$ does not contain a vertex of $p$, then*

$$\lambda(T[W, Y] \cup w_Y) \leqq 2\alpha(Y) - 1.$$

*Proof.* Proposition 9.11.2 implies that $w_Y - Y$ contains at most two main vertices of type 3. Let $Y_s$ denote the terminus of $T[W, Y] \cup w_Y$, i.e., the end vertex of $w_Y$ with $Y_s \neq Y$. Obviously, $w_Y - Y$ has at most $\alpha_2(Y_s)$ main vertices of type 2.

Case 1. If $w_Y - Y$ does not contain a triple path of $p$, then by Proposition 9.10.4 it does not contain three consecutive main vertices of type 1. Hence between two consecutive main vertices of $T$ of types 2 and 3 there are at most two main vertices of type 1 and at most two intermediate vertices. Consequently, $w_Y - Y$ contains at most two main vertices of type 3, at most $\alpha_2(Y_s)$ main vertices of type 2, at most $2(\alpha_2(Y_s) + 3)$ main vertices of type 1 and at most $2(\alpha_2(Y_s) + 3)$ intermediate vertices. Hence

$$\lambda(w_Y) \leqq 2 + \alpha_2(Y_s) + 4(\alpha_2(Y_s) + 3) = 5\alpha_2(Y_s) + 14. \tag{1}$$

By Corollary 9.10.3 it follows with $\alpha_3(Y_s) \leqq \alpha_3(Y) + 2$ and $\lambda(T[W, Y]) = 2\alpha_1(Y) + \alpha_2(Y) + \alpha_3(Y) = \alpha(Y) + \alpha_1(Y)$ that

$$\alpha_2(Y_s) \leqq \alpha_3(Y_s) \leqq \alpha_3(Y) + 2 \leqq (\alpha(Y) - \alpha_1(Y)) + 2. \tag{2}$$

Evidently,

$$\lambda(T[W, Y]) = \alpha(Y) + \alpha_1(Y). \tag{3}$$

By (1), (2) and (3) we conclude that

$$\lambda(w_Y) \leqq 5\alpha_2(Y_s) + 14 \leqq 5(\alpha(Y) - \alpha_1(Y)) + 24, \tag{4}$$

$$\lambda(T[W, Y] \cup w_Y) \leqq (\alpha(Y) + \alpha_1(Y)) + 5(\alpha(Y) - \alpha_1(Y)) + 24 \leqq 6\alpha(Y) + 24. \tag{5}$$

Case 2. If $w_Y - Y$ does not contain a chord of $p$, then by Proposition 9.10.4 it does not contain two consecutive vertices of type 1. By the same arguments as used in case 1 we arrive at the assertion.

Case 3. If $w_Y$ has no vertex of $p$, then $Y \neq W$, and on $w_Y$ there are not vertices of types 1 and 3. Consequently,

$$\lambda(w_Y) = \alpha_2(Y_s) - \alpha_2(Y) \leqq \alpha_2(Y_s) - 1 \tag{6}$$

(note that $\alpha_2(Y) \geqq 1$, because $Y$ is not a vertex of type 1 or 3). By Corollary 9.10.3 it follows with $\alpha_3(Y_s) \leqq \alpha_3(Y)$ that

(7) $$\alpha_2(Y_s) \leqq \alpha_3(Y_s) \leqq \alpha_3(Y) \leqq \alpha(Y) - \alpha_1(Y).$$

By (3), (6) and (7) we conclude that

(8) $$\lambda(T[W, Y] \cup w_Y) = (\alpha(Y) + \alpha_1(Y)) + (\alpha(Y) - \alpha_1(Y) - 1) = 2\alpha(Y) - 1. \square$$

**Proposition 9.11.4.** *Let $Y$ be a main vertex of $T$. If*

$$\alpha(Y) \leqq \left\lfloor \frac{t-6}{12} \right\rfloor - 4, \quad t \geqq 54,$$

*then $w_Y$ contains a triple path of $p$. If*

$$\alpha(Y) \leqq \left\lfloor \frac{t-2}{8} \right\rfloor - 4, \quad t \geqq 34,$$

*then $w_Y$ contains a chord of $p$. If*

$$\alpha(Y) \leqq \lfloor t/4 \rfloor - 1, \quad t \geqq 5,$$

*then $w_Y$ contains a vertex of $p$.*

*Proof.* By Proposition 9.11.2 it follows that $\lambda(T[W, Y] \cup w_Y) \geqq h - 1$, $h = \lfloor t/2 \rfloor - 1$. Assume that Proposition 9.11.4 is not true. Then Proposition 9.11.3 implies that

1) if $\alpha(Y) \leqq \left\lfloor \frac{t-6}{12} \right\rfloor - 4$ and $w_Y$ does not contain a triple path of $p$, then

$$\lambda(T[W, Y] \cup w_Y) \leqq 6\alpha(Y) + 24 \leqq \frac{t-6}{2} < \lfloor t/2 \rfloor - 2 = h - 1;$$

2) if $\alpha(Y) \leqq \left\lfloor \frac{t-2}{8} \right\rfloor - 4$ and $w_Y$ does not contain a chord of $p$, then

$$\lambda(T[W, Y] \cup w_Y) \leqq 4\alpha(Y) + 14 \leqq \frac{t-2}{2} - 2 < h - 1;$$

3) if $\alpha(Y) \leqq \lfloor t/4 \rfloor - 1$ and $w_Y$ does not contain a vertex of $p$, then $Y \neq W$ and

$$\lambda(T[W, Y] \cup w_Y) \leqq 2\alpha(Y) - 1 < h - 1.$$

In all three cases we arrive at a contradiction! Consequently, the assertions of Proposition 9.11.4 are valid. $\square$

## 9.12. The number of vertices $Y$ of $T$ with given $\alpha(Y) = \alpha$

In section 9.11 we have proved that for each main vertex $V$ of $T$ having a sufficiently small distance from $W$ the branch $T_V$ contains a triple path of $p$. Hence in this section 9.12 we shall determine a *branching subtree* $T_\alpha$ of $T$ which contains all main vertices of $T$ that have the property mentioned above. The branches $T_Y$, where $Y$ is an end vertex of the subtree $T_\alpha$, are pairwise disjoint, and each of them contains a triple path of $p$. Counting the end vertices of $T_\alpha$, we obtain that $T$ has many (internally) disjoint triple paths of $p$. We start with the proofs.

Let $\alpha$ be an integer with $1 \leqq \alpha \leqq 2\lfloor (t-6)/8 \rfloor$, $t \geqq 14$, and let $T_\alpha$ denote the subtree of $T$ induced by the vertices $Y$ with $\alpha(Y) \leqq \alpha$. We consider two cases.

Case 1. $1 \leqq \alpha \leqq \lfloor (t-6)/8 \rfloor$, $t \geqq 14$. Since two vertices of $T_\alpha$ have distance at most $4\alpha \leqq 4\lfloor (t-6/8 \rfloor < \lfloor t/2 \rfloor - 2$, by property $(Q_m)$ at most $m$ main vertices $W_1^*, \ldots, W_s^*$, $s \leqq m$, of $T$ have a neighbour in $H$.

If $W_i^* \neq W$, let $W_i'$ be the neighbour of $W$ in $T$ such that $W_i^* \in T_{W_i'}$, $1 \leqq i \leqq s$.

Let $\bar{T}_\alpha =_{\text{def}} T_\alpha - \bigcup_i T_{W_i'}$, and let $\bar{F}_\alpha$ be the subgraph of $\bar{T}_\alpha$ induced by all main vertices and intermediate vertices of type 1. By $\bar{D}_\alpha$ we denote the component of $\bar{F}_\alpha$ that contains $W$. Obviously, $\bar{D}_\alpha \subseteqq D(p)$. If $r \geqq \max\{3, m+1\}$ or $r \geqq \max\{3, m+1\} + 1$, then in $\bar{T}_\alpha$ and $\bar{D}_\alpha$, respectively, the vertices have the following valency:

**Proposition 9.12.1.** *Let $1 \leqq \alpha \leqq \lfloor (t-6)/8 \rfloor$, $t \geqq 14$. Then the valency of each intermediate vertex of $\bar{T}_\alpha$ and $\bar{D}_\alpha$ is equal to 2.*

*If $r \geqq \max\{3, m+1\}$, then $W$ and the main vertices $V$ of $\bar{T}_\alpha$ have the valency*

$$v(\bar{T}_\alpha : V) \begin{cases} \geqq r - m & \text{if } V = W \\ \geqq r & \text{if } \alpha(V) < \alpha, \\ = 1 & \text{if } \alpha(V) = \alpha. \end{cases}$$

*If $r \geqq \max\{3, m+1\} + 1$, then $W$ and the main vertices of $\bar{D}_\alpha$ have the valency*

$$v(\bar{D}_\alpha : V) = \begin{cases} v(\bar{T}_\alpha : V) - 1 & \text{if } \alpha(V) < \alpha, \\ 1 & \text{if } \alpha(V) = \alpha. \end{cases}$$

Case 2. $\lfloor (t-6)/8 \rfloor + 1 \leqq \alpha \leqq 2\lfloor (t-6)/8 \rfloor$, $t \geqq 14$. The tree $\bar{T}_{\lfloor (t-6)/8 \rfloor}$ satisfies Proposition 9.12.1. Let $R$ denote the set of all vertices $Y$ of $\bar{T}_{\lfloor (t-6)/8 \rfloor}$ with $\alpha(Y) = \lfloor (t-6)/8 \rfloor$, that are all end vertices of this tree. By construction of $\bar{T}_{\lfloor (t-6)/8 \rfloor}$ no vertex of $R$ has a neighbour in $H$.

Let $Y$ be an arbitrarily chosen vertex of $R$, and let $T_{\alpha,Y}$ be the branch of $T_\alpha$ with root $Y$. By $R_Y$ we denote the set of all main vertices $Z$ of $T_{\alpha,Y}$ with $\alpha(Z) = \lfloor (t-6)/8 \rfloor + 1$. Since any two vertices of $T_{\alpha,Y}$ have distance at most $4\alpha \leqq 4\lfloor (t-6)/8 \rfloor < \lfloor t-2 \rfloor - 2$, by property $(Q_m)$ at most $m$ main vertices $Y_1^*, \ldots, Y_s^*$, $s \leqq m$, of $T_{\alpha,Y}$ have a neighbour in $H$.

Since by construction of $\bar{T}_{\lfloor (t-6)/8 \rfloor}$ the vertex $Y$ has not a neighbour in $H$, without loss of generality $Y_1^*, \ldots, Y_q^* \notin R_Y$ and $Y_{q+1}^*, \ldots, Y_s^* \in R_Y$. Then for each $1 \leqq i \leqq q$ there is a vertex $Z_i \in R_Y$ such that $Y_i^* \in T_{Z_i} - \{Z_i\}$.

By $Z_i'$ we denote the neighbour of $Z_i$ in $T_{Z_i}$ such that $Y_i^* \in T_{Z_i'}$, $1 \leqq i \leqq q$. Let $\bar{T}_{\alpha,Y} =_{\text{def}} T_{\alpha,Y} - \bigcup_{i=1}^{q} T_{Z_i'}$ for all $Y \in R$ ($q = 0$ is possible for some $Y$), and let

$$\bar{T}_\alpha =_{\text{def}} \bar{T}_{\lfloor t-6)/8 \rfloor} \cup \bigcup_{Y \in R} \bar{T}_{\alpha,Y}.$$

If $r \geqq \max\{3, m+1\}$ or $r \geqq \max\{3, m+1\} + 1$, then in $\bar{T}_\alpha$ or $\bar{D}_\alpha$, respectively, the vertices have the following valency:

**Proposition 9.12.2.** *Let $\lfloor (t-6)/8 \rfloor + 1 \leqq \alpha \leqq 2\lfloor (t-6)/8 \rfloor$, $t \geqq 14$. Then the valency of each intermediate vertex of $\bar{T}_\alpha$ and $\bar{D}_\alpha$ is equal to 2.*

*If $r \geqq \max\{3, m+1\}$, then $W$ and the main vertices $V$ of $\bar{T}_\alpha$ have the valency*

$$v(\bar{T}_\alpha : V) \begin{cases} \geqq r-m & \text{if } V = W \text{ or } V \neq W \text{ and } \alpha(V) = \lfloor (t-6)/8 \rfloor + 1, \\ \geqq r & \text{if } \alpha(V) < \alpha \text{ and } V \neq W, \alpha(V) \neq \lfloor (t-6)/8 \rfloor + 1, \\ = 1 & \text{if } \alpha(V) = \alpha. \end{cases}$$

*Moreover, if $R'_Y$ denotes the set of all main vertices $V$ of $\bar{T}_{\alpha, Y}$ with $\alpha(V) = \lfloor (t-6)/8 \rfloor + 2$, then*

$$|R'_Y| \geqq (r-2)(r-1).$$

*If $r \geqq \max\{3, m+1\} + 1$, then $W$ and the main vertices of $\bar{T}_\alpha$ have the valency*

$$v(\bar{D}_\alpha : V) = \begin{cases} v(\bar{T}_\alpha : V) - 1 & \text{if } \alpha(V) < \alpha, \\ 1 & \text{if } \alpha(V) = \alpha. \end{cases}$$

*Moreover, if $R''_Y$ denotes the set of all main vertices $V$ of $\bar{D}_{\alpha, Y}$ with $\alpha(V) = \lfloor (t-6)/8 \rfloor + 2$, then*

$$|R''_Y| \geqq (r-3)(r-2). \quad \square$$

Simple arguments show that Propositions 9.12.1 and 9.12.2 imply the following result:

**Proposition 9.12.3.** *Let $R^T_\alpha$ and $R^D_\alpha$ denote the set of main vertices of $T_\alpha$ and $D_\alpha$ with $\alpha(X) = \alpha$, respectively. If $1 \leqq \alpha \leqq \lfloor (t-6)/8 \rfloor$, $t \geqq 14$, then*

$$|R^T_\alpha| \geqq (r-m)(r-1)^{\alpha-1} \quad \text{for } r \geqq \max\{3, m+1\},$$
$$|R^D_\alpha| \geqq (r-m-1)(r-2)^{\alpha-1} \quad \text{for } r \geqq \max\{3, m+1\} + 1.$$

*If $\lfloor (t-6)/8 \rfloor + 1 \leqq \alpha \leqq 2\lfloor (t-6)/8 \rfloor$, $t \geqq 14$, then*

$$|R^T_\alpha| \geqq (r-m)(r-2)(r-1)^{\alpha-2} \quad \text{for } r \geqq \max\{3, m+1\},$$
$$|R^D_\alpha| \geqq (r-m-1)(r-3)(r-2)^{\alpha-2} \quad \text{for } r \geqq \max\{3, m+1\} + 1.$$

## 9.13. Construction of a cycle $C$ in $G\langle p \cup T\rangle$

In sections 9.9-9.12 the existence of many (internally) disjoint triple paths of $p$ has been proved. Here in section 9.13 we prove that almost all of them are triple paths of some cycle $C$ of $\mathfrak{B} - H$. Hence this cycle $C$ contains many (internally) disjoint triple paths of $C$ which proves the validity of Lemma 9.6.1.

Let $\mathfrak{E}(\Gamma)$ denote the edge set of the graph $\Gamma$, and let $B$ be that vertex of $p \cap T$ that is the only vertex of $p[A, B]$ which belongs to $T$.

**Proposition 9.13.1.**

(i) *$B$ is an intermediate vertex of $T$ of type 1.*

(ii) *Let $\bar{B}$ denote the main vertex of $T$ of type 1 with*

$$[B, \bar{B}] \subseteqq T[W, \bar{B}] \quad \text{and} \quad p_{\bar{B}} =_{\text{def}} p \triangle T[W, \bar{B}].$$

*Then $C =_{\text{def}} p_{\bar{B}}[B, \bar{B}] \cup (\bar{B}, B)$ is a cycle with the following property: all vertices, edges and diagonals of $p$ which are in $T - \mathfrak{E}(T[W, \bar{B}])$ are also vertices, edges and diagonals of $C$.*

*Proof.* First we shall prove that

(1) $\lambda(p[A, B]) \leqq \lambda(p) + 1 - t.$

*Proof of (1).* By Proposition 9.11.2 there is a $W$-walk $w_W$ such that $w_W - \{W\}$ has at most two main vertices of type 3. Then by Corollary 9.10.3 the path $w_W - \{W\}$ contains at most two vertices of type 2. Since by Proposition 9.11.2 the $W$-walk $w_W$ has length at least $h - 1 = \lfloor t/2 \rfloor - 2 \geqq 5$, the $W$-walk $w_W$ contains an intermediate vertex $Q'$ of $T$ of type 1. Let $Q$ be the neighbour of $Q'$ with $[Q, Q'] \subseteqq T[W, Q']$. Then $Q$ is a main vertex of $T$, and $p_Q =_{\text{def}} p \triangle T[W, Q]$ is an $A$-path with second end vertex $Q$ and diagonal $(Q, Q')$. The cycle $p_Q[Q', Q] \cup (Q, Q')$ has length at least $t$. Hence $\lambda(p_Q[Q', Q]) \geqq t - 1$. Therefore we have $\lambda(p[A, B]) = \lambda(p_Q[A, B]) \leqq \lambda(p_Q[A, Q']) = \lambda(p_Q) - \lambda(p_Q[Q', Q]) \leqq \lambda(p) + 1 - t$. □

Next we shall prove the following statement:

(2) *$B$ is an intermediate vertex of type* 1.

*Proof of (2).* Let $V$ be an arbitrarily chosen main vertex of $T$ belonging to $p$. Obviously, $\alpha_3(V) \leqq h \leqq t/2$. By assertion (i) of Proposition 9.10.2 and assertion (1) we have

$$\begin{aligned}\lambda(p_V) &= \lambda(p \triangle T[W, V]) = \lambda(p) - \alpha_3(V) + \alpha_2(V) \\ &\geqq \lambda(p) - \alpha_3(V) \geqq \lambda(p) - \frac{t}{2} > \lambda(p) + 1 - t \geqq \lambda(p[A, B]).\end{aligned}$$

Forming $p_V = p \triangle T[W, V]$, the path $p[A, B]$ remains unchanged, i.e., $p_V[A, B] = p[A, B]$. Since $p_V = p_V[A, V]$ is longer than $p[A, B] = p_V[A, B]$, the vertices $V$ and $B$ are distinct. Consequently, we have $V \neq B$ for all main vertices $V$ of $T$. Thus $B$ is an intermediate vertex of $T$. □

Since $p_{\bar{B}} = p \triangle T[W, \bar{B}]$, all vertices, edges and diagonals of $p$ which do not meet a vertex of $T[W, \bar{B}]$ are also vertices, edges and diagonals of $C$, respectively. □

Now we will complete the proof of Lemma 9.6.1.

*Proof of Lemma 9.6.1.* By Proposition 9.13.1 every vertex, every diagonal and every triple path of $p$ which belong to a $Y$-walk $w_Y$ with $w_Y \cap T[W, \bar{B}] = \emptyset$ is also a vertex, a diagonal or a triple path of $C$, respectively.

Let $\alpha = \lfloor (t-6)/12 \rfloor - 4$ ($\alpha = \lfloor (t-2)/8 \rfloor - 4$ or $\alpha = \lfloor t/4 \rfloor - 3$). By Proposition 9.11.4 every $Y$-walk, $Y \in R_\alpha^T$, contains a triple path (diagonal or vertex) of $p$. Since at most one $Y'$-walk, $Y' \in R_\alpha^T$, contains vertices of $T[W, \bar{B}]$, at least $|R_\alpha^T| - 1$ of these triple paths (diagonals or vertices) of $p$ also are triple paths (diagonals or vertices) of $C$.

By Proposition 9.12.3 the assertions (i), (ii) and for $r = 3$ also assertion (iii) can be deduced. Let $\alpha = 2\lfloor (t-6)/8 \rfloor - 2$. By the construction $C_S$ of $D(p)$ every main vertex $Y$ of $D(p)$ with $\alpha(Y) = \alpha$ is the initial vertex of at least $(r-2)$ $Y$-walks $w_Y$ of $D_Y$. Hence the first, the second and the third main vertices of $w_Y$ are of type 1, and $w_Y$ contains a triple path of $p$. Consequently, $D$ has at least $(r-2)\,|R_\alpha^D|$ internally disjoint triple paths of $p$ and $D$ has at least $(r-2)\,|R_\alpha^D| - 1$ internally disjoint triple paths of $C$. By Proposition 9.12.3 this also implies the validity of assertion (iii) in the case $r \geqq 4$. □

# Chapter 10

# Extremal problems related to long cycles with many diagonals

## 10.0. Introduction

P. Turán, 1941, 1954, raised and solved the following problem: how many edges guarantee that a graph with this number of edges (and $n$ given vertices) has a complete subgraph with $k$ vertices? This type of question has been extensively investigated since then (see H. Sachs, 1970a). The problem is to find the maximum number of edges among graphs on $n$ vertices not containing certain subgraphs. Still, these subgraphs may only be prescribed up to isomorphism, up to homeomorphism (if, for example, cycles are excluded) or in some other way (if, for example, cycles with diagonals are excluded), and this leads to investigations with quite different methods. The reader will find many such problems in all graph theory books and a large variety of solved and unsolved problems in the papers of P. Erdős (see P. Erdős, 1973) and B. Bollobás, 1978a. There is also a good survey in the paper of M. Simonovits, 1983.

One feature common to all of these extremal problems is that they can all be translated from questions involving the number of edges into ones involving, instead, the degrees of the vertices. There are several well-known tricks for interpreting information from one formulation in terms of the other.

A first and well-known result is that of K. Zarankiewicz, 1953, 1954, who raised and solved the following problem: which minimum degree of a graph guarantees that a graph with this minimum degree (and $n$ given vertices) has a complete subgraph on $k$ vertices? This problem is the formulation of Turán's problem involving the minimum degree of the graph instead the maximum number of edges.

We may ask: how many edges guarantee that a graph $G^n$ of order $n$ with a given number of edges has a cycle or an odd cycle or an even cycle? The following result can easily be proved ($e(G^n)$ denotes the number of edges of $G^n$):

(i) *if* $e(G^n) \geqq n$, *then* $G^n$ *contains a cycle;*

(ii) *if* $e(G^n) \geqq \frac{3}{2}n - 1$, *then* $G^n$ *contains an even cycle;*

(iii) *if* $G^n$ *is 2-connected and* $e(G^n) \geqq n + 1$, *then* $G^n$ *contains an even cycle;*

(iv) *if* $e(G^n) > \left\lfloor \frac{n^2}{4} \right\rfloor$, *then* $G^n$ *contains an odd cycle.*

These lower bounds are best possible.

Assertion (iv) is a consequence of Turán's theorem. Moreover, J. A. Bondy, 1971, proved hat these graphs contain cycles of all lengths at least 3, i.e., they are pancyclic.

B. Bollobás, 1978a, presents more sufficient conditions for a graph to contain a cycle of length $l$ modulo $k$, where $l$, $k$ with $l \leqq k$ are integers. J.A.Bondy and M.Simonovits, 1974, dealt with sufficient conditions for the existence of cycles of given even length.

In Chapters 7-9 we determined the length of longest cycles and the maximum number of diagonals in cycles depending on the degree of the vertices. Here we translate some of these results into the corresponding assertions involving the number of edges.

The first result in this direction is due to P. Erdős and T. Gallai, 1959, who proved that every graph of order $n$ and more than $\frac{l}{2}(n-1)$ edges contains a cycle of length greater than $l$. This lower bound is attained for infinitely many values of $n$, but not for all. The precise lower bound for all $n$ has been found independently by D. R. Woodall, 1976, and G.N.Kopylov, 1977. They also independently determined the corresponding precise lower bound in the case when the graphs under consideration are 2-connected.

With respect to even cycles a theorem similar to the Erdős-Gallai theorem can be proved: every graph of order $n$ with more than $\frac{l+1}{2}(n-1)$ edges, $l \geqq 3$, contains an *even* cycle of length greater than $l$. Taking into consideration also the girth of the graphs we have shown that every graph of order $n$ with girth at least $t \geqq 3$ and more than $(r-1)(n-1)$ edges, $r \geqq 3$, contains a cycle of length at least $(r-1)^{t/4-7}$. All these results are included in section 10.1.

Sections 10.2 and 10.3 are devoted to similar problems related to the existence of long cycles through given edges and the existence of long *odd* cycles, respectively. A first result related to the maximum number of diagonals in cycles is the one of L. Pósa (see L. Lovász, 1979). With the aid of a theorem of J. Czipszer (Theorem 8.3.1) he proved that every graph $G$ of order $n$ with at least $2n-3$ edges, $n \geqq 4$, contains a cycle with one diagonal and if $G$ has at least $(r-1)(n-1)$ edges, $n \geqq r+1$, then $G$ contains a cycle with at least $r-2$ diagonals. The graph $I_2 + I_{n-2}$ shows that the bound $2n-3$ is best possible; in fact, if $n \geqq 6$, then $I_2 + I_{n-2}$ is the unique extremal graph (see Cambridge Mathematical Tripos 1974, Part II, Paper I, Nr. 10).

C. Thomassen and B. Toft, 1978, proved that every graph of order $n$ with more than $2n-2$ edges contains an even cycle with at least one diagonal, and if $G$ is also non-bipartite, it also contains an odd cycle with at least one diagonal.

All these results will be discussed and stronger versions will be proved. We do this in the following way: each assertion on the existence of a long cycle or a long even or odd cycle is supplemented by an assertion on the existence of a cycle or of an even or odd cycle with many diagonals.

The proofs are based on the results of Chapters 7-9 and involve sometimes the investigation of bridges of cycles.

## 10.1. Long cycles and long even cycles with many diagonals

We start with the result that every graph of order $n$ with girth at least $t \geqq 3$ and more than $(r-1)(n-1)$ edges, $r \geqq 3$, contains a cycle of length at least $(r-1)^{t/4-7}$ (Theorem 10.1.1). For small $t$ this bound makes no sense.

For $t \geqq 25$ and $l \geqq 2^{t/4-7}$ this result can be reformulated as follows: every graph of order $n$, girth $t$ and with at least $\left\lceil (2l)^{\frac{4}{t-24}} \right\rceil (n-1)$ edges contains a cycle of length at least $l$ (Corollary 10.1.2).

Since for small $t$ this bound is very bad, we present the well-known theorem of P. Erdős and T. Gallai, 1959, that every graph of order $n$ with more than $\frac{l}{2}(n-1)$ edges, $l \geqq 3$, contains a cycle of length greater than $l$ (Theorem 10.1.4(i)). This theorem was the first result on the length of longest cycles in graphs of order $n$ involving the number of edges.

The lower bound in the Erdős-Gallai theorem is attained for infinitely many values of $n$, but not for all. The precise lower bound for all $n$ has been found independently by D. R. Woodall, 1976, and G. N. Kopylov, 1977 (Theorem 10.1.5). They also independently determined the corresponding lower bound in the case when the graphs under consideration are 2-connected (Theorem 10.1.6).

With respect to even cycles a theorem similar to the Erdős-Gallai theorem can be proved: every graph of order $n$ with more than $\frac{l+1}{2}(n-1)$ edges, $l \geqq 3$, contains an *even* cycle of length greater than $l$. In our result stated at the very beginning of this section involving the girth of the graphs, the cycle obtained can be chosen to be even.

A similar program will be carried out for the number of diagonals of an (even) cycle instead of the length of such a cycle (Theorems 10.1.1, 10.1.4 (ii), 10.1.7 (ii), 10.1.8 and 10.1.9). The first theorem of this section is devoted to the existence of long even cycles with many diagonals depending on the girth of the graphs considered. This result will be essential only for graphs without small cycles, i.e., for graphs with high girth.

We defined above: $\gamma(r, t) = 2^{t/12-8}$ for $r = 3$, $t \geqq 3$, and $\gamma(r, t) = \frac{1}{2}(r-2)^{t/4-8}$ for $r \geqq 4$, $t \geqq 3$.

**Theorem 10.1.1.** *Let $G$ be a graph of girth at least $t$ with $n$ vertices and more than $(r-1)(n-1)$ edges, $r \geqq 3$. Then $G$ contains both an even cycle of length at least $\frac{1}{2}(r-1)^{t/4-6}$ and an even cycle with at least $\gamma(r, t)$ diagonals.*

The disjoint union of some copies of $G(2r-2, t)$ (definitions see page 183) shows that Theorem 10.1.1 cannot be essentially improved. This is a consequence of the following fact: if $n$ is the order of such a graph, then its edge number is $(r-1)n$ and each cycle has length at most $(2r-3)^{t+1}$ and at most $(2r-3)^{t+1}$ diagonals (cf. Proposition 9.1.1).

For small $n$ Theorem 10.1.1 is a trivial statement, because for small $n$ there is no graph with $n$ vertices and more than $(r-1)(n-1)$ edges. Thus the assertions of Theorem 10.1.1 are essential only for all $n \geqq n_0(r, t)$, where $n_0(r, t)$ is an appropriate constant.

Theorem 10.1.1 implies the following two corollaries.

**Corollary 10.1.2.** *Let $l$ and $t$ be integers with $l \geqq t \geqq 25$ and $\left\lceil \sqrt[t-24]{(2l)^4} \right\rceil \geqq 2$. Then every graph $G$ of girth at least $t$ with $n$ vertices and more than $\left\lceil \sqrt[t-24]{(2l)^4} \right\rceil (n-1)$ edges contains an even cycle of length at least $l$.* □

**Corollary 10.1.3.** *Let $d$ and $t$ be integers with $t \geqq 33$ and $\left\lceil \sqrt[t-32]{(2d)^4} \right\rceil \geqq 3$. Then every graph $G$ of girth at least $t$ with $n$ vertices and more than $\left\lceil \sqrt[t-32]{(2d)^4} \right\rceil (n-1)$ edges contains an even cycle with at least $d$ diagonals.* □

*Proof of Theorem 10.1.1.* The theorem will be proved by demonstrating the following result:

(A) *Let $t \geqq 3$, $r \geqq 3$ be two integers. Every graph $G$ of order $k$ and girth $t$ containing no even cycle of length at least $\frac{1}{2}(r-1)^{t/4-6}$ or no even cycle with at least $\gamma(r, t)$ diagonals has at most $(r-1)(k-1)$ edges.*

*Proof of* (A). The proof is performed by induction on $k$. The assertion (A) is obviously true for $2 \leqq k \leqq t$. We assume that (A) is true for $2 \leqq k \leqq n$. Now let $G$ be a graph of order $n+1\ (\geqq t+1)$ satisfying the hypotheses of (A). By Theorem 9.4.2 the graph $G$ contains a vertex $V$ of valency $v(G:V) \leqq r-1$. By induction $e(G-V) \leqq (r-1)(n-1)$, and consequently $e(G) \leqq e(G-V) + (r-1) \leqq (r-1)((n+1)-1)$. □

The proof of (A) is based on Theorem 9.4.2 which provides the existence of a vertex $V$ of valency at most $r-1$. If $t=3$, a result of G. A. Dirac, 1952a, and Theorem 9.2.1 of R. P. Gupta, J. Kahn and N. Robertson, 1980, also assert the existence of a vertex of valency at most $r-1$: every graph with minimum degree at least $r$ contains a cycle of length at least $r+1$ and a cycle with at least $\frac{1}{2}(r+1)(r-2)$ diagonals. Therefore, if a graph $G$ contains no cycle of length at least $r+1$ and no cycle with at least $\frac{1}{2}(r+1)(r-2)$ diagonals, then it has a vertex of valency at most $r-1$.

Next, in the same way as in (A), it can be proved that

*every graph $G$ of order $k$ containing no cycle of length greater than $l$ has at most $(l-2)(k-1)$ edges, and every graph $G$ of order $k$ containing no cycle with at least $\frac{1}{2}(r+2)(r-1)$ diagonals has at most $(r-1)(k-1)$ edges.*

(Note that Theorem 10.1.1 provides only the existence of an even cycle with at least one diagonal.)

Both results can be improved by using Theorem 7.2.1 of G. A. Dirac, 1952a, and Theorem 9.2.2 of P. Ash, 1985a: every 2-connected graph of order at least $2r$ with minimum degree at least $r$ contains a cycle of length at least $2r$ and a cycle with at least $r(r-2)$ diagonals.

In the proofs the assumed 2-connectivity has to be taken into account, hence the proofs are a little more complicated than the proof of assertion (A).

We formulate the improved version.

**Theorem 10.1.4.** *Let $n \geqq l \geqq 3$, $r \geqq 3$ be integers.*

(i) *Every graph $G$ of order $n$ containing no cycle of length greater than $l$ has at most $\frac{l}{2}(n-1)$ edges.*

(ii) *Every graph $G$ of order $n$ containing no cycle with at least $r(r-2)$ diagonals has at most $(r-1)(n-1)$ edges.*

Assertion (i) is the famous theorem of P. Erdős and T. Gallai, 1959. Assertion (ii) generalizes the result of L. Pósa stating that every graph $G$ of order $n$ with more than $(r-1)(n-1)$ edges contains a cycle with at least $r-2$ diagonals (see L. Lovász, 1979).

*Proof of Theorem 10.1.4.* The proof is performed by induction on $n$. The assertions (i) and (ii) are obviously true for $2 \leqq n \leqq l$ and $2 \leqq n \leqq 2r-2$, respectively. We assume that (i) and (ii) are true for $2 \leqq n \leqq k$.

Now let $G$ be a graph of order $k+1$ ($\geqq l+1$ or $\geqq r+1$, respectively) satisfying the hypotheses of (i) or (ii).

Case 1. Suppose that $G$ is not 2-connected. Then $G = G_1 \cup G_2$, where $G_1$ and $G_2$ have at most one vertex in common. By induction $e(G_i) \leqq \frac{l}{2}(k_i - 1)$ or $e(G_i) \leqq (r-1)(k_i - 1)$, respectively, where $k_i$ denotes the order of $G_i$, $i = 1, 2$. Hence $k+1 \leqq k_1 + k_2 \leqq k+2$ and $e(G) = e(G_1) + e(G_2) \leqq \frac{l}{2}(k_1 + k_2 - 2) \leqq \frac{l}{2}((k+1)-1)$ or $e(G) \leqq (r-1)((k+1)-1)$.

Case 2. Let $G$ be 2-connected. The order of $G$ is at least $l+1$ or at least $2r-1$, respectively. In case (ii) we assume first that the order of $G$ is at least $2r$. By Theorem 7.2.1 (G. A. Dirac, 1952a) and Theorem 9.2.2 (P. Ash, 1985a) one vertex $V$ of $G$ has valency at most $\lfloor l/2 \rfloor \leqq l/2$ or at most $r-1$, respectively. By induction $e(G-V) \leqq \frac{l}{2}(n-1)$ or $e(G-V) \leqq (r-1)(n-1)$ and $e(G-V) + \frac{l}{2} \leqq \frac{l}{2}((n+1)-1)$ or $e(G-V) + (r-1) \leqq (r-1)((n+1)-1)$, respectively.

With respect to assertion (ii) it remains to prove the case $n = 2r-1$. If one vertex of $G$ has valency at most $r-1$, the proof can be performed as described above. If the minimum degree of $G$ is at least $r$, then by Dirac's Theorem 7.2.1 the graph $G$ has a Hamiltonian cycle with fewer than $r(r-2)$ diagonals. Hence $e(G) \leqq n + r(r-2) = 2r-1+r(r-2) = (r-1)(r+1) \leqq (r-1)(n-1)$. □

The bound in Theorem 10.1.4 (i) is sharp for $n = s(l-1)+1$. It is attained by connected graphs, each block of which is a $K_l$.

Next we define a more general class. Let $l$, $n$, $t$, $s$ be positive integers with $l \geqq 3$ and $n = t(l-1)+1+s$ with $1 \leqq s \leqq l-1$. Then $\Delta(n, l)$ denotes the class of all connected graphs consisting of $t+1$ blocks, $t$ of them being complete $l$-graphs and one of them a complete $(s+1)$-graph. If $l$ is an even number and $s = \frac{l}{2}$ or $s = \frac{l}{2}+1$, then $K_{l/2} + I_{n-l/2}$ is also in $\Delta(n, l)$. An example with $n = 15$, $l = 4$ and $s = 2$ is depicted in Fig. 10.1.1.

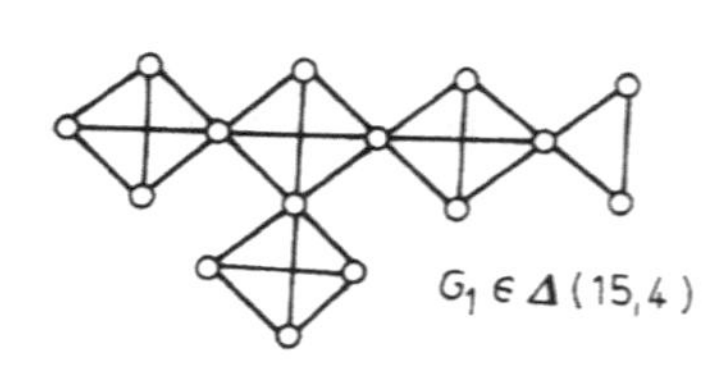


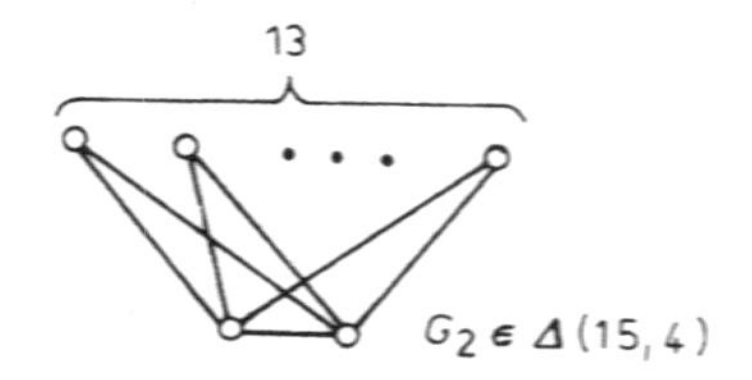


Fig. 10.1.1

If $s = l-1$, then the graphs of $\Delta(n, l)$ have $\frac{l}{2}(n-1)$ edges, i.e., the bound of Theorem 10.1.4(i) is best possible in this case. P. Erdős and T. Gallai, 1959, conjectured that the graphs of $\Delta(n, l)$ are the extremal graphs for all $1 \leqq s \leqq l-1$.

In the following years some authors reproved Theorem 10.1.4(i). So the theorem and parts of the conjecture were proved by J. A. Bondy, 1971 (exact solution for $n = l + 1$ and for $n \leqq l + c\sqrt{l}$, where $c$ is a suitable constant), D. R. Woodall, 1972 (exact solution for $l + 1 \leqq n \leqq 2l - 3$) and M. Lewin, 1975, D. R. Woodall, 1976, and G. N. Kopylov, 1977, have independently proved the conjecture of P. Erdős and T. Gallai in the affirmative.

**Theorem 10.1.5.** *Let $n = t(l-1) + 1 + s$ with $1 \leqq s \leqq l - 1$. Then every graph $G$ of order $n$ with at least $t\binom{l}{2} + \binom{s+1}{2}$ edges is a graph of $\Delta(n, l)$ (and has precisely $t\binom{l}{2} + \binom{s+1}{2}$ edges) or contains a cycle of length greater than $l$.*

In 1976 I found another proof of Theorem 10.1.5 (unpublished). In the case when $G$ is 2-connected and the circumference $l$ is odd, J. A. Bondy, 1971, proved that every 2-connected graph of order $n$ has at most $\frac{l-1}{2} n$ edges. The proofs of D. R. Woodall, 1976, and N. G. Kopylov, 1977, of Theorem 10.1.5 are based on the exact solution of the corresponding problem for 2-connected graphs.

It is convenient to formulate the following theorem in a form slightly different from the other theorems of this chapter.

**Theorem 10.1.6.** *The number $e(G)$ of edges in a 2-connected graph of circumference $l$ and of order at least $l + 1$ is*

$$e(G) \leqq \max\{e(K_{l/2} + I_{n-l/2}),\ e(K_2 + (K_{l-3} \cup I_{n-l+1}))\} \quad \text{if } l \text{ is even,}$$
$$e(G) \leqq \max\{e(K_{(l-1)/2} + (K_2 \cup I_{n-(l+3)/2})),\ e(K_2 + (K_{l-3} \cup I_{n-l+1}))\} \quad \text{if } l \text{ is odd.}$$

*The graphs in the brackets are the only graphs for which equality can be attained.*

D. R. Woodall, 1976, and N. G. Kopylov, 1977, have independently proved this theorem.

The bound in Theorem 10.1.4(ii) is sharp for $r = 3$ and $n = 3s + 1$. It is attained by connected graphs each block of which is a $K_4$. For general $r \geqq 3$ the bound can be improved. By arguments used in this chapter the bound $(r-1)(n-1)$ can be replaced by $(r-1)(n-1) - c(r)$, where $c(r)$ is a suitable polynomial in $r$ with $c(r) \geqq 0$ for $r \geqq 3$. I have made no attempt to determine $c(r)$.

If we require the cycles of Theorem 10.1.4 to be *even*, then a similar theorem can be proved using Theorem 7.2.1 of G. A. Dirac, 1952a, Theorem 7.6.1 of H.-J. Voss and C. Zuluaga, 1977, and Theorem 9.2.4 of H.-J. Voss, 1985a; every 2-connected graph of order at least $2r$ with minimum degree at least $r$ contains an *even* cycle of length at least $2r$ and an *even* cycle with at least $\frac{1}{2}\binom{r-1}{2}$ diagonals. Repeating the proof of Theorem 10.1.4 with these arguments a proof of the following result is obtained.

**Theorem 10.1.7.** *Let $n \geqq l \geqq 3$ be integers.*

(i) *Every graph of order $n$ containing no even cycle of length greater than $l$, $l$ even, has at most $\frac{l+1}{2}(n-1)$ edges.*

(ii) *Every graph of order $n$ containing no even cycle with more than $\frac{1}{2}\binom{r-1}{2}$ diagonals has at most $(r-1)(n-1)$ edges.*

The bound in assertion (i) of Theorem 10.1.7 is sharp for $n = ls + 1$, $l$ even. It is attained by the connected graphs each block of which is a $K_{l+1}$, $l$ even.

The bound in assertion (ii) of Theorem 10.1.7 is far from being sharp. It should be improved.

For $r = 3$ Theorem 10.1.7(ii) implies that each graph of order $n$ with more than $2n - 2$ edges has an even cycle with at least one diagonal. This was proved by C. Thomassen and B. Toft, 1978. This result can be improved (H.-J. Voss, 1982a).

**Theorem 10.1.8.** *Let G be a graph with n vertices* ($n \geqq 4$) *and at least* $2n - 2$ *edges. Then G contains an even cycle with at least two diagonals, or G is a connected graph having exactly* $2n - 2$ *edges and each block of it is isomorphic to* $W_4$.

The proof of Theorem 10.1.8 will be accomplished by demonstrating the following result.

(B) *Every graph G of order k containing no even cycle with at least two diagonals has at most* $2k - 2$ *edges. If G has precisely* $2k - 2$ *edges, then G is a connected graph and each of its blocks is isomorphic to* $W_4$.

*Proof of* (B). The proof is performed by induction on $k$. The assertion (B) is obviously true for $2 \leqq k \leqq 4$. We assume (B) to be true for $2 \leqq k \leqq n$. Now let $G$ be a graph of order $n + 1$ ($\geqq 5$) satisfying the hypotheses of (B).

Case 1. Suppose that $G$ is not 2-connected. Then $G$ has $s \geqq 2$ blocks $\mathfrak{B}_1, \mathfrak{B}_2, \ldots, \mathfrak{B}_s$ which form $p \geqq 1$ components with $s - p$ cut vertices. By the induction hypothesis $e(\mathfrak{B}_i) \leqq 2n_i - 2$, where $n_i$ is the order of $\mathfrak{B}_i$, and equality is attained if and only if $\mathfrak{B}_i \cong W_4$. Consequently,

$$n(G) = n + 1 = \sum_{i=1}^{s} n_i - (s - p)$$

and

$$\begin{aligned} e(G) &= \sum_{i=1}^{s} e(\mathfrak{B}_i) \leqq \sum_{i=1}^{s} (2n_i - 2) = 2\left(\sum_{i=1}^{s} n_i - s\right) \\ &= 2\left(\sum_{i=1}^{s} n_i - s + p\right) - 2p = 2(n+1) - 2p \\ &\leqq 2(n+1) - 2. \end{aligned}$$

Equality is obtained if and only if $G$ is connected (i.e., $p = 1$) and $\mathfrak{B}_i \cong W_4$ for all $1 \leqq i \leqq s$.

Case 2. Let $G$ be 2-connected. By Theorem 8.4.2 the graph $G \cong W_4$, or $G$ contains a vertex $V$ of valency $v(G:V) = 2$. Then $n(G - V) = n$ and, by the induction hypothesis, $e(G - V) \leqq 2n - 2$. Hence

$$e(G) = e(G - V) + v(G:V) \leqq 2n - 2 + 2 = 2(n+1) + 2.$$

The bound $e(G) = 2(n+1) - 2$ is attained if any only if $e(G - V) = 2n - 2$. Then, by the induction hypothesis, $G - V$ is connected and each block is isomorphic to $W_4$. In this case it is easy to show that $G$ contains an even cycle with at least two diagonals. Contradiction! □

Theorem 10.1.8 states the existence of even cycles with certain properties in arbitrary graphs of order $n$ with at least $2n - 2$ edges. For 2-connected graphs a stronger result will be stated; it corresponds to the Theorem 10.3.2 on odd cycles.

**Theorem 10.1.9.** *Let $G$ be a 2-connected graph with $n$ vertices ($n \geqq 4$) and at least $2n-3$ edges. Let $G \not\cong W_4$. Then $G$ contains an even cycle with at least two diagonals, or $G$ has exactly $2n-3$ edges and is a graph of one of the types depicted in Fig. 10.1.2.* □

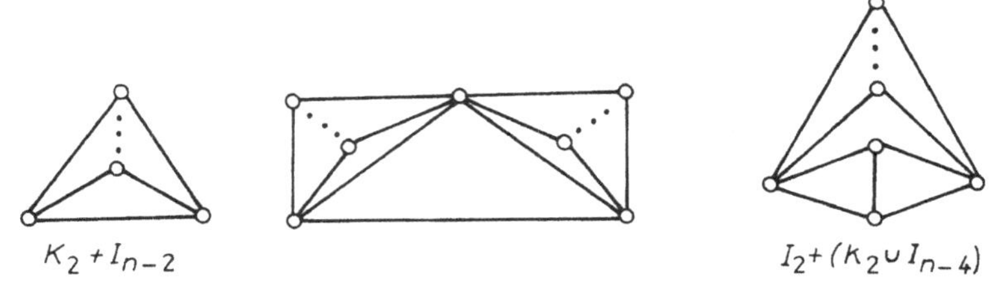


Fig. 10.1.2

The proof of this theorem is omitted. It can be proved in a way similar to Theorem 10.3.2 (the proof of Theorem 10.3.2 is a little more complicated; hence we shall prove this theorem).

With the aid of Theorem 8.4.4 it can be proved that

*if the graph $G$ of Theorem 10.1.9 has no 3-cycles or 4-cycles, then $G$ contains an even cycle with at least three diagonals.*

From Theorem 8.3.1 of J. Czipszer, 1963, and its proof we know that if $\delta(G) \geqq r \geqq 3$, then $G$ contains a cycle with at least $r-2$ diagonals incident with the same vertex. This implies (B. Bollobás, 1978a) the following result:

**Theorem 10.1.10.** *Let $k \geqq 1$ be an integer and denote by $g_k(n)$ the maximum size of a graph of order $n$ that does not contain a cycle with $k$ diagonals incident with the same vertex. Then*

(i) $g_k(n+1) \leqq g_k(n) + k + 1, \quad n \geqq 1;$

(ii) *if $n \geqq k+1$, then*

$$(k+1)\,n - (k+1)^2 \leqq g_k(n) \leqq (k+1)\,n - \binom{k+2}{2};$$

(iii) *furthermore, if $n \geqq 2k+2$ and $k = 1, 2$ or $3$, then*

$$g_k(n) = (k+1)\,n - (k+1)^2.$$

*Proof.* Inequality (i) follows from the observation preceding this theorem (Theorem 8.3.1 of J. Czipszer, 1963). Since $g_k(k+1) = \binom{k+1}{2} = (k+1)^2 - \binom{k+2}{2}$, (i) implies the right inequality of (ii). The validity of the left inequality of (ii) is shown by the graph $I_{k+1} + I_{n-(k+1)}$. In order to prove (iii) one has only to check that $g_1(4) = 4$, $g_2(6) = 9$ and $g_3(8) = 16$.

P. Erdős made the conjecture that if $n \geqq 2(k+1)$, then $g_k(n)$ is in fact always equal to $(k+1)\,n - (k+1)^2$, the lower bound in (ii). This was disproved by M. Lewin (see B. Bollobás, 1978a). B. Bollobás, 1978a, made the conjecture that it is rather likely that there is a function $n(k)$ for which $n \geqq n(k)$ implies that $g_k(n) = (k+1)n - (k+1)^2$ does hold.

## 10.2. Long cycles through a given edge with many diagonals

In Theorem 10.1.1 the existence of long cycles with many diagonals has been stated. If the graphs are also 2-*connected*, then through each edge such a cycle can be found.

**Theorem 10.2.1.** *Let $G$ be a 2-connected graph of girth at least $t$ with $n$ vertices and more than $(r-1)(n-1)$ edges, $r \geqq 3$. Then for each edge $e$ of $G \cup \bar{G}$ the graph $G \cup \{e\}$ contains both a cycle of length at least $\frac{1}{2}(r-1)^{t/4-6}$ through $e$ and a cycle with at least $\gamma(r, t)$ diagonals through $e$.*

$\gamma(r, t)$ has been defined on page 209.

For infinitely many natural numbers $n$ an upper bound is provided by the graphs $\underline{H}_t(T_{2r-2})$ described in section 9.1.

Let $d(T_{2r-2})$ again denote the diameter of $T_{2r-2}$. If $n$ denotes the order of $\underline{H}_t(T_{2r-2})$, then its edge-number is at least $(r-1)n$ and each cycle of $\underline{H}_t(T_{2r-2})$ has length at most $2d(T_{2r-2}) + 2(2r-3)^{t+1}$ and at most $2(2r-3)^{t+1}$ diagonals (compare with Theorem 9.1.2).

Theorem 10.2.1 will be proved by demonstrating the following result (let $\varepsilon(r, t) =_{\text{def}} \frac{1}{2}(r-1)^{t/4-6}$):

(C) *Let $t \geqq 3$, $r \geqq 3$ be two integers. In every 2-connected graph $G$ of order $k$ and girth $t$ containing two distinct vertices $A$ and $B$ such that no $A,B$-path has length at least $\varepsilon(r, t)$ or no $A,B$-path has at least $\gamma(r, t)$ diagonals the number of edges is at most $(r-1)(k-1)$.*

*Proof of* (C). The proof is performed by induction on $k$. The assertion (C) is obviously true for $2 \leqq k \leqq t$. We assume (C) to be true for $2 \leqq k \leqq n$. Now let $G$ be a 2-connected graph of order $n+1$ ($\geqq t+1$) satisfying the hypotheses of (C). By Theorem 9.3.1 the graph $G$ contains a vertex $V$, $V \notin \{A, B\}$, of valency $v(G:V)$ with $2 \leqq v(G:V) \leqq r-1$.

Let $\mathfrak{B}_1, \ldots, \mathfrak{B}_s$, $s \geqq 1$, denote the blocks of $G - V$. Since $G$ is 2-connected, the vertices $A$ and $B$ are joined by two disjoint paths $v_i$ and $w_i$ to $\mathfrak{B}_i$. Let $A$, $A_i$ and $B$, $B_i$ denote the end vertices of $v_i$ and $w_i$, respectively. (Note that if $A \in \mathfrak{B}_i$, then $v_i$ has length 0 and $A_i = A$.) Each $A_i,B_i$-path $p_i$ of $\mathfrak{B}_i$ forms with $v_i$ and $w_i$ an $A,B$-path $v_i \cup p_i \cup w_i$ of $G$. Hence each $A_i,B_i$-path of $\mathfrak{B}_i$ has length at most $\varepsilon(r, t) - 1$ or each $A_i,B_i$-path of $\mathfrak{B}_i$ has at most $\gamma(r, t) - 1$ diagonals, respectively. Consequently, by the induction hypothesis, $e(\mathfrak{B}_i) \leqq (r-1)(n_i - 1)$ for all $1 \leqq i \leqq s$, where $n_i$ denotes the number of vertices of $\mathfrak{B}_i$. Hence

$$n + 1 = n(G) = \sum_{i=1}^{s} n_i - (s-1) + 1 = \sum_{i=1}^{s} n_i - s + 2$$

and

$$\begin{aligned} e(G) &= \sum_{i=1}^{s} e(\mathfrak{B}_i) + v(G:V) \leqq \sum_{i=1}^{s} (r-1)(n_i - 1) + (r-1) \\ &= (r-1)\left(\sum_{i=1}^{s} n_i - s + 1\right) = (r-1)((n+1)-1). \ \square \end{aligned}$$

Theorem 10.2.1 remains true if $\varepsilon(r, 3)$ is replaced by $r+1$ and $\gamma(r, 3)$ is replaced by $\frac{1}{2}\binom{r-1}{2}$. In the proof of Theorem 10.2.1 the only change is that a vertex $V$ of valency at most $r-1$ is found by applying Theorem 7.5.3 or Theorem 9.2.3, respectively. Thus we obtain the following theorem.

**Theorem 10.2.2.** *Let $n \geqq l \geqq 3$, $r \geqq 3$ be integers.*

(i) *Every 2-connected graph of order n containing an edge e which lies on no cycle of length at least l has at most $(l-2)(n-1)$ edges.*

(ii) *Every 2-connected graph of order n containing an edge e which lies on no cycle with at least $\frac{1}{2}\binom{r-1}{2}$ diagonals has at most $(r-1)(n-1)$ edges.* □

Assertion (i) of Theorem 10.2.2 can be improved:

**Theorem 10.2.3.** *Let G be a 2-connected graph of order n containing an edge $e = (A, B)$ which lies on no cycle of length at least $2r-1$. Then G has at most $\frac{2r-1}{2}(n-2)+1$ edges.*

If $n = s(2r-4)+1$, then the bound $\frac{2r-1}{2}(n-2)+1$ is attained by the graphs $K_2 + sK_{2r-4}$, $s \geqq 1$.

*Proof of Theorem 10.2.3.* The proof is performed by induction on $n$. The assertion is obviously true for $2 \leqq n \leqq 2r-2$ because

$$e(G) \leqq \frac{n(n-1)}{2} = \frac{n+1}{2}(n-2)+1 \leqq \frac{2r-1}{2}(n-2)+1.$$

We assume that Theorem 10.2.3 is true for $2 \leqq n \leqq k$. Now let $G$ be a graph of order $k+1\,(\geqq 2r-1)$ satisfying the hypotheses of the theorem.

Case 1. Suppose that $G$ is not 3-connected. Hence $G = G_1 \cup G_2$ is such that 1) $G_1$, $G_2$ have precisely two vertices, say $P$ and $Q$, and no edge in common; 2) if $(P, Q)$ is an edge of $G$, then $(P, Q)$ is in $G_1$. The subgraph $G_1$ can be chosen so that $G_1$ is a block. Then $G_2 \cup (P, Q)$ is also a block ($(P, Q) \notin G_2$). Since $(A, B)$ is an edge of $G$, the vertices $A$ and $B$ both belong to $G_1$ or both belong to $G_2$.

Without loss of generality, let $A, B \in G_1$. Since no $A,B$-path of the 2-connected graph $G_1$ has length at least $2r-2$, by induction hypothesis

$$e(G_1) \leqq \frac{2r-1}{2}(k_1-2)+1,$$

where $k_1$ denotes the order of $G_1$. Let $k_2$ be the order of $G_2$. Since $G_1$ is 2-connected, $G_1$ contains two disjoint paths $v_1$, $v_2$ joining $A$, $B$ to $P$, $Q$. Each $P,Q$-path $p$ of $G_2 \cup (P, Q)$ together with $v_1$, $v_2$ form an $A,B$-path of $G$. Therefore, each $P,Q$-path of $G_2 \cup (P, Q)$ has a length less than $2r-2$. Consequently, by the induction hypothesis,

$$e(G_2 \cup (P, Q)) \leqq \frac{2r-1}{2}(k_2-2)+1,$$

$$e(G_2) \leqq \frac{2r-1}{2}(k_2-2).$$

With $k+1=k_1+k_2-2$ we obtain that

$$e(G) \leqq \frac{2r-1}{2}((k_1+k_2-2)-2)+1=\frac{2r-1}{2}((k+1)-2)+1.$$

Case 2. Let $G$ be 3-connected. By Theorem 7.5.3 the graph $G$ contains a vertex $V$ with $3 \leqq v(G:V) \leqq r-1$. Then $G-V$ is 2-connected and contains no $A,B$-path of length at least $2r-2$. By induction $e(G-V) \leqq \frac{2r-1}{2}(k-2)+1$ and

$$e(G)=e(G-V)+v(G:V)<\frac{2r-1}{2}((k+1)-2)+1. \square$$

We reformulate Theorem 10.2.3 as follows:

**Corollary 10.2.4.** *Let $n \geqq 4$, $r \geqq 3$ be integers. Let $G$ be a 2-connected graph with $n$ vertices and more than $\frac{2r-1}{2}(n-2)+1$ edges. Then each pair of vertices $A$, $B$ of $G$ is joined by an $A,B$-path of length at least $2r-2$.* $\square$

For $t=3$ and $r=3$ Theorem 10.2.1 states that there is a cycle through each edge with at least one diagonal. This result can be improved.

**Theorem 10.2.5.** *Let $G$ be a 2-connected graph with $n$ vertices ($n \geqq 3$) and at least $2n-3$ edges. Then each pair of vertices $A$, $B$ of $G$ is joined by an $A,B$-path with at least two diagonals different from $(A, B)$, or $G$ is the following graph (with exactly $2n-3$ edges): $G$ consists of two vertices $\bar{A}$ and $\bar{B}$, the edge $(\bar{A}, \bar{B})$, and bridges of $\bar{A}$, $\bar{B}$ which are of type $D_1$, $D_2$ and/or $D_3$* (see Fig. 10.2.1).

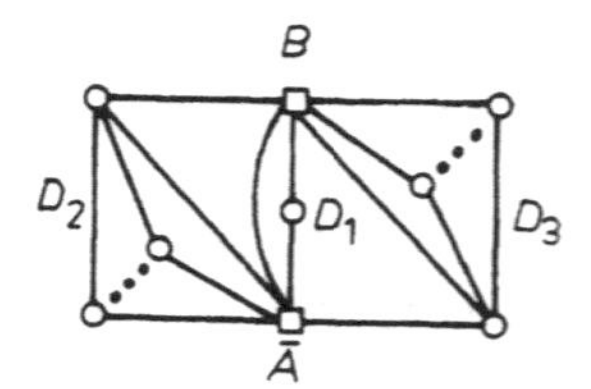


Fig. 10.2.1

Note that the bound $2n-2$ is the same as that in Dirac's well-known theorem that every graph with $n$ vertices and at least $2n-2$ edges contains a subdivision of $K_4$.

Theorem 10.2.5 will be proved by demonstrating the following result.

(D) *Let $G$ be a block of order $k$ containing two vertices $A$, $B$ such that each $A,B$-path has at most one diagonal different from $(A, B)$. Then $G$ has at most $2k-3$ edges. If the number of edges of $G$ is exactly $2k-3$, then $G$ consists of the two vertices $A$ and $B$, the edge $(A, B)$, and bridges of $\{A, B\}$ which are of type $D_1$, $D_2$ and/or $D_3$* (see Fig. 10.2.1).

The graphs of (D) with $2k-3$ edges will be referred to as extremal graphs of (D).

*Proof of* (D). The proof of (D) can be performed by exactly the same arguments as the proof of (C). The only change is that, by the induction hypothesis, $e(\mathfrak{B}_i) \leqq 2n_i - 3$ and

$e(\mathfrak{B}_i) = 2n_i - 3$ if and only if $\mathfrak{B}_i$ is an extremal graph of (D). By Theorem 8.3.7 the vertex $V$ has precisely two neighbours $V_1$ and $V_2$. Consequently,

$$n + 1 = n(G) = \sum_{i=1}^{s} n_i - s + 2$$

(note that $p = 1$) and

$$\begin{aligned} e(G) &= \sum_{i=1}^{s} e(\mathfrak{B}_i) + v(G:V) \leqq \sum_{i=1}^{s} (2n_i - 3) + 2 \\ &= 2\left(\sum_{i=1}^{s} n_i - s + 1\right) - s = 2((n+1) - 1) - s \\ &\leqq 2(n+1) - 3 . \end{aligned}$$

Equality holds if and only if $s = 1$ and $G - V$ is an extremal graph of (D). Adding $V_1VV_2$ to $G - V$ so that the hypotheses of (D) are not violated results in an extremal graph of (D). □

By Theorem 8.6.2 a 2-connected graph $G$ is minimally 2-connected if and only if does not contain a cycle with a diagonal. Theorem 10.2.5 and assertion (D) imply that $G$ has at most $2n - 3$ edges, where $n$ denotes the order of $G$. Since all extremal graphs of (D) of order at least 4 contain a cycle with one diagonal, this bound can be reduced.

**Corollary 10.2.6.** *For $n \geqq 4$ every minimally 2-connected graph of order $n$ has at most $2n - 4$ edges.* □

This bound is sharp, as the graphs $I_2 + I_{n-2}$ show. We note that Corollary 10.2.6 can also be proved with the aid of Theorem 8.3.4.

## 10.3. Long odd cycles with many diagonals in non-bipartite graphs

In Theorem 10.2.1 the existence of large *even* cycles with many diagonals has been stated. If these graphs are also 2-*connected* and *non-bipartite*, then *odd* cycles with many diagonals can be found.

**Theorem 10.3.1.** *Let $G$ be a 2-connected non-bipartite graph of girth at least $t$ with $n$ vertices and more than $(r-1)(n-1)$ edges, $r \geqq 3$. Then $G$ contains both an odd cycle of length at least $\frac{1}{2}(r-1)^{t/4-6}$ and an odd cycle with at least $\gamma(r, t)$ diagonals.*

Again the graphs $\underline{H}_t(T_{2r-2})$ provide the same upper bounds as derived above for the even cycles.

Theorem 10.3.1 will be proved by demonstrating the following result.

(E) *Let $t \geqq 3$, $r \geqq 3$ be two integers. Every 2-connected non-bipartite graph $G$ of order $k$ and girth $t$ containing no odd cycle of length at least $\varepsilon(r, t)$ or no odd cycle with at least $\gamma(r, t)$ diagonals has at most $(r-1)(k-1)$ edges.*

*Proof of* (E). The proof is performed by induction on $k$. The assertion (E) is obviously true for $2 \leqq k \leqq t$. We assume that (E) is true for $2 \leqq k \leqq n$.

Now let $G$ be a graph of order $n+1$ ($\geqq t+1$) satisfying the hypotheses of (E). By Theorem 9.4.1 the graph $G$ contains a vertex $V$ of valency $v(G:V)$ with $2 \leqq v(G:V) \leqq r-1$. Let $\mathfrak{B}_1, \ldots, \mathfrak{B}_s$ denote the blocks of $G-V$.

Case 1. Let $G-V$ be bipartite. Since $G$ is non-bipartite, each vertex of $G$ is contained in an odd cycle. Hence for each $1 \leqq i \leqq s$ there is a $\mathfrak{B}_i,\mathfrak{B}_i$-path $v_i$ of $G$ with ends $A_i$, $B_i$ such that each $A_i,B_i$-path of $\mathfrak{B}_i$ together with $v_i$ forms an odd cycle. Obviously, $V \in v_i$. Therefore, each $A_i,B_i$-path of $\mathfrak{B}_i$ has length at most $\varepsilon(r,t)-1$ or each $A_i,B_i$-path of $\mathfrak{B}_i$ has at most $\gamma(r,t)-1$ diagonals. By assertion (C), $e(\mathfrak{B}_i) \leqq (r-1)(n_i-1)$. The proof of case 1 can be accomplished in the same way as the proof of (C).

Case 2. Without loss of generality, let $\mathfrak{B}_1$ be non-bipartite. By the induction hypothesis, $e(\mathfrak{B}_1) \leqq (r-1)(n_1-1)$. Since $G$ is 2-connected, two disjoint paths $v_i$, $w_i$ join $\mathfrak{B}_1$ and $\mathfrak{B}_i$, $2 \leqq i \leqq s$, to ends $A_1^i$, $B_1^i$ in $\mathfrak{B}_1$ and $A_i$, $B_i$ in $\mathfrak{B}_i$, respectively.

Let $p_1^i$ and $\bar{p}_1^i$ denote two $A_1^i$, $B_1^i$-paths of $\mathfrak{B}_1$ having different parities. Then each $A_i,B_i$-path of $\mathfrak{B}_i$ together with either $p_1^i$ or $\bar{p}_1^i$ forms an odd cycle. Hence each $A_i,B_i$-path of $\mathfrak{B}_i$ has length at most $\varepsilon(r,t)-1$ or each $A_i,B_i$-path of $\mathfrak{B}_i$ has at most $\gamma(r,t)-1$ diagonals.

Assertion (C) implies that $e(\mathfrak{B}_i) \leqq (r-1)(n_i-1)$, $2 \leqq i \leqq s$. Hence $e(\mathfrak{B}_i) \leqq (r-1)(n_i-1)$ for all $1 \leqq i \leqq s$. Thus the proof of case 2 can also be accomplished in the same way as the proof of assertion (C). □

For $t=3$, $r=3$ Theorem 10.3.1 states the existence of an odd cycle with at least one diagonal. This result can be improved.

**Theorem 10.3.2.** *Let $G$ be a 2-connected non-bipartite graph with $n$ vertices ($n \geqq 5$) and at least $2n-3$ edges. Then $G$ contains an odd cycle with at least two diagonals, or $G$ has exactly $2n-3$ edges and is either a prism graph or a graph of one of the types depicted in Fig. 10.3.1.*

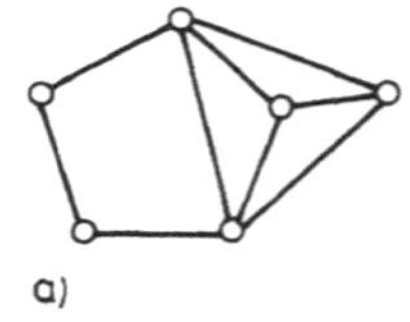
a)

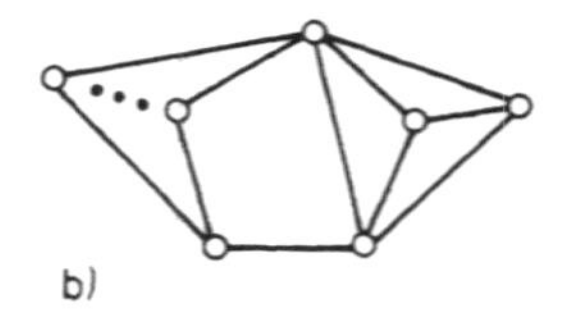
b)

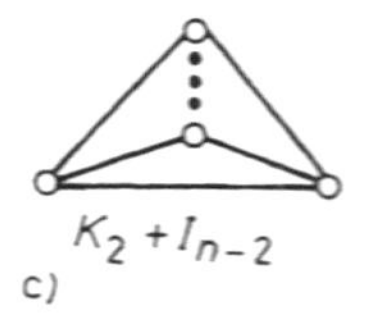

c)

Fig. 10.3.1

With the aid of Theorem 8.3.4 it can be proved that

*if the graph $G$ has no 3-cycles and no 4-cycles, then $G$ contains even an odd cycle with at least three diagonals.* □

Theorem 10.3.2 will be proved by demonstrating the following result.

(F) *Let $G \not\equiv W_3$ be a non-bipartite block of order $k$ containing no odd cycle with two diagonals. Then $G$ has at most $2k-3$ edges. If the number of edges of $G$ is exactly $2k-3$, then $G$ is a prism graph or a graph of one of the types depicted in Fig. 10.3.1.*

The graphs of (F) with $2k-3$ edges will be referred to as extremal graphs of (F).

*Proof of* (F). The proof of (F) can be performed by similar arguments as in the proof of (E). Here we only discuss the differences in the proofs of assertions (E) and (F).

Suppose that $G \not\equiv W_3$ and that $G$ is not a prism graph. By Theorem 8.3.2 the vertex $V$ has degree 2. Let $\mathfrak{B}_1, \ldots, \mathfrak{B}_s$ again denote the blocks of $G-V$.

Case 1. Let $G - V$ be bipartite. By assertion (D) the edge number $e(\mathfrak{B}_i)$ is at most $2n_i - 3$, and $e(\mathfrak{B}_i) = 2n_i - 3$ if and only if $\mathfrak{B}_i$ is an extremal graph of (D). Consequently,

$$n + 1 = n(G) = \sum_{i=1}^{s} n_i - s + 2$$

(note that $G - V$ is connected) and

$$\begin{aligned} e(G) &= \sum_{i=1}^{s} e(\mathfrak{B}_i) + v(G:V) \leqq \sum_{i=1}^{s} (2n_i - 3) + 2 \\ &= 2\left(\sum_{i=1}^{s} n_i - s + 1\right) - s = 2((n+1) - 1) - s \\ &\leqq 2(n+1) - 3. \end{aligned}$$

Equality holds if and only if $s = 1$ and $\mathfrak{B}_1$ is an extremal graph of (D) such that $A_1$, $B_1$ coincide with $\bar{A}$, $\bar{B}$. The vertex $V$ is joined to $\bar{A}$, $\bar{B}$ by edges, i.e., $v_1 = \bar{A}V\bar{B}$. Since $D_\varrho \cup \bar{A}V\bar{B}\bar{A}$, $\varrho = 2, 3$, contains a 5-cycle with two diagonals, $G - V \cong K_2 + I_{n-2}$, and consequently $G \cong K_2 + I_{n-1}$ (see Fig. 10.3.1).

Case 2. Without loss of generality, let $\mathfrak{B}_1$ be non-bipartite. By the induction hypothesis $\mathfrak{B}_1 \cong W_3$ or $e(\mathfrak{B}_1) \leqq 2n_1 - 3$. Since $G$ is 2-connected, two paths $v_i$, $w_i$ join $\mathfrak{B}_1$ and $\mathfrak{B}_i$, $2 \leqq i \leqq s$, to ends $A_1$, $B_1$ in $\mathfrak{B}_1$ and $A_i$, $B_i$ in $\mathfrak{B}_i$. (Note that since $v(G:V) = 2$, the graph $G - V$ is a chain of blocks, and therefore $A_1, B_1$ are the same vertices for all $2 \leqq i \leqq s$.)

Let $p_1$ and $\bar{p}_1$ denote two $A_1,B_1$-paths of $\mathfrak{B}_1$ having different parities. Then each $A_i,B_i$-path of $\mathfrak{B}_i$ together with $v_i$, $w_i$ and either $p_1$ or $\bar{p}_1$ form an odd cycle. Hence each $A_i,B_i$- path of $B_i$ has at most one diagonal.

Assertion (D) implies that $e(\mathfrak{B}_i) \leqq 2n_i - 3$ for all $2 \leqq i \leqq s$. Consequently,

$$n + 1 = n(G) = \sum_{i=1}^{s} n_i - s + 2$$

and

$$\begin{aligned} e(G) &= \sum_{i=1}^{s} e(\mathfrak{B}_i) + v(G:V) \leqq e(\mathfrak{B}_1) + \sum_{i=2}^{s} (2n_i - 3) + 2 \\ &= (e(\mathfrak{B}_1) - (2n_1 - 3)) + \sum_{i=1}^{s} (2n_i - 3) + 2 \\ &\leqq (e(\mathfrak{B}_1) - (2n_1 - 3)) + 2\left(\sum_{i=1}^{s} n_i - s + 1\right) - s \\ &= (e(\mathfrak{B}_1) - (2n_1 - 3)) + (2(n+1) - 3) - (s - 1). \end{aligned}$$

If $\mathfrak{B}_1 \ncong W_3$, then $e(\mathfrak{B}_1) \leqq 2n_1 - 3$ and

$$e(G) \leqq 2(n+1) - 3.$$

Equality holds if and only if $s = 1$ and $G - V = \mathfrak{B}_1$ is an extremal graph of (F). Consequently, $G$ is also an extremal graph of (F). If $\mathfrak{B}_1 \cong W_3$, then $e(\mathfrak{B}_1) = 2n_1 - 2$. Obviously, $\mathfrak{B}_1$ contains $A_1,B_1$-paths of lengths 3 and 2 having three and one diagonals, respectively. Therefore, each $A_1,B_1$-path of $G - (\mathfrak{B}_1 - \{A_1, B_1\})$ has no diagonal different from $(A_1, B_1)$, and its length is odd. Hence $V$ has at most one neighbour in $\mathfrak{B}_1$, and consequently $s \geqq 2$. The inequalities $e(\mathfrak{B}_1) = 2n_1 - 2$ and $s \geqq 2$ imply that

$$e(G) \leqq (e(\mathfrak{B}_1) - (2n_1 - 3)) + (2(n+1) - 3) - (s-1) = 2(n+1) - 3 - (s-2)$$
$$\leqq 2(n+1) - 3.$$

Equality holds if and only if $s = 2$ and $\mathfrak{B}_2$ is an extremal graph of (D) containing no $\bar{A},\bar{B}$-path with a diagonal. Consequently, $\mathfrak{B}_2 \cong K_2$ and $G$ is the union of a $W_3$ and a path of length 3 whose only common elements are the end vertices of the path (see Fig. 10.3.1a). □

# Chapter 11

# Longest cycles in graphs of given maximum degree

## 11.0. Introduction

G. A. Dirac already observed that the circumference in the class of all 2-connected graphs $G$ with given maximum degree $\Delta$ tends to infinity as the number of vertices does so. Since these graphs obviously do not have $\Delta + 1$ isomorphic bridges, Dirac's observation is also a consequence of Theorem 4.2.3.

The 2-connectedness of $G$ is an essential requirement even in the case when $G$ is $\Delta$-regular.

Dirac's result implies that if the circumference is a given integer $l$ (i.e., we consider the class $H_\Delta(l)$ of all 2-connected graphs with given maximum degree $\Delta$ having circumference $l$), then the maximum possible order $n_\Delta(l)$ of these graphs can be determined. Conversely, if the order of those graphs is a fixed integer $n$, then the maximum integer $c_\Delta(n)$ can be determined such that each 2-connected graph with maximum degree $\Delta$ and given order $n$ has a cycle of length at least $c_\Delta(n)$.

In this chapter we shall deal with both $n_\Delta(l)$ and $c_\Delta(n)$.

### 1° Graphs with given circumference

By definition, $n_\Delta(l)$ is the smallest integer $k$ such that every 2-connected graph with maximum degree $\Delta$ and order greater than $n_\Delta(l)$ contains a cycle of length greater than $l$. We shall show that

$$(\Delta - 1)^{\lfloor l/4 \rfloor - 1} \leqq n_\Delta(l) \leqq l\,(\Delta - 1)^{\lfloor l/4 \rfloor}.$$

In a forthcoming paper I shall present the improved upper bound $n_\Delta(l) \leqq c\,(\Delta - 1)^{\lfloor l/4 \rfloor}$, where $c$ is an appropriate constant.

P. Erdős and H. Sachs, 1963, and others intensively investigated $\Delta$-regular graphs ($\Delta \geqq 3$) with given girth $t$ having the minimum order among all these graphs (see H. Walther and H.-J. Voss, 1974).

Correspondingly, one may ask for properties of all 2-connected $\Delta$-regular graphs of given circumference $l$ having the maximum order. Let $\bar{H}_\Delta(l)$ denote the class of all these graphs and let $\bar{n}_\Delta(l)$ denote this maximum order. Already in the years 1967 and 1968 H. Walther and R. Lang determined lower bounds for $\bar{n}_\Delta(l)$ (H. Walther, 1967, and R. Lang and H. Walther, 1968):

$$\bar{n}_\Delta(l) \geqq 3\,(\Delta - 1)^{\lfloor l/4 - \Delta/2 \rfloor - 1}.$$

In this chapter for all even $l$ I shall determine the extremal graphs of $\bar{H}_3(l)$, and consequently the number $\bar{n}_3(l)$ itself (Theorems 11.3.3 and 11.3.4; see also H.-J. Voss, 1975). These extremal graphs are just the graphs of H. Walther and R. Lang.

Our proofs are essentially based on an intensive study of a longest cycle and its bridges of the graph in question.

## 2° Graphs with given order $n$

By definition, $c_\Delta(n)$ denotes the largest integer $k$ such that every 2-connected graph on $n$ vertices with maximum degree $\Delta$ contains a cycle of length at least $k$. Let $\bar{c}_\Delta(n)$ denote the same for $\Delta$-regular graphs.

J. A. Bondy and R. C. Entringer, 1980, proved that for $\Delta \geqq 3$ and $n \geqq \Delta + 2$

$$4\log_{\Delta-1} n - 4\log_{\Delta-1}\log_{\Delta-1} n - 20 < c_\Delta(n) < 4\log_{\Delta-1} n + 4.$$

Recently I have been able to improve the lower bound by showing that

$$c_\Delta(n) > 4\log_{\Delta-1} n - c',$$

where $c'$ is an appropriate constant.

For $\Delta$-regular graphs on $n \geqq 2\Delta$ vertices the well-known theorem of G. A. Dirac, 1952a, says that

$$\bar{c}_\Delta(n) \geqq 2\Delta.$$

Combining G. A. Dirac's theorem with the result of J. A. Bondy and R. C. Entringer we obtain for all $n \geqq 2\Delta$ that

$$\bar{c}_\Delta(n) \geqq \max\{2\Delta, 4\log_{\Delta-1} n - 4\log_{\Delta-1}\log_{\Delta-1} n - 20\}.$$

J. A. Bondy and R. C. Entringer, 1980, obtained upper bounds by constructing appropriate 2-connected $\Delta$-regular graphs on $n$ vertices for $\Delta \geqq 3$ and $n \geqq 2 + \Delta(\Delta + 1)$:

$$\bar{c}_\Delta(n) \leqq 4\log_{\Delta-1} n + 2\Delta.$$

With my improved lower bound we have

$$-c' < \bar{c}_\Delta(n) - 4\log_{\Delta-1} n < 2\Delta$$

with constant $c'$. I conjecture that

$$\bar{c}_\Delta(n) \geqq 4\log_{\Delta-1} n + 2\Delta + c'',$$

where $c''$ is an appropriate constant.

The last result means that for $\Delta \geqq 3$ and $n \geqq 2 + \Delta(\Delta + 1)$ the bounds of $\bar{c}_\Delta(n)$ are fairly close to each other. However, the bounds are rather weak for large values of $\Delta$ $\left(\Delta \geqq \sqrt{n}\right)$. For example, B. Jackson, 1980a, has proved that for $\Delta \geqq n/3$,

$$\bar{c}_\Delta(n) = n,$$

and J. A. Bondy, 1978, has conjectured that for $\Delta \geqq n/k$, where $k \geqq 3$ and $n$ is sufficiently large,

$$\bar{c}_\Delta(n) \geqq \frac{2n}{k-1}.$$

In this chapter we shall determine the precise value of $\bar{c}_3(n)$ for all even integers $n$ (note that every cubic graph has an even number of vertices). The result is presented in section 11.4.

Since for even $n \geqq 14$ the 2-connected graphs at which the bound $\bar{c}_3(n)$ is attained are planar, we have

$$\tilde{c}_3(n) = \bar{c}_3(n) \quad \text{for all even } n \geqq 14,$$

where $\tilde{c}_3(n)$ denotes the largest integer $k$ such that every 2-connected cubic planar graph on $n$ vertices contains a cycle of length at least $k$. The results on $\bar{c}_3(n)$ and $\tilde{c}_3(n)$ are presented in section 11.4.

Recently B. Bollobás and G. Brightwell, 1989, obtained similar results for some class of graphs with no subgraphs of minimum degree $r \geqq 3$.

## 3° Graphs of higher connectivity

What can be said about graphs which are at least 3-connected? How long a cycle must there be in a $k$-connected $r$-regular graph on $n$ vertices, $3 \leqq k \leqq r$?

For planar graphs in the case $k = r = 3$ this question goes back to P. G. Tait, 1880, who conjectured that any planar 3-connected 3-regular graph is Hamiltonian. W. T. Tutte, 1946, disproved this conjecture by finding a counterexample on 46 vertices.

Using Tutte's example, B. Grünbaum and T. S. Motzkin, 1962, constructed an infinite family of 3-connected 3-regular planar graphs such that the length of a longest cycle in each member of the family is at most $n^c$, where $c = 1 - 2^{-17}$ and $n$ is the number of vertices. The exponent $c$ was subsequently reduced by H. Walther, 1967 and 1969a, and by B. Grünbaum and H. Walther, 1973.

H. Walther, B. Grünbaum and others constructed infinite families of graphs with specified properties such that the length of a longest cycle in each member of the family does not exceed $n^c$, $c < 1$. A survey (up to 1972) on such investigations of the so-called *shortness exponent* can be found in H. Walther and H.-J. Voss, 1974.

All these results imply that

$$\bar{c}_3^3(n) \leqq n^c, \quad c \text{ constant}, \quad c < 1,$$

for infinitely many integers $n$, where $\bar{c}_\Delta^3(n)$ is the analogue of $\bar{c}_\Delta(n)$ for 3-connected graphs.

What is the lower bound for $\bar{c}_3^3(n)$?

The transition from 2-connected graphs to graphs of higher connectivity has a striking effect on the problems treated in 2°. J. A. Bondy and M. Simonovits, 1980, have proved, for example, that

$$e^{c_1\sqrt{\log n}} \leqq \bar{c}_3^3(n) \leqq n^{\log 8/\log 9}.$$

This improves a result of D. Barnette, 1966, that

$$\bar{c}_3^3(n) \geqq 3\log_2 n - 10.$$

(At the Colloquium on Graph Theory at Tihany M. Schäuble, 1966, also proposed to investigate $\bar{c}_\Delta^3(n)$ and to find the extremal graphs.) J. A. Bondy and M. Simonovits, 1980, made the conjecture that

$$\bar{c}_3^3(n) > n^c$$

for some constant $c < 1$; this was recently proved by B. Jackson in the affirmative. Moreover, B. Jackson, 1986, proved a stronger theorem:

*Any two edges of a 3-connected cubic graph of order n are in a cycle of length greater than $n^c$, where c can be chosen as $c = \log_2(1 + \sqrt{5}) - 1$.*

J. A. Bondy and M. Simonovits, 1980, state that their methods may also be used to obtain analogous results about 3-connected graphs with prescribed maximum degree $\Delta > 3$:

$$e^{c_2 \sqrt{\log n}} < c_\Delta^3(n) < n^{\log 2/\log(\Delta - 1)},$$

where $c_\Delta^3(n)$ is the analogue of $c_\Delta(n)$ for 3-connected graphs.

Another conjecture due to R. Häggkvist, 1976 (see B. Jackson, 1979, 1980a) concerns $\bar{c}_\Delta^k(n)$, the analogue of $\bar{c}_\Delta(n)$ for $k$-connected graphs, and asserts that, for $\Delta \geqq k + 2$ and $n \leqq \Delta(k + 1)$,

$$\bar{c}_\Delta^k(n) = n.$$

Results concerning this conjecture are discussed in section 7.7.

Let $\tilde{c}^3(n)$ denote the largest integer $k$ such that every 3-connected planar graph on $n$ vertices contains a cycle of length at least $k$. Then

$$e^{\sqrt{n}/6} \leqq \tilde{c}^3(n) \leqq 9n^{\ln 2/\ln 3}.$$

The lower and the upper bounds are due to L. Clark, 1985, and J. W. Moon and L. Moser, 1963, respectively. A first lower bound

$$\tilde{c}^3(n) \geqq 2\sqrt{\log_2 n - 5}$$

was found by D. W. Barnette, 1966.

We conclude this section with the remark that by Tutte's famous theorem every 4-connected planar graph has a Hamiltonian cycle. Consequently, $\tilde{c}^4(n) = n$.

## 11.1. On the number of vertices of 2-connected graphs with given maximum degree and circumference

Here we shall deal with the class $H_\Delta(l)$ of all 2-connected graphs with given maximum degree $\Delta$ and circumference $l$. We shall determine an upper bound on the number of vertices.

**Theorem 11.1.1.** *Every 2-connected graph of maximum degree $\Delta$ and circumference $l$ has*

$$n \leqq n_\Delta(l) \leqq l\,(\Delta - 1)^{\lfloor l/4 \rfloor}$$

*vertices.*

*Proof.* Let $G$ be a 2-connected graph with circumference $l$ and maximum degree $\Delta$. Let $L$ denote a longest cycle of $G$. Since $G$ is 2-connected, each vertex $V$ of $G - L$ is joined by two paths $p_1, p_2$ to $L$, where the only common vertex of $p_1$ and $p_2$ is $V$. The path $p_1 \cup p_2$ is an $L,L$-path. Hence it has length at most $\lfloor l/2 \rfloor$. Consequently, one of the paths has length at most $\lfloor l/4 \rfloor$.

Let $A_i$ denote the set of all vertices of $G$ having distance $i$ from $L$. Then

$$|A_0| = l,\ |A_1| \leqq l(\Delta - 2),\ \ldots,\ |A_{\lfloor l/4 \rfloor}| \leqq l(\Delta - 2)(\Delta - 1)^{\lfloor l/4 \rfloor - 1}$$

and

$$|A_i| = 0 \text{ for all } i \geqq \lfloor l/4 \rfloor.$$

Hence $G$ has order $n$, where

$$n \leqq l\,(1 + (\Delta - 2) + (\Delta - 2)(\Delta - 1) + \ \ldots\ + (\Delta - 2)(\Delta - 1)^{\lfloor l/4 \rfloor - 1}) = l\,(\Delta - 1)^{\lfloor l/4 \rfloor}. \ \square$$

Recently I was able to improve Theorem 11.1.1 by using the methods which will be presented in this chapter.

**Theorem 11.1.2.** *Every 2-connected graph with maximum degree $\Delta$ and circumference $l$ has at most*

$$n_\Delta(l) \leqq c(\Delta - 1)^{\lfloor l/4 \rfloor}$$

*vertices, where $c$ is an appropriate constant.* □

Examples show that the bound of Theorem 11.1.2 cannot be essentially improved.

Let $T^h$ be a tree with root $W$ of height $h$, $h \geqq 1$, and end vertices $Y_1, \ldots, Y_s$, where $W$ is an inner vertex, each inner vertex having valency $\Delta$ and the distance between $W$ and $Y_i$ being $h$ for all $1 \leqq i \leqq s$. Consequently, $s = \Delta(\Delta - 1)^{h-1}$. Let $\bar{T}^h$ be a tree such that $T^h$ and $\bar{T}^h$ are disjoint and there is an isomorphism $\varphi$ from $T^h$ onto $\bar{T}^h$. We identify $Y_i$ and $\bar{Y}_i$ for all $1 \leqq i \leqq s$, where $\bar{Y}_i = \varphi(Y_i)$. The resulting graph $\Gamma_\Delta(4h)$ is 2-connected, has circumference $l = 4h$ and maximum degree $\Delta$ (see Fig. 11.1.1a and e). Its order is

$$2 + \frac{1}{\Delta - 2}\{\Delta^2(\Delta - 1)^{\lfloor l/4 \rfloor - 1} - 2\Delta\}.$$

**Proposition 11.1.3.** *The graph $\Gamma_\Delta(4h)$ is a 2-connected graph of circumference $l = 4h$ and maximum degree $\Delta$ having*

$$2 + \frac{1}{\Delta - 2}\{\Delta^2(\Delta - 1)^{l/4 - 1} - 2\Delta\} \geqq (\Delta - 1)^{l/4}$$

*vertices.* □

The graphs $\Gamma_\Delta(4h)$, $h = 1, 2, \ldots$, are not regular. Next we introduce a sequence of $\Delta$-regular graphs $\bar{\Gamma}_\Delta(4h + 2\Delta)$. Let $G_1, \ldots, G_s$ be $s$ disjoint copies of $K^-_{\Delta+1}$. We identify one vertex of valency $\Delta - 1$ of $G_i$ with $Y_i$ and the other vertex of valency $\Delta - 1$ with $\bar{Y}_i$ for all $1 \leqq i \leqq s$. The resulting graph $\bar{\Gamma}_\Delta(4h + 2\Delta)$ is $\Delta$-regular, 2-connected and has circumference $l = 4h + 2\Delta$. Its order is

$$2 + \frac{1}{\Delta - 2}(\Delta^2(\Delta - 1)^{(l - 2\Delta)/4} - 2\Delta).$$

**Proposition 11.1.4.** *The graph $\bar{\Gamma}_\Delta(4h + 2\Delta)$ is a $\Delta$-regular, 2-connected graph of circumference $l = 4h + 2\Delta$ having*

$$2 + \frac{1}{\Delta - 2}(\Delta^2(\Delta - 1)^{(l - 2\Delta)/4} - 2\Delta) \geqq (\Delta - 1)^{(l - 2\Delta)/4}$$

*vertices.*

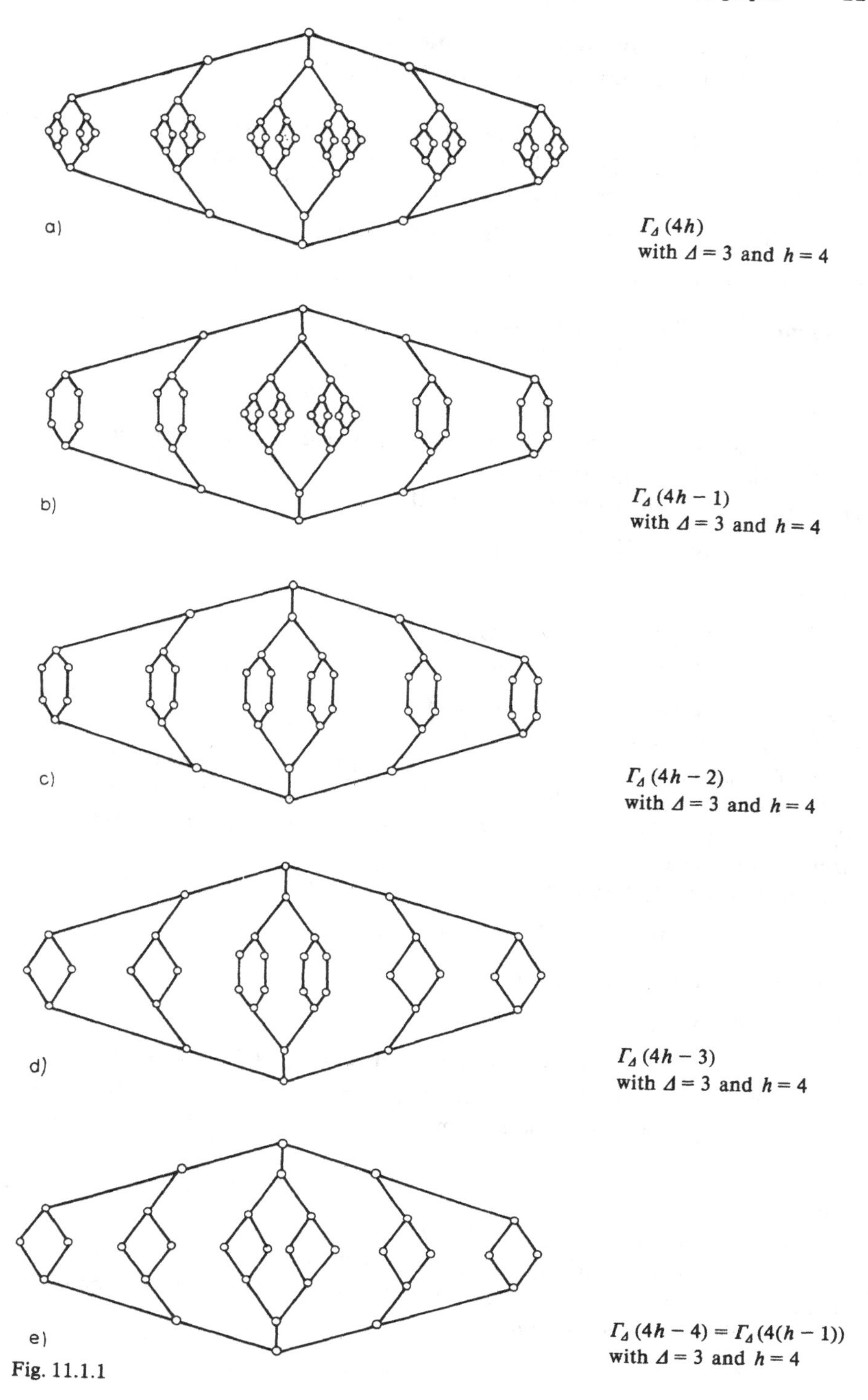
a)
$\Gamma_\Delta(4h)$
with $\Delta = 3$ and $h = 4$
b)
$\Gamma_\Delta(4h-1)$
with $\Delta = 3$ and $h = 4$
c)
$\Gamma_\Delta(4h-2)$
with $\Delta = 3$ and $h = 4$
d)
$\Gamma_\Delta(4h-3)$
with $\Delta = 3$ and $h = 4$
e)
$\Gamma_\Delta(4h-4) = \Gamma_\Delta(4(h-1))$
with $\Delta = 3$ and $h = 4$

Fig. 11.1.1

## 11.2. On the circumference of 2-connected graphs with given maximum degree and order

By definition, $c_\Delta(n)$ is the largest integer $k$ such that every 2-connected simple graph on $n$ vertices with maximum degree $\Delta$ contains a cycle of length at least $k$.

J. A. Bondy and R. C. Entringer, 1980, proved for $\Delta \geqq 3$ and $n \geqq \Delta + 2$ that

$$4 \log_{\Delta-1} n - 4 \log_{\Delta-1} \log_{\Delta-1} n - 20 < c_\Delta(n) < 4 \log_{\Delta-1} n + 4 .$$

We shall prove this theorem with a slightly improved lower bound.

**Theorem 11.2.1.** *For every pair of integers $\Delta \geqq 3$ and $n \geqq \Delta + 2$, we have*

$$4 \log_{\Delta-1} n - 4 \log_{\Delta-1} \log_{\Delta-1} n - 8 < c_\Delta(n) < 4 \log_{\Delta-1} n + 4 .$$

*Proof.*

(A) *The lower bound.* By Theorem 11.1.1 we know that every 2-connected graph $G$ of circumference $l$ and maximum degree $\Delta$ has $n \leqq l(\Delta - 1)^{\lfloor l/4 \rfloor}$ vertices. If $l \geqq 4 \log_{\Delta-1} n$, then Theorem 11.2.1 is true with respect to $G$. If $l < 4 \log_{\Delta-1} n$, then

$$n \leqq l(\Delta - 1)^{\lfloor l/4 \rfloor} < 4(\log_{\Delta-1} n)(\Delta - 1)^{l/4} .$$

Therefore,

$$\begin{aligned} l &\geqq 4 \log_{\Delta-1} n - 4 \log_{\Delta-1} \log_{\Delta-1} n - 4 \log_{\Delta-1} 4 \\ &\geqq 4 \log_{\Delta-1} n - 4 \log_{\Delta-1} \log_{\Delta-1} n - 8 . \end{aligned}$$

The inequality is true for all 2-connected graphs with maximum degree $\Delta \geqq 3$. This proves the validity of the lower bound of $c_\Delta(n)$.

(B) *The upper bound.* For each $n \geqq \Delta + 2$ we construct a 2-connected graph on $n$ vertices with maximum degree $\Delta$. We start with the sequence of graphs $\Gamma_\Delta(4), \ldots, \Gamma_\Delta(4h-4), \Gamma_\Delta(4h), \ldots$ of order $n(\Gamma_\Delta(4)), \ldots, n(\Gamma_\Delta(4h-4)), n(\Gamma_\Delta(4h)), \ldots$, respectively. These graphs have been defined above.

Next for each $h \geqq 2$ we construct a sequence of 2-connected graphs

$$\Gamma_0^h = \Gamma_\Delta(4h), \Gamma_1^h, \ldots, \Gamma_{s'}^h = \Gamma_\Delta(4h-4)$$

with maximum degree $\Delta$ such that

$$n(\Gamma_i^h) = n(\Gamma_{i-1}^h) - 1 \quad \text{for all} \quad 0 \leqq i \leqq s' - 1 .$$

We start with $\Gamma_0^h = \Gamma_\Delta(4h)$. The notation defined above will be used here.

Step 1. For $1 \leqq i \leqq s$ we delete the vertex $Y_i$ from $\Gamma_{i-1}^h$ and if the two neighbours of $Y_i$ are not already adjacent, then we join them by a new edge $y_i$. The resulting graph is denoted by $\Gamma_i^h$.

Step 2. For $s + 1 \leqq i \leqq s' = s + \dfrac{s}{\Delta - 1}$ let $\Gamma_i^h$ be obtained from $\Gamma_{i-1}^h$ by contracting one of the new edges $y_j$. Obviously, $\Gamma_{s'}^h = \Gamma_\Delta(4h-4)$. The sequence $\Gamma_0^h, \Gamma_1^h, \ldots, \Gamma_{s'}^h$ consists of 2-connected graphs of maximum degree $\Delta$ with $n(\Gamma_i^h) = n(\Gamma_{i-1}^h) - 1$ for all $1 \leqq i \leqq 2s$. Thus, for each $n$ with $n(\Gamma_\Delta(4h-4)) < n \leqq n(\Gamma_\Delta(4h))$ we have such a graph of order $n$ with circumference $l$, $4h - 4 < l \leqq 4h$.

By Proposition 11.1.3,

$$2+\frac{1}{\Delta-2}\{\Delta^2(\Delta-1)^{h-2}-2\Delta\}<n\leqq 2+\frac{1}{\Delta-2}\{\Delta^2(\Delta-1)^{h-1}-2\Delta\}.$$

This implies that

$$n>(\Delta-1)^{h-1}\geqq(\Delta-1)^{l/4-1} \quad \text{and} \quad l<4\log_{\Delta-1}n+4.$$

Consequently,

$$c_\Delta(n)\leqq 4\log_{\Delta-1}n+4 \quad \text{for all} \quad n\geqq\Delta+2,\ \Delta\geqq 3.\ \square$$

With respect to the proof of Theorem 11.2.1 there are two corollaries: if in section (A) of this proof we apply Theorem 11.1.2, then an improved lower bound is obtained.

**Corollary 11.2.2.** $c_\Delta(n)\geqq 4\log_{\Delta-1}n-c'$, *where* $c'$ *is a constant.*

**Corollary 11.2.3.** $c(\Gamma_s^h)=4h-2$.

With Corollary 11.2.2 we define $\Gamma_\Delta(4h-2)=_{\text{def}}\Gamma_s^h$ (see Fig. 11.1.1c).

Next we are interested in 2-connected graphs of prescribed maximum degree and given odd circumference having order as large as possible.

Let $e$ be an edge of $T^h$ incident with $W$. Then $e$ and $\varphi(e)$ form an edge cut of $\Gamma_\Delta(4h)$.

Let $K$ denote the component of $\Gamma_\Delta(4h)-\{e,\varphi(e)\}$ not containing $W$. If we apply step 1 of section (B) of the proof of Theorem 11.2.1 to all vertices $Y_i$ of $\Gamma_\Delta(4h)-K$, then we obtain a 2-connected graph $\Gamma_\Delta(4h-1)$ of maximum degree $\Delta$ and circumference $4h-1$ (see Fig. 11.1.1b).

Let $K'$ denote the component of $\Gamma_\Delta(4h-2)-\{e,\varphi(e)\}$ not containing $W$. If we apply step 2 of section (B) of the proof of Theorem 11.2.1 to all edges $y_j$ of $\Gamma_\Delta(4h-2)-K'$, then we obtain a 2-connected graph $\Gamma_\Delta(4h-3)$ of maximum degree $\Delta$ and circumference $4h-3$ (see Fig. 11.1.1d).

## 11.3. The class $\bar{H}_3(l)$ of all 2-connected cubic graphs with circumference $l$

For convenience, let $H(l)=_{\text{def}}\bar{H}_3(l)$ and $n(l)=_{\text{def}}\bar{n}_3(l)$ in this and the following sections.

For $l\leqq 9$ the class $H(l)$ can easily be determined.

**Theorem 11.3.1.** $H(4)$, $H(6)$ *and* $H(8)$ *consist of all 2-connected cubic graphs of orders* 4, 6 *and* 8, *respectively;* $H(3)=H(5)=H(7)=\emptyset$. $\square$

By Theorem 11.3.1 the graphs of $H(2k)$, $2\leqq k\leqq 4$, are Hamiltonian.

Let $P_{12}$ denote the graph obtained from the Petersen graph when one vertex of it is replaced by a triangle. The following theorem will be useful.

**Theorem 11.3.2.** *Every cubic graph of order* 10 *is either a Petersen graph or a Hamiltonian graph.* $H(9)$ *consists of the Petersen graph. Every cubic graph of order* 12 *is either a graph* $P_{12}$ *or a Hamiltonian graph.* $H(11)$ *consists of the graph* $P_{12}$. $\square$

In proving Theorems 11.3.1 and 11.3.2 and the assertion $n(13) \neq 16$ I essentially applied Theorems 3.3.1, 3.3.4 and 3.3.6 (reformulated as Lemma 11.5.1 for cubic graphs).

Let $D(10)$ and $D(12)$ be the graphs depicted in Fig. 11.3.1d and e. $D(2j+4)$ is obtained from $D(2j)$ in the following way: in $D(2j)$ each configuration depicted in Fig. 11.3.1a is replaced by the configuration in Fig. 11.3.1b (see also $D(14)$ and $D(16)$ in Fig. 11.3.1f and g).

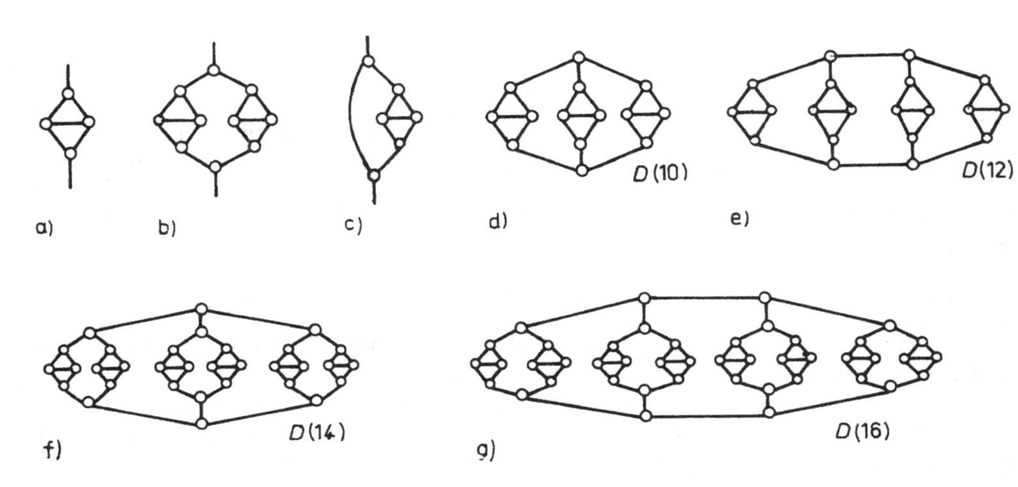


Fig. 11.3.1

Let $H(\leqq l)$ denote the class of all 2-connected cubic graphs with circumference at most $l$. In section 11.6, I shall prove the main theorem concerning this class:

**Theorem 11.3.3.** *Let $j \geqq 5$ be an integer. If $G \in H(\leqq 2j+1)$, then $G$ is isomorphic with $D(2j)$ or $P_{12}$ (in the case $j = 5$) or $G$ has order at most $n(D(2j)) - 4$.*

With the aid of Theorems 11.3.1 and 11.3.3 we can determine the maximum order $n(l)$ of $H(l)$ for all even $l$.

**Theorem 11.3.4.** $n(4) = 4$ *and* $n(6) = 6$. *For all even $l \geqq 8$ the maximum order of $H(l)$ is*

$$n(l) = -4 + 3 \cdot 2^{l/4} \begin{cases} 1 & \text{if} \quad 4 \mid l, \\ \dfrac{3}{4}\sqrt{2} & \text{if} \quad 4 \nmid l. \ \square \end{cases}$$

Next we deal with the case when $l$ is odd. Obviously, $c(D(2j)) = 2j$. Consequently, $n(11) = 12$ and $n(2j+1) \leqq n(2j) - 4$ for all $j \geqq 6$. This implies $n(13) \leqq n(12) - 4 = 16$.

For all $j \geqq 7$, I have constructed graphs of $H(2j+1)$ of order $n(2j) - 12$ (H.-J. Voss, 1982c). Therefore $n(2j+1) \geqq n(2j) - 12$.

Let $P_{14}$ denote the graph obtained from the Petersen graph by replacing one edge by the configuration depicted in Fig. 11.3.1a. Since $P_{14} \in H(13)$, we have $n(13) \geqq 14$. With the aid of Theorem 3.3.1, 3.3.4 and 3.3.6 (reformulated for cubic graphs as Lemma 11.5.1) it can easily be shown that $n(13) \neq 16$. Hence $n(13) = 14$.

These observations and Theorem 11.3.1 imply the following result:

**Theorem 11.3.5.** *$n(3)$, $n(5)$ and $n(7)$ do not exist. $n(9) = 10$, $n(11) = 12$ and $n(13) = 14$. For all integers $j \geqq 7$ the maximum order of $H(2j+1)$ satisfies the inequalities*

$$n(2j) - 12 \leqq n\,(2j+1) \leqq n(2j) - 4. \ \square$$

Let $\tilde{H}(l)$ denote the class of all 2-connected cubic planar graphs with circumference $l$. Let $\tilde{n}(l)$ denote the maximum order in $\tilde{H}(l)$.

Since $D(2j)$ is planar, Theorem 1.3.3 implies the following statement.

**Theorem 11.3.6.** *For all even $l \geqq 4$ we have*

$$\tilde{n}(l) = n(l). \ \square$$

## 11.4. The class $K(n)$ of all 2-connected cubic graphs of order $n$

Let $K(n)$ denote the class of all 2-connected cubic graphs of order $n$. Then $\bar{c}_3(n)$ is the minimum circumference in $K(n)$; for convenience we write $c(n)$ instead of $\bar{c}_3(n)$. Since every cubic graph has an even number of vertices, $c(n)$ is only defined for all even integers $n \geqq 4$.

First we shall prove the following result:

**Theorem 11.4.1.** *$c(14) = 10$. For all integers $j \geqq 5$ and all even integers $n$ with $n(2j) < n \leqq n\,(2j+2)$ the minimum circumference of $K(n)$ is*

$$c(n) = \begin{cases} 2j+2 & \text{if} \quad n \neq n(2j+2) - 2, \\ 2j+4 & \text{if} \quad n = n(2j+2) - 2. \end{cases}$$

A weaker theorem has already been proved in H.-J. Voss, 1982c).

*Proof.* Let $j \geqq 5$ be an integer. For $j \geqq 6$ Theorem 11.3.3 implies that if $G \in H(\leqq 2j+1)$, then $G$ is isomorphic to $D(2j)$ and has $n(2j)$ vertices or $G$ has order at most $n(D(2j)) - 4 = n(2j) - 4$. Consequently, if $n > n(2j)$ or $n = n(2j) - 2$, then the inequality $c(n) \geqq 2j+2$ holds.

For $j = 5$ Theorem 11.3.3 implies that if $G \in H(\leqq 11)$, then $G$ is isomorphic to $D(10)$ and has $n(10) = 14$ vertices or $G$ has order at most $n(D(10)) - 2 = n(10) - 2$. Consequently, if $n > n(10) = 14$, then $c(n) \geqq 12$ and if $n = n(10) = 14$, then $c(n) = 10$.

These results can be reformulated as follows:

$$c(14) = 10$$

and

$$c(n) \geqq 2j+2 \quad \text{for all} \quad n(2j) < n \leqq n\,(2j+2);$$

moreover

$$c(n) \geqq 2j+4 \quad \text{for} \quad n = n\,(2j+2) - 2.$$

The proof will be completed by the construction of a graph $H_n \in K(n)$ for which the lower bound is attained, i.e. $c(n) = c(H_n)$ (H.-J. Voss, 1982c). Let $m =_{\text{def}} n\,(2j+2)$, $j \geqq 5$. Then

$$H_m =_{\text{def}} D\,(2j+2);$$

$H_{m-4}$ is obtained from $H_m$ by replacing a configuration depicted in Fig. 11.3.1b by the configuration depicted in Fig. 11.3.1c;

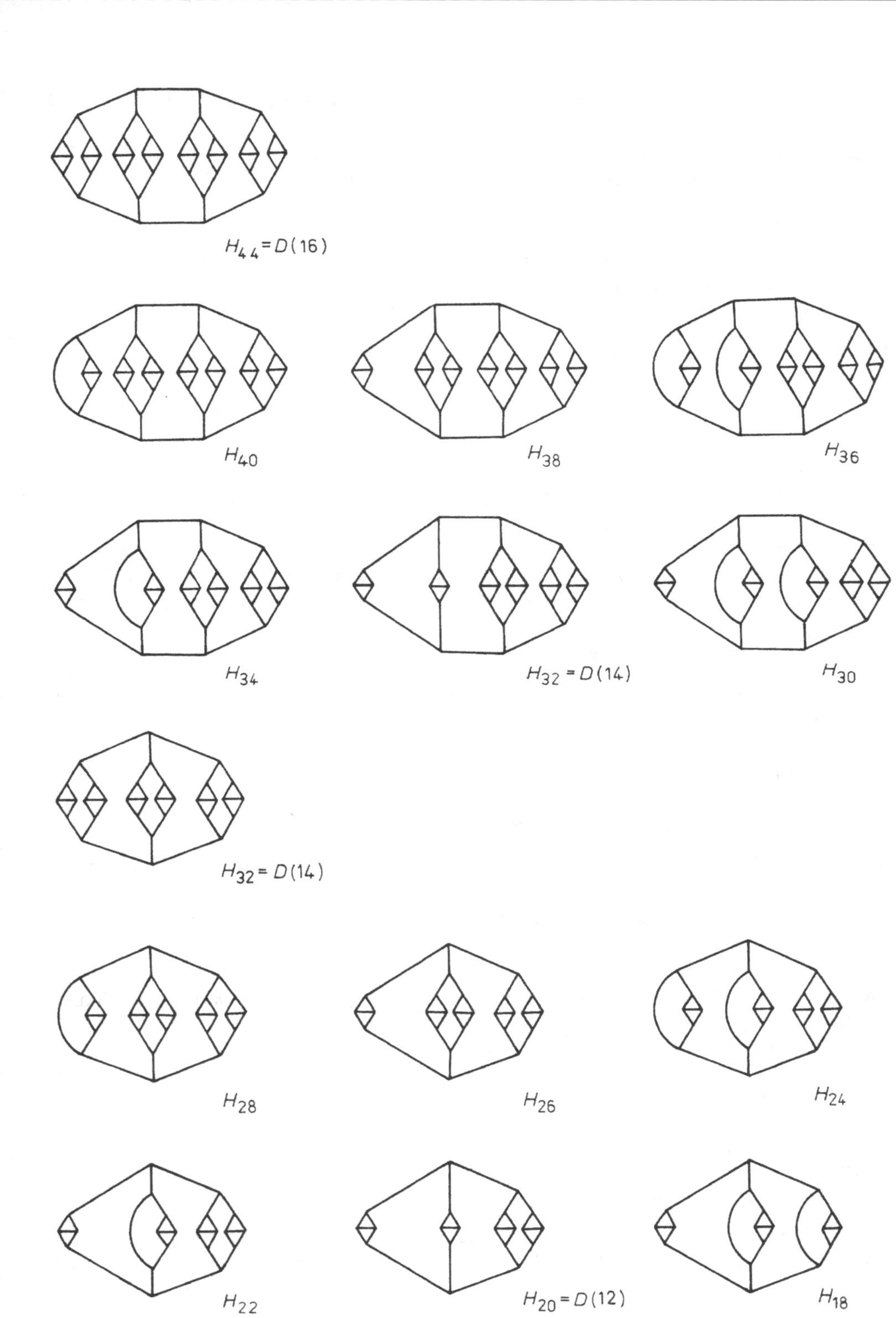
$H_{44} = D(16)$
$H_{40}$
$H_{38}$
$H_{36}$
$H_{34}$
$H_{32} = D(14)$
$H_{30}$
$H_{32} = D(14)$
$H_{28}$
$H_{26}$
$H_{24}$
$H_{22}$
$H_{20} = D(12)$
$H_{18}$

Fig. 11.4.1

$H_{m-6}$ is obtained from $H_m$ by replacing a configuration depicted in Fig. 11.3.1b by the configuration depicted in Fig. 11.3.1a;

$H_{m-8}$ is obtained from $H_{m-4}$ by replacing a configuration depicted in Fig. 11.3.1b by the configuration depicted in Fig. 11.3.1c.

For all even $n$ with $n(2j) - 2 \leqq n \leqq n(2j+2) - 4 = m - 4$ graphs $H_n$ are obtained from the graphs $H_{m-4}$, $H_{m-6}$ and $H_{m-8}$ by replacing zero, one or more configurations depicted in Fig. 11.3.1b by the configuration depicted in Fig. 11.3.1a.

For $n = n(2j)$ the graph $D(2j)$ is obtained again.

By construction, all graphs $H_n$ with

$$n(2j) - 2 \leqq n \leqq n(2j+2), \qquad n \notin \{n(2j),\, n(2j+2) - 2\},$$

have circumference $2j+2$. In Fig. 11.4.1 graphs $H_n$ are depicted for all even $n$, $18 \leqq n \leqq 44$, $n \neq 42$. □

For real $\alpha$, let $\lceil \alpha \rceil$ and $\lfloor \alpha \rfloor$ denote the smallest integer not less than $\alpha$ and the largest integer not greater than $\alpha$, respectively. Let

$$\varphi(n) =_{\text{def}} \min \left\{ 4 \left\lceil \log_2 \frac{n+4}{3} \right\rceil, 4 \left\lceil \log_2 \frac{4(n+4)}{9} \right\rceil - 2 \right\}$$

and

$$K =_{\text{def}} \{ n(2j) - 2 \mid j = 6, 7, \ldots \}.$$

**Theorem 11.4.2.** $c(4) = 4$, $c(6) = 6$, $c(8) = 8$, $c(10) = 9$ *and* $c(12) = 11$. *For all even* $n \geqq 14$ *the minimum circumference of* $K(n)$ *is*

$$c(n) = \begin{cases} \varphi(n) & \text{if } n \notin K, \\ \varphi(n) + 2 & \text{if } n \in K. \end{cases} \; \square$$

*Proof.* By Theorem 11.3.1 all 2-connected cubic graphs of order $2k$ form the class $H(2k)$ for $k = 2, 3, 4$. Since all these graphs are Hamiltonian, $c(4) = 4$, $c(6) = 6$ and $c(8) = 8$.

Theorem 11.3.2 implies that $c(10) = 9$ and $c(12) = 11$. The validity of the general case $n \geqq 14$ follows from Theorem 11.4.1. □

In Fig. 11.4.2 the function $c(n)$ is depicted. The points $(2j, c(2j))$ and $(2j+2, c(2j+2))$ are joined by a straight line segment for all $j \geqq 2$.

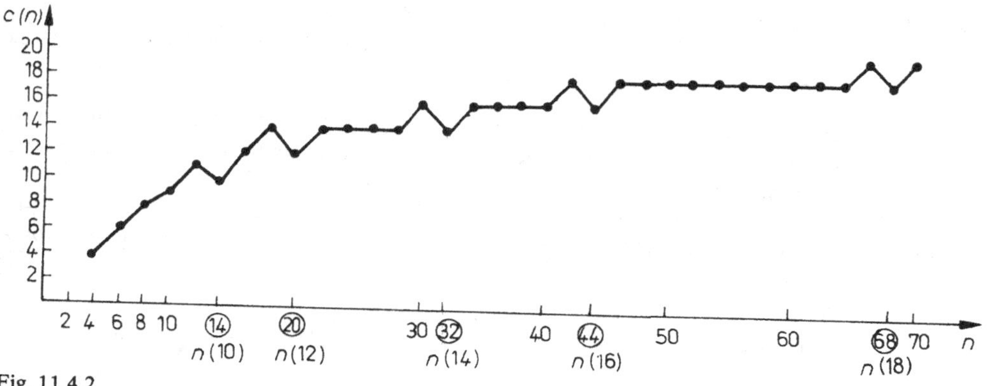


Fig. 11.4.2

The validity of the equality sign in Theorem 11.4.1 has been proved by constructing a 2-connected cubic graph $H_n$ of each order $n \geqq 14$. These graphs have been derived from $\{D(2j+2) \mid j \geqq 5\}$.

Correspondingly, from the set of graphs $\{\bar{\Gamma}_\Delta(4h+2\Delta) \mid h \geqq 1\}$ (as defined in section 11.1) 2-connected $\Delta$-regular graphs $\bar{H}_{\Delta,n}$ of order $n$ for each even $\Delta \geqq 4$ and each $n \geqq 2+\Delta(\Delta+1)$ and for each odd $\Delta \geqq 5$ and each *even* $n \geqq 2+\Delta(\Delta+1)$ can be constructed. This implies the result of J. A. Bondy and R. C. Entringer, 1980:

**Theorem 11.4.3.** *For each $\Delta \geqq 3$ and each $n \geqq 2+\Delta(\Delta+1)$ we have*

$$\bar{c}_\Delta(n) < 4\log_{\Delta-1} n + 2\Delta. \ \square$$

Let $\tilde{K}(n)$ denote the class of all 2-connected cubic planar graphs of order $n$. Then $\tilde{c}_3(n)$ is the minimum circumference in $\tilde{K}(n)$. For even $n \geqq 14$ the graph $H_n \in K(n)$ is planar, and consequently $H_n \in \tilde{K}(n)$. This implies that the following theorem is valid.

**Theorem 11.4.4.** *For all even $n \geqq 14$ we have*

$$\tilde{c}_3(n) = \bar{c}_3(n). \ \square$$

## 11.5. Preliminaries

If $C$ is a cycle with the two vertices $A$ and $B$, then $C[A, B]$ and $C[B, A]$ denote the two arcs of $C$ with ends $A$ and $B$. If $\pi$ is a path or a cycle, then $\lambda(\pi)$ is the length of it.

The proof of Theorem 11.3.3 will be accomplished by an intensive study of bridges of a longest cycle.

A bridge with $\alpha$ vertices of attachment is said to be an *$\alpha$-bridge.*

By Lemma 1.1.1 the kernel of a bridge $B$ is connected. Therefore, each bridge contains a tree whose end vertices are precisely the vertices of attachment of $B$. The number of edges of a largest such tree in $B$ is called the *length* $\lambda(B)$ of $B$.

In Chapter 3 we proved the following lemma (formulated here for cubic graphs):

**Lemma 11.5.1.** *Let $L$ be a longest cycle of a cubic graph $G$ and $l$ its length. Then*

(i) $\lambda(B) \leqq \lfloor l/2 \rfloor$ *for each bridge $B$ of $L$;*

(ii) $\lambda(B_1) + \lambda(B_2) \leqq \lfloor l/2 \rfloor$ *for each pair of different overlapping bridges $B_1$ and $B_2$;*

(iii) $\lambda(B_1) + \lambda(B_2) + \lambda(B_3) \leqq \lfloor l/2 \rfloor$ *for each triple of bridges $B_1$, $B_2$, $B_3$ which form an induced path of length 2 in the overlap graph $O(G:L)$.*

Next we define the bridges $M(2j-1)$ as follows: $M(1)$, $M(3)$ and $M(5)$ are depicted in Fig. 11.5.1 (small rectangles represent the vertices of attachment). For $j \geqq 3$ the bridge $M(2j+1)$ is obtained from $M(2j-1)$ in the following way: in $M(2j-1)$ each configuration depicted in Fig. 11.3.1a is replaced by the configuration depicted in Fig. 11.3.1b (the configuration of Fig. 11.3.1b is the bridge $M(7)$ without the vertices of attachment). Obviously, $\lambda(M(2j-1)) = 2j-1$.

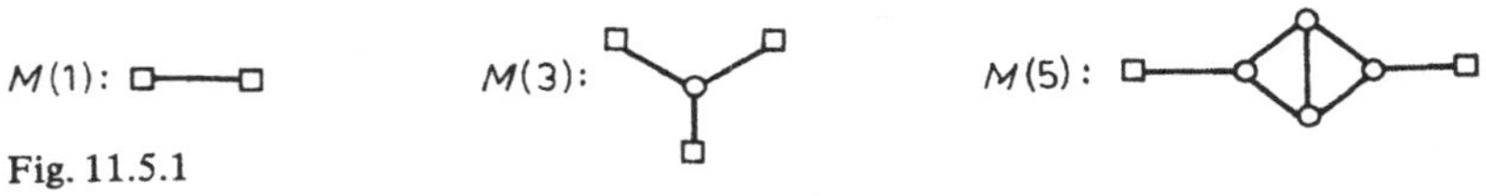

Fig. 11.5.1

The following lemmas can easily be proved:

**Lemma 11.5.2.**

(i) $n(M(1)) = 2$, $n(M(3)) = 4$ *and* $n(M(2j-1)) = 3 \cdot 2^{j-2}$ *for all* $j \geqq 3$;

(ii) $2n(M(2j-1)) = n(M(2j+1))$ *for all* $j \neq 2$;

(iii) $n(M(2j-1)) + 2 \leqq n(M(2j+1))$ *for all* $j \geqq 1$;

(iv) $n(M(2j-1)) + 6 \leqq n(M(2j+1))$ *for all* $j \geqq 3$. □

For convenience we put $M(2j) =_{\text{def}} M(2j-1)$. Then the length of $M(2j)$ is $\lambda(M(2j)) = \lambda(M(2j-1))$. For $j \neq 2$ the bridges $M(2j-1)$ and $M(2j)$ are 2-bridges.

**Lemma 11.5.3.**

(i) $n(D(4j-2)) = 3n(M(2j-1)) - 4$ *for all* $j \geqq 3$;

(ii) $n(D(4j)) = 4n(M(2j-1)) - 4$ *for all* $j \geqq 3$. □

## 11.6. The proof of Theorem 11.3.3

For convenience we reformulate this theorem as Theorem 11.6.1.

**Theorem 11.6.1.** *Let $j \geqq 5$ be an integer. If $G \in H(\leqq 2j+1)$, then $G$ is isomorphic to $D(2j)$ or $P_{12}$ (in the case $j = 5$) or $G$ has order at most $n(D(2j)) - 4$.*

*Proof of Theorem 11.6.1.* Assume that Theorem 11.6.1 is not true. Let $m$ be the smallest integer $j \geqq 5$ such that Theorem 11.6.1 is not true; i.e. there exists a graph $G \in H(\leqq 2m+1)$ which has at least $n(D(2m)) - 2$ vertices and is not isomorphic to $D(2m)$.

For all cubic graphs it is obvious that

(1) *all vertices of attachment and all inner vertices of a bridge $B$ with respect to a cycle in a cubic graph have degree* 1 *and* 3 *in $B$, respectively.* □

Therefore, in the following we shall consider only bridges with these properties.

(2) *Let $B$ be a 2-bridge in a cubic graph. Then each longest cycle $C$ of $B$ has length $\lambda(C) \leqq 2\lambda(B) - 4$.*

*Let $P$, $Q$ be the two vertices of attachment of $B$ and $P'$, $Q'$ the only neighbour of $P$, $Q$ in $B$, respectively.*

*If $\lambda(C) = 2\lambda(B) - 4$, then $C$ goes through $P'$ and $Q'$ and the two arcs of $C$ between $P'$ and $Q'$ have equal length $\lambda(B) - 2$.*

*Proof of (2).* Let $B$ be a 2-bridge with vertices of attachment $P$ and $Q$. By (1) the vertices $P$ and $Q$ have degree 1 in $B$. Let $P'$ and $Q'$ be the unique neighbour of $P$ and $Q$, respectively. Let $C$ be any longest cycle of $B$. This is also a cycle of $B - \{P, Q\}$. Since $G$ is 2-connected, there are two disjoint paths $w_1$ and $w_2$ joining $P'$ and $Q'$ to $C$, respectively. Let $a$ denote the larger arc of $C$ between the end vertices of $w_1$ and $w_2$ on $C$. Obviously, $\lambda(a) \geqq \frac{1}{2}\lambda(C)$. Then $p =_{\text{def}} [P, P'] \cup w_1 \cup a \cup w_2 \cup [Q', Q]$ is a $P,Q$-path of $B$ of length

$$\lambda(p) \geqq 1 + \lambda(w_1) + \lambda(a) + \lambda(w_2) + 1.$$

By the definition of the length of $B$, $\lambda(p) \leqq \lambda(B)$. Hence $\frac{1}{2}\lambda(C) + 2 \leqq \lambda(p) \leqq \lambda(B)$ and $\lambda(C) \leqq 2\lambda(B) - 4$.

If $\lambda(C) = 2\lambda(B) - 4$, then $C$ goes through $P'$ and $Q'$ and $\lambda(a) = \lambda(B) - 2$. Therefore, the second $P', Q'$-arc of $C$ has also length $\lambda(B) - 2$. □

(3) *Every 2-bridge $B$ with $\lambda(B) \leqq m + 2$ is isomorphic to $M(\lambda(B))$ or has order at most $n(M(\lambda(B))) - 4$.*

From an analysis of all bridges of length at most 6 we have the following result:

**Remark 11.6.2.** *$M(1)$ and $M(5)$ are the only 2-bridges of lengths at most 6.* □

*Proof of (3).* Assertion (3) will be proved by induction on $\lambda(B)$. Remark 11.6.2 implies the validity of (3) for all 2-bridges $M$ with $\lambda(M) \leqq 6$. Assume that (3) is true for all 2-bridges $M$ of length $\lambda(M) < 2p \leqq m + 2$, $p \geqq 4$, or $\lambda(M) < 2p + 1 \leqq m + 2$, $p \geqq 3$. Let $B$ be a 2-bridge of length $\lambda(B) = 2p$ (or $\lambda(B) = 2p + 1$). Note that the case $\lambda(B) = 2p + 1$ will be dealt with in brackets.

Let $P$, $Q$ be the vertices of attachment of $B$, having degree 1 by (1), and let $P'$, $Q'$ be the neighbours of $P, Q$, respectively. By (2) the circumference $c(B)$ is at most $2\lambda(B) - 4$. We consider two cases:

Case 1. Assume that $c(B) \leqq 2\lambda(B) - 5$. Since $\lambda(B) = 2p \leqq m + 2$ (or $\lambda(B) = 2p + 1 \leqq m + 2$), the circumference of $B$ is $c(B) \leqq 4p - 5 \leqq 2m - 1$ (or $c(B) \leqq 4p - 3 \leqq 2m - 1$).

Case 1.1. Assume that $p \geqq 4$. Let $B$ and $M(2p - 3)$ be two disjoint 2-bridges. A new 2-connected cubic graph $\Gamma$ is formed by deleting all vertices of attachment of $B$ and $M(2p - 3)$ and adding two new edges as depicted in Fig. 11.6.1. Obviously,

$$c(\Gamma) \leqq \max\{c(B), c(M(2p - 3)), \lambda(B) + \lambda(M(2p - 3)) - 2\}.$$

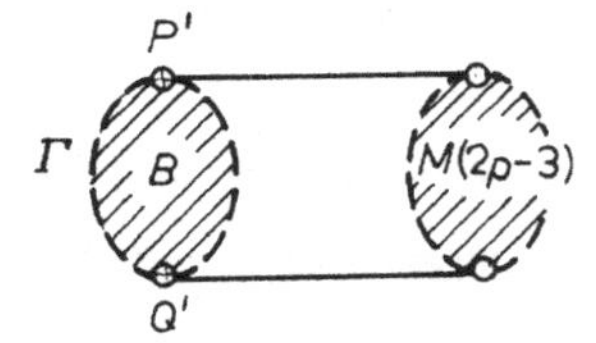


Fig. 11.6.1

Since $c(B) \leqq 4p - 5$, $c(M(2p - 3)) = 4p - 10$, $\lambda(B) = 2p$ and $\lambda(M(2p - 3)) = 2p - 3$ it follows that $c(\Gamma) \leqq 4p - 5$, where $4p - 5 \geqq 11$. (Since $c(B) \leqq 4p - 3$, $c(M(2p - 3)) = 4p - 10$, $\lambda(B) = 2p + 1$ and $\lambda(M(2p - 3)) = 2p - 3$ it follows that $c(\Gamma) \leqq 4p - 3$, where $4p - 3 \geqq 13$.)

Since $m$ is minimal, the assertion of Theorem 11.6.1 is valid for all $5 \leqq j \leqq m - 1$. Hence $\Gamma \cong D(4p - 6)$, or $\Gamma \cong P_{12}$ if $p = 4$, or $n(\Gamma) \leqq n(D(4p - 6)) - 4$ (or $n(\Gamma) \leqq n(D(4p - 4))$).

If $\Gamma \cong D(4p - 6)$, then $\Gamma$ and $D(4p - 6)$ have the same structure.

We "delete" $M(2p - 3)$ in $D(4p - 6)$ and obtain $B \cong M(2p - 1)$. By the definition $M(2p) =_{\text{def}} M(2p - 1)$ this implies that $B \cong M(2p)$. The case $\Gamma \cong P_{12}$ is impossible, because $\Gamma$ has an edge-cut of size 2 and $P_{12}$ has none.

If $n(\Gamma) \leqq n(D(4p-6)) - 4$, then since

$$n(\Gamma) = n(B) + n(M(2p-3)) - 4 \quad \text{and} \quad n(D(4p-6)) = 3n(M(2p-3)) - 4$$

(by Lemma 11.5.3) we obtain $n(B) \leqq 2n(M(2p-3)) - 4$. By Lemma 11.5.2 (ii) it follows that

$$n(B) \leqq 2n(M(2p-3)) - 4 = n(M(2p-1)) - 4 = n(M(\lambda(B))) - 4.$$

(Since $n(\Gamma) \leqq n(D(4p-4))$ and $n(D(4p-4)) = 4n(M(2p-3)) - 4$ by Lemma 11.5.3, we have $n(B) \leqq 3n(M(2p-3))$. Lemma 11.5.2 implies that

$$\begin{aligned} n(B) &\leqq 3n(M(2p-3)) = n(M(2p-3)) + n(M(2p-1)) \\ &\leqq (n(M(2p-1)) - 6) + n(M(2p-1)) = n(M(2p+1)) - 6 \\ &< n(M(\lambda(B))) - 4.) \end{aligned}$$

Case 1.2. Assume that $p = 3$. Then $\lambda(B) = 2p + 1 = 7$ and $c(B) \leqq 2\lambda(B) - 5 \leqq 9$. If the edge $(P', Q')$ belongs to $B$, then $B - \{P, Q, (P', Q')\}$ is a bridge with respect to $(\{P', Q'\}, \emptyset)$ of length 5. Consequently, $n(B) \leqq n(M(7)) - 4$.

Suppose that the edge $(P', Q')$ does not belong to $B$. Let

$$\Gamma =_{\text{def}} (B - \{P, Q\}) \cup \{(P', Q')\}.$$

Obviously, $c(\Gamma) \leqq 9$.

Theorem 11.3.1 implies that if $c(B) \leqq 6$, then $n(\Gamma) \leqq 6$ and $n(B) \leqq 8 \leqq n(M(7)) - 4$. Theorems 11.3.1 and 11.3.2 imply that if $c(B) \geqq 6$, then $\Gamma$ is isomorphic to the Petersen graph or $\Gamma$ is a Hamiltonian graph of order 8. In both cases we arrive at $\lambda(B) > 7$. Contradiction!

Case 2. Assume that $c(B) = 2\lambda(B) - 4$. Then, by (2), each longest cycle $C$ of $B$ passes through $P'$ and $Q'$ and both $P', Q'$-arcs of $C$ have equal length $\lambda(B) - 2$. Let them be denoted by $a$ and $b$.

($\alpha$) *There is no path joining an inner vertex of $a$ to an inner vertex of $b$.*

*Proof of* ($\alpha$). Assume that there is a path $w$ joining an inner vertex $A$ of $a$ to an inner vertex $B$ of $b$. Then

$$p_1 = [P, P'] \cup a[P', A] \cup w \cup b[B, Q'] \cup [Q', Q]$$

and

$$p_2 = [P, P'] \cup b[P', B] \cup w \cup a[A, Q'] \cup [Q', Q]$$

are two $P, Q$-paths of $B$ which cover $(P, P')$, $(Q, Q')$ and $w$ twice and $C$ once. Hence $\lambda(p_1) + \lambda(p_2) \geqq 2 + 2 + 2 + \lambda(C)$. By the definition of $\lambda(B)$ the length $\lambda(p_i) \leqq \lambda(B)$, $i = 1, 2$. Hence $\lambda(C) + 6 \leqq \lambda(p_1) + \lambda(p_2) \leqq 2\lambda(B)$. Consequently, $\lambda(C) \leqq 2\lambda(B) - 6$, contradicting the assumption $\lambda(C) = 2\lambda(B) - 4$! □

By ($\alpha$), the graph $B - \{P, Q\}$ is the union of the subgraph $H' = (\{P', Q'\}, \emptyset)$ and two of its bridges $B_1$, $B_2$ of length $\lambda(B) - 2$. By induction $B_i \cong M(\lambda(B) - 2)$ or $n(B_i) \leqq n(M(\lambda(B) - 2)) - 4$. Consequently, $B \cong M(\lambda(B))$ or $n(B) \leqq n(M(\lambda(B))) - 4$. Thus the proof of (3) is complete. □

Let $L$ be a largest cycle of the graph $G$. A bridge $B$ of $L$ is called *large (small)* if $\lambda(B) > m/2$ ($\lambda(B) \leqq m/2$). Assertion (ii) of Lemma 11.5.1 implies that

(4) *large bridges avoid each other.* □

Assertion (4) and assertion (iii) of Lemma 11.5.1 imply that

(5) *each small bridge overlaps at most one large bridge.* □

We shall show that

(6) *if $B$ is any $\alpha$-bridge of length $2j-1$ or $2j$, then $n(B) \leqq n(M(2j-1))$.*

*Proof of (6).* For a bridge $B$ with $\alpha \geqq 3$ vertices of attachment the assertion (6) will be proved by constructing a 2-bridge $B'$ with $n(B') = n(B) - 2\delta$ and $\lambda(B') \leqq \lambda(B) - 2\delta$ for some integer $\delta$, $0 \leqq \delta \leqq \alpha - 2$. From this result we shall infer that $n(B) \leqq n(M(2j-1))$, i.e. the assertion of (6).

Let $A_1, \ldots, A_\alpha$ denote the vertices of attachment of $B$. If $\alpha = 2$, then $B' =_{\text{def}} B$. Let $\alpha \geqq 3$. We consider the subgraph $B_1^1$ of $B$ induced by the union of all blocks of $B$ containing at least one vertex which is not on an $A_i$,$A_j$-path of the subgraph $B - \{A_1\}$, $2 \leqq i < j \leqq \alpha$. Obviously, these blocks contain only one vertex $P_1$ which is on an $A_i$,$A_j$-path of $B - \{A_1\}$. Since $G$ is 2-connected, each vertex of $B_1^1$ is on an $A_1$,$P_1$-path of $B_1^1$ (see Fig. 11.6.2).

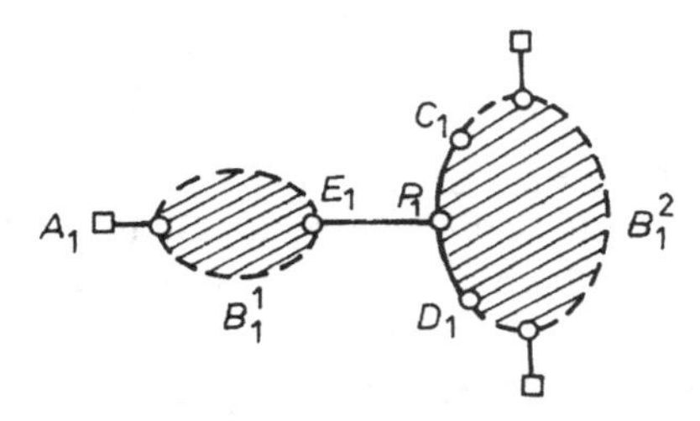


Fig. 11.6.2

Let $B_1^2 =_{\text{def}} B - (B_1^1 - \{P_1\})$. Since, by assertion (1), $v(G:P_1) = 3$ and $P_1$ is on an $A_i$,$A_j$-path of $B_1^2$, it follows that $v(B_1^2 : P_1) = 2$ and $v(B_1^1 : P_1) = 1$. Let $C_1$, $D_1$ and $E_1$ denote the neighbours of $P_1$ such that $E_1 \in B_1^1$ and $C_1$, $D_1 \in B_1^2$. Next the bridge $\overline{B}_1^2$ will be defined. We consider two cases:

Case 1. If $(C_1, D_1) \notin B_1^2$, then

$$\overline{B}_1^2 =_{\text{def}} (B_1^2 - \{P_1\}) \cup \{(C_1, D_1)\}.$$

It can easily be proved that

$$(\alpha) \qquad n(B_1^1) + n(\overline{B}_1^2) = n(B) \quad \text{and} \quad \lambda(B_1^1) + \lambda(\overline{B}_1^2) \leqq \lambda(B) - 1.$$

Case 2. If $(C_1, D_1) \in B_1^2$, then

$$\overline{B}_1^2 =_{\text{def}} (B_1^2 - \{(C_1, D_1)\}) \cup \{P_1', (P_1', P_1), (P_1', C_1), (P_1', D_1)\},$$

where $P_1'$ is a vertex not in $G$. It can easily be proved that

$$(\beta) \qquad n(B_1^1) + n(\overline{B}_1^2) = n(B) + 2 \quad \text{and} \quad \lambda(B_1^1) + \lambda(\overline{B}_1^2) \leqq \lambda(B) + 1.$$

In both cases we identify the vertex $E_1$ of $B_1^1 - \{P_1\}$ with the vertex $A_\alpha$ of $\overline{B}_1^2$. This new graph is denoted by $B_1$.

In $B_1$ the new vertex has degree 3. Therefore, $A_1, \ldots, A_{\alpha-1}$ have valency 1 and all other vertices have degree 3. Consequently, $B_1$ is a new $(\alpha-1)$-bridge with the vertices of attachment $A_1, A_2, \ldots, A_{\alpha-1}$. The bridge $B_1$ satisfies:

$$(\gamma) \qquad \begin{cases} n(B_1) = n(B) - 2 \quad \text{and} \quad \lambda(B_1) \leqq \lambda(B) - 2 \quad \text{in case } 1, \\ n(B_1) = n(B) \quad \text{and} \quad \lambda(B_1) \leqq \lambda(B) \quad \text{in case } 2. \end{cases}$$

By repeatedly applying this method we obtain a 2-bridge $B'$. By $(\gamma)$ we have

$$n(B') = n(B) - 2\delta \quad \text{and} \quad \lambda(B') \leqq \lambda(B) - 2\delta,$$

where $0 \leqq \delta \leqq \alpha - 2$. This, assertion (3), the assertion (iii) of Lemma 11.5.2, and the definition $M(2j) =_{\text{def}} M(2j-1)$ imply that

$$\begin{aligned} n(B) &= n(B') + 2\delta \leqq n(M(\lambda(B'))) + 2\delta \\ &\leqq n(M(\lambda(B) - 2\delta)) + 2\delta \leqq n(M(\lambda(B))) \\ &\leqq n(M(2j-1)). \ \square \end{aligned}$$

Let $B$ denote a bridge of $L$ with $k = k(B)$ vertices of attachment $A_1, A_2, \ldots, A_k$, which are on $L$ in this cyclic order.

If $k(B) \geqq 3$, then $L[A_i, A_{i+1}]$ denotes the segment of $L$ with ends $A_i$ and $A_{i+1}$. The other arc of $L$ is denoted by $L[A_{i+1}, A_i]$.

If $k(B) = 2$, then $L[A_1, A_2]$ denotes the smaller segment of $L$ with ends $A_1$ and $A_2$.

Let the indices be chosen so that $\lambda(L[A_k, A_1])$ is the largest segment of $L$ with respect to $B$, i.e.

$$\lambda(L[A_k, A_1]) \geqq \lambda(L[A_i, A_{i+1}]) \quad \text{for all} \quad 1 \leqq i \leqq k-1.$$

We define

$$\varrho(B) =_{\text{def}} \sum_{i=1}^{k-1} \lambda(L[A_i, A_{i+1}]) = \lambda(L[A_1, A_k]).$$

A large bridge $B$ is said to be a *P-bridge* or a *$\bar{P}$-bridge* if $\varrho(B) \leqq m$ or $\varrho(B) > m$, respectively. If $B$ is a $\bar{P}$-bridge, then $\lambda(L[A_k, A_1]) \leqq m$. Then, by (4), no other (large) $\bar{P}$-bridge overlaps $B$. This implies that

(7) *$L$ has at most one $\bar{P}$-bridge.* $\square$

(8) *$\varrho(B) \geqq \lambda(B) + (k-2)$ for each bridge $B$ with $k$ vertices of attachment.*

*Proof of (8).* The proof can be accomplished by investigating a subtree $T$ in $B$ of length $\lambda(B)$ with ends $A_1, A_2, \ldots, A_k$. Obviously,

$$\lambda(L[A_i, A_{i+1}]) \geqq \lambda(T[A_i, A_{i+1}]) \quad \text{for all} \quad 1 \leqq i \leqq k-1.$$

Thus

$$\begin{aligned} \varrho(B) &= \sum_{i=1}^{k-1} \lambda(L[A_i, A_{i+1}]) \geqq \sum_{i=1}^{k-1} \lambda(T[A_i, A_{i+1}]) \\ &\geqq \varrho(B) + (k-2); \end{aligned}$$

the latter inequality is valid because in $\sum_{i=1}^{k-1} \lambda(T[A_i, A_{i+1}])$ each edge of $T$ is counted at least once and each edge of $T$ incident with $A_2, \ldots, A_{k-1}$ is counted at least twice (see also Chapter 3). $\square$

(9) *Let B denote a bridge of L avoiding all other bridges. Then* $\lambda(L[A_{k(B)}, A_1]) \geqq m+3$ *and* $\varrho(B) \leqq m-2$.

*Proof of (9).* Assume that assertion (9) is not true, i.e. $\lambda(L[A_i, A_{i+1}]) \leqq m+2$ for all $1 \leqq i \leqq k$ (indices taken modulo $k$). We delete the kernel $K$ of bridge $B$. The graph $G-K$ is a "ring" of bridges $B_i$, where $B_i$ is a bridge with respect to $\{A_i, A_{i+1}\}$ which contains $L[A_i, A_{i+1}]$. Consequently, $\lambda(B) \leqq m$, $\lambda(B_i) \leqq \lambda(L[A_{k(B)}, A_1]) \leqq m+2$ and $\sum_{i=1}^{k} \lambda(B_i) \leqq 2m+1$. With the aid of (3) and Lemma 11.5.2, 11.5.3 it can easily be proved that $G \cong D(2m)$ or $n(G) \leqq n(D(2m)) - 4$. □

(10) *G has no bridge of L of length m and G has at most one bridge of length* $m-1$.

*Proof of (10).* By assertion (ii) of Lemma 11.5.1 each bridge of $L$ of length at least $m$ avoids all other bridges. Therefore, by (8) and (9) the graph $G$ has no bridge of $L$ of length $m$. Next, assume that $G$ has two bridges $B$ and $B'$ of $L$ of length $m-1$. By (4) the bridges $B$ and $B'$ do not overlap. Obviously,

$$\varrho(B) + \varrho(B') + 2 \leqq c(B) \leqq 2m+1.$$

Let $B$ and $B'$ have $\alpha$ and $\alpha'$ vertices of attachment, respectively. Then, by (8), it follows that

$$\begin{aligned} 2m+1 &\geqq \varrho(B) + \varrho(B') + 2 \\ &\geqq \lambda(B) + \lambda(B') + (\alpha-2) + (\alpha'-2) + 2 \\ &= 2m+1+(\alpha+\alpha'-5) \end{aligned}$$

and

$$\alpha + \alpha' \leqq 5.$$

Therefore, we have two cases: $\alpha = 3$, $\alpha' = 2$, and $\alpha = \alpha' = 2$. With the help of assertions (ii) and (iii) of Lemma 11.5.1 and the assertions (8) and (4) it can easily be shown that $B$ or $B'$ is a 2-bridge which avoids all other bridges. By (9) the length

$$\lambda(B) \leqq \varrho(B) \leqq m-2.$$

This contradicts the assumption that $\lambda(B) = m-1$! □

Let the indices of the vertices of attachment $A_1^{(j)}, \ldots, A_{k_j}^{(j)}$ of the bridges $B_j$ be chosen so that $\varrho(B_j) = \lambda L[A_1^{(j)}, A_{k_j}^{(j)}]$.

In the set of all $P$-bridges of $L$ we introduce the relation "parallel" as follows: two $P$-bridges $B_1$, $B_2$ are said to be *parallel* if $B_1 = B_2$ or

$$L[A_1^{(1)}, A_{k_1}^{(1)}] \cap L[A_1^{(2)}, A_{k_2}^{(2)}] \neq \emptyset \quad \text{or} \quad L[A_1^{(1)}, A_{k_1}^{(1)}] \cap L[A_1^{(2)}, A_{k_2}^{(2)}] = \emptyset$$

and there is no third $P$-bridge $B_3$ such that

$$L[A_1^{(i)}, A_{k_i}^{(i)}] \cap L[A_1^{(j)}, A_{k_j}^{(j)}] = \emptyset \quad \text{for all} \quad 1 \leqq i < j \leqq 3.$$

With the help of (4) it can easily be proved that the relation "parallel" is an equivalence relation in the set of all $P$-bridges. It follows that

(11) *either all P-bridges are parallel or there are three classes of parallel P-bridges.* □

If $\bar{B}$ is a $\bar{P}$-bridge, then $\lambda(L[\bar{A}_j, \bar{A}_{j+1}]) \leqq m$ for all $1 \leqq j \leqq k(\bar{B})$. This implies that

(12) *if $G$ contains a $\bar{P}$-bridge of $L$, then all $P$-bridges of $L$ have length at most $m-2$.* □

We denote the large bridges of $L$ by $B_1, B_2, \ldots, B_q$ so that

$$\varrho(B_1) \geqq \varrho(B_2) \geqq \ldots \geqq \varrho(B_q).$$

Assertion (ii) of Lemma 11.5.1 and assertions (8)–(12) imply that

(13) *$\lambda(B_1) \leqq m$ and $\lambda(B_i) \leqq \varrho(B_i) \leqq m - i + 1$ for all $2 \leqq i \leqq q \leqq \frac{m+1}{2}$, where $\lambda(B_1) < m$ and $\lambda(B_2) < m$.* □

Obviously, if $G$ has a $\bar{P}$-bridge $M$ of $L$, then $B_1 = M$.

(14) *Small bridges $B'_1, B'_2, \ldots, B'_q$ can be determined so that*

$$n(B_i) + n(B'_i) \leqq n(M(m-i+1)) + n(M(i-1))$$

*for all $1 \leqq i \leqq q$, where*

$$n(B_1) + n(B'_1) \leqq n(M(m-2)) + n(M(1)) \quad \text{if } m \text{ is odd}$$

*and*

$$n(B_2) + n(B'_2) \leqq n(M(m-3)) + n(M(3)) \quad \text{if } m \text{ is even.}$$

*Proof of (14).*

1) If $B_i$ overlaps a small bridge $B'_i$, then $\lambda(B_i) + \lambda(B'_i) \leqq m$ (see assertion (ii) of Lemma 11.5.1). Lemma 11.5.2 and assertion (13) imply that the pair $[B_i, B'_i]$ satisfies (14).

2) Let $B_i$ avoid all other bridges. By (9) the bridge $B_i$ is a $P$-bridge with $\varrho(B_i) \leqq m-2$. Assertion (13) implies that $\varrho(B_i) \leqq m - i + 1$.

Let $\bar{A}_1$ and $\bar{A}_k$ denote the neighbour of $A_1$ and $A_k$ on $L[A_k, A_1]$, respectively. The graph $G - \{(A_1, \bar{A}_1), (A_k, \bar{A}_k)\}$ consists of two components $K_1, K_2$, where $K_1 \supseteqq L[A_1, A_k]$. The bridge

$$B' =_{\text{def}} K_1 \cup \{\bar{A}_1, \bar{A}_k, (A_1, \bar{A}_1), (A_k, \bar{A}_k)\}$$

is a 2-bridge with respect to the subgraph $(\{\bar{A}_1, \bar{A}_k\}, \emptyset)$ of length

$$\lambda(B') = \varrho(B_i) + 2$$
$$\leqq \min\{m - i + 3, m\}.$$

Obviously, $B_i \subseteqq B'$. Assertion (3) implies that $B' \cong M(\varrho(B_i) + 2)$ or

$$n(B') \leqq n(M(\varrho(B_i) + 2)) - 4.$$

Since $n(G) \geqq n(D(2m)) - 2$, we have only to consider the first case. Consequently, $B'$ contains the bridge $B_i \cong M(\varrho(B_i))$ and the uniquely determined bridge $B'_i \cong M(m - \varrho(B_i))$. If $i \geqq 3$, then with Lemma 11.5.2(ii)

$$n(B_i) + n(B'_i) = n(M(\varrho(B_i))) + n(M(m - \varrho(B_i)))$$
$$\leqq n(M(m-i+1)) + n(M(i-1)).$$

Now let $i \in \{1, 2\}$. Since, by (9), $\lambda(B_i) \leqq m - 2$, the pair $[B_i, B'_i]$ satisfies (14). □

It is possible that only small bridges remain. They are also arranged in pairs. Thus the system (14) can be supplied so that the following result holds:

(15) *The bridges of L can be denoted by* $B_1, B_2, \ldots, B_j, \ldots$ *so that*

$$\lambda(B_1) \geqq \lambda(B_2) \geqq \ldots \geqq \lambda(B_j) \geqq \ldots$$

*and*

$$\lambda(B_i) \leqq m - i + 1 \quad \textit{for all } i, \quad 1 \leqq i \leqq \frac{m+1}{2}$$

*where*

$$\lambda(B_1) < m \quad \textit{and} \quad \lambda(B_2) < m - 1 .$$

In order to complete the proof of the theorem when $m = 5$ or 6 the following remark is very useful.

**Remark 11.6.3.** *If* $m = 5$ *or* 6, *then L has only bridges of the type depicted in Fig. 11.6.3.* □

Fig. 11.6.3

With the aid of Remark 11.6.3 and assertions (15) and (9) it can be proved that

(16) *if* $m = 5$ *or* 6, *then* $G \cong P_{12}$ *or* $n(G) \leqq n(D(2m)) - 4$. □

This completes the proof of the theorem when $m = 5$ or 6.

Now suppose that $m \geqq 7$. With the help of (3), the definition $M(2j) =_{\text{def}} M(2j-1)$, and (iv) of Lemma 11.5.2 we can assign to each bridge $B_i$ an inequality such that by (14) the sum over all right-hand sides of the inequalities is an upper bound for the number of vertices of $G$.

(17) *If* $m \geqq 7$ *is an odd integer, then we may assign to* $B_i$ *the following inequalities:*

$$\begin{aligned} n(B_1) &\leqq n(M(m-2)) \\ &\leqq n(M(m)) - 6, \end{aligned}$$

$$\left.\begin{array}{l} n(B_{2j}) \\ n(B_{2j+1}) \end{array}\right\} \leqq n(M(m-2j)) \quad \textit{for all } j = 1, 2, \ldots$$

(18) *If* $m \geqq 8$ *is an even integer, then we may assign to* $B_i$ *the following inequalities:*

$$n(B_1) \leqq n(M(m-1)),$$

$$\begin{aligned} n(B_2) &\leqq n(M(m-3)) \\ &\leqq n(M(m-1)) - 6, \end{aligned}$$

$$\left.\begin{array}{l} n(B_{2j-1}) \\ n(B_{2j}) \end{array}\right\} \leqq n(M(m-2j+1)) \quad \textit{for all } j = 2, 3, \ldots$$

By (17) and (18) the proof can be completed in the following way: if $L$ has at most one bridge $M(3)$, then

$$\begin{aligned} n(G) &\leqq 1 + \sum' n(B_j) + n(M(3)) \\ &= 1 + \sum' n(B_j) + n(M(1)) + 2 \\ &\leqq 1 + n(D(2m)) - 6 + 2 = n(D(2m)) - 3, \end{aligned}$$

where the sum $\sum'$ is taken over all bridges $B_j$ except the bridge $M(3)$.

If $L$ has precisely two bridges $M(3)$, then they are replaced by three bridges of type $M(1)$. Consequently,

$$\begin{aligned} n(G) &\leqq 1 + \sum'' n(B_j) + n(M(3)) + n(M(3)) \\ &\leqq 1 + \sum'' n(B_j) + n(M(1)) + n(M(1)) + n(M(1)) + 2 \\ &\leqq 1 + n(D(2m)) - 6 + 2 = n(D(2m)) - 3, \end{aligned}$$

where the sum $\sum''$ is taken over all bridges $B_j$ except the two bridges $M(3)$.

Since $n(G)$ and $n(D(2m))$ are even integers, $n(G) \leqq n(D(2m)) - 4$. This contradicts the assumption that $n(G) \geqq n(D(2m)) - 2$.

Thus the proof of Theorem 11.6.1 is complete. □

# Bibliography


Aigner, M., 1984: Graphentheorie. Eine Entwicklung aus dem 4-Farben Problem. B. G. Teubner, Stuttgart, 1984.

Ainouche, A., and N. Christofides, 1985: Conditions for the existence of Hamiltonian circuits in graphs based on vertex degrees. J. London Math. Soc. (2) **32** (1985), 385-391.

Aldred, R. E. L., D. A. Holton, and C. Thomassen, 1985: Cycles through four edges in 3-connected cubic graphs, Graphs Combin. **1** (1985)1, 7-11.

Alon, N., 1986: The longest cycle of a graph with a large minimal degree. J. Graph Theory **10** (1986), 123-127.

Amar, D., I. Fournier, A. Germa, and R. Häggkvist, 1984: Covering of vertices of a simple graph with given connectivity and stability number. In: Convexity and Graph Theory (Jerusalem, 1981), North-Holland Math. Stud. 87, North-Holland Publ. Comp., Amsterdam – New York, 1984, p. 43-45.

Andrasfai, B., 1967: On critical graphs. In: Theory of Graphs, Intern. Sympos. held at Rome 1966 (1967), p. 9-19.

Appel, K., and W. Haken, 1976: Every planar map is four-colorable. Bull. Amer. Math. Soc. **82** (1976)5, 711-712.

Appel, K., and W. Haken, 1977: Every planar map is four-colorable. Illinois J. Math. **21** (1977), 84, 429-567.

Appel, K., and W. Haken, 1989: Every planar map is four colourable. In: Contemporary Mathematics, AMS Vol. 98, Providence, Rhode Island, 1989.

Archdeacon, D., and Ph. Huneke, 1989: A Kuratowski theorem for nonorientable surfaces. J. Combin. Theory B **46** (1989), 173-231.

Ash, P., 1985a: The maximum number of diagonals of a cycle in a block. Discrete Math. **55** (1985), 305-309.

Ash, P., 1985b: Ph. D. Thesis, Goldsmiths' College, London, England.

Ash, P., and B. Jackson, 1984: Dominating cycles in bipartite graphs. In: Progress in Graph Theory (Waterloo, Ont., 1982), Academic Press, Toronto, Ont., 1984.

Aung, Min, 1988: On the circumference of triangle-free and regular graphs. Dissertation D 83, TU Berlin, 1988.

Aung, Min, 1989a: Circumference of a regular graph. J. Graph Theory **13** (1989)2, 149-155.

Aung, Min, 1989b: Longest cycles in triangle-free graphs. J. Combin. Theory B **47** (1989), 171-186.

Barnette, D. W., 1966: Trees in polyhedral graphs. Canad. J. Math. **18** (1966), 731-736.

Barnette, D. W. (1968): Long paths imply long circuits in 3-polytopes. (Preprint.)

Barnette, D. W., and B. Grünbaum, 1969: On Steinitz's theorem concerning convex 3-polytopes and on some properties of planar graphs. In: The Many Facets of Graph Theory, Lecture Notes in Math. 110, Springer-Verlag, Berlin – Heidelberg – New York, 1969, p. 27-40.

Bauer, D., 1985: A note on degree conditions for Hamiltonian cycles in line graphs. Congr. Numerantium **49** (1985), 11-18.

Bauer, D., H. A. Jung, and E. Schmeichel, 1989: On 2-connected graphs with small circumference. Preprint, 1989.

Bauer, D., and E. Schmeichel, 1986: Long cycles in tough graphs. Stevens Research Reports in Math. 8612. Stevens Institute of Technology, Hoboken, New Jersey 07030, 1986.

Bauer, D., and E. Schmeichel, 1987: A proof of an extension of Ore's theorem. Ars Combinatorica **24B** (1987), 93-99.

Bauer, D., H. J. Veldman, A. Morgana, and E. Schmeichel, 1989/90: Long cycles in graphs with large degree sums. Discrete Math. **79** (1989/90), 59-70.

Beineke, L. W. → White, A. T.

Benhocine, A., and A. P. Wojda, 1990: Graphs containing every matching in a cycle. Discrete Math., to appear.

Berge, C., 1960: Les problèmes de coloration en théorie des graphes. Publ. Inst. Statist. Univ. Paris **9** (1960), 123-160.

Berge, C., 1961: Färbung von Graphen, deren sämtliche bzw. deren ungerade Kreise starr sind. Wiss. Z. Martin-Luther-Univ. Halle-Wittenberg, Math.-Nat. Reihe, **10** (1961)1, 114-115.

Berge, C., 1970: Graphes et Hypergraphes. Monographies Universitaires de Mathématiques, Dunod, Paris, 1970.

Berman, A. K., 1983: Proof of a conjecture of Häggkvist on cycles and independent edges. Discrete Math. **46** (1983), 9-13.

Bermond, J. C., 1975: Thesis, University of Paris XI, Orsay 1975.

Bermond, J. C., 1976: On Hamiltonian walks. In: Proc. Fifth British Combinatorial Conference (edited by C. St. J. A. Nash-Williams and J. Sheehan), Utilitas Math., Winnipeg, 1976, p. 41-51.

Bermond, J. C., 1978: Hamiltonian graphs. In: Selected Topics in Graph Theory (edited by L. W. Beineke and R. J. Wilson), Academic Press, New York, 1978, p. 127-167.

Bermond, J. C., and C. Thomassen, 1981: Cycles in digraphs—a survey. J. Graph Theory **5** (1981), 1-43.

Bigalke, A., and H. A. Jung, 1979: Über Hamiltonsche Kreise und unabhängige Ecken in Graphen. Mh. Math. **88** (1979), 195-210.

Bodendiek, R., and K. Wagner, 1985: Über den Kuratowskischen Satz für beliebige Flächen. Math. Nachr. **124** (1985), 239-251.

Bodendiek, R., and K. Wagner, 1989: Solution to König's graph embedding problem. Math. Nachr. **140** (1989), 215-272.

Bodendiek, R. → Širáň, J.

Bollobás, B., 1977a: Cycles modulo *k*. Bull. London. Math. Soc. **9** (1977), 97-98.

Bollobás, B., 1977b: Semi-topological subgraphs. Discrete Math. **20** (1977), 83-85.

Bollobás, B., 1978a: Extremal Graph Theory. Academic Press, London-New York-San Francisco, 1978.

Bollobás, B., 1978b: Nested cycles in graphs. Colloques internationaux CNRS 260—Problèmes Combinatoires et Théorie des Graphes, Paris, 1978, p. 49-50.

Bollobás, B., und G. Brightwell, 1989: Long cycles in graphs with no subgraphs of minimal degree 3. Discrete Math. 75 (1989) 1/2/3, 47-53.

Bollobás, B., and A. M. Hobbs, 1978: Hamilton cycles in regular graphs. In: Advances in Graph Theory (edited by B. Bollobás), North-Holland Publ. Comp., Amsterdam, 1978, p. 138-158.

Bondy, J. A., 1970: Large cycles in graphs. In: Proc. Louisiana Conf. on Combinatorics, Graph Theory and Computing (edited by R. C. Mullin, K. B. Reid, and D. P. Roselle), 1970, p. 47-60.

Bondy, J. A., 1971: Large cycles in graphs. Discrete Math. **1** (1971), 121-132.

Bondy, J. A., 1978: Hamilton cycles in graphs and digraphs. Congr. Numerantium **21** (1978), 3-28.

Bondy, J. A., 1980a: Longest paths and cycles in graphs of high degree. Dept. Combinatorics and Optimization. Research Report Cort. 80-16 (1980). Univ. Waterloo.

Bondy, J. A., 1980b: Integrity in graph theory. In: The Theory and Applications of Graphs (edited by G. Chartrand and others), Proc. Fourth Intern. Conf. Theory and Appl. of Graphs, May 6-9, 1980, Western Univ., Kalamazoo, p. 117-125.

Bondy, J. A., and R. C. Entringer, 1980: Longest cycles in 2-connected graphs with prescribed maximum degree. Canad. J. Math. **32** (1980), 987-992.

Bondy, J. A., and G. Fan, 1987: A sufficient condition for dominating cycles. Discrete Math. **67** (1987)2, 205-208.

Bondy, J. A., J. B.-A. Hartman, and S. C. Locke, 1981: A new proof of a theorem of Dirac. Congr. Numerantium **32** (1981), 131-136.

Bondy, J. A., and R. L. Hemminger, 1977: Graph reconstruction—a survey. J. Graph Theory **1** (1977), 227-268.

Bondy, J. A., and B. Jackson, 1985: Long paths between specified vertices of a block. In: Cycles in Graphs (edited by B. R. Alspach and C. D. Godsil), Annals of Discrete Math. **27** (1985), 195-200.

Bondy, J. A., and M. Kouider, 1988: Hamilton cycles in regular 2-connected graphs. J. Combin. Theory B **44** (1988), 177-186.

Bondy, J. A., and S. C. Locke, 1980: Relative lengths of paths and cycles in $k$-connected graphs. In: Proceedings First Canada-France Combin. Colloquium, Annals of Discrete Math. **8** (1980), 253-259.

Bondy, J. A., and S. C. Locke, 1981: Relative lengths of paths and cycles in 3-connected graphs. Discrete Math. **33** (1981), 111-122.

Bondy, J. A., and L. Lovász, 1981: Cycles through specified vertices of a graph. Combinatorica **1** (1981)1, 117-140.

Bondy, J. A., and U. S. R. Murty, 1976: Graph Theory with Applications. Macmillan Press, London and Basingstroke, 1976.

Bondy, J. A., and M. D. Plummer, 1975: Separating cycles in graphs. In: Proc. Fifth British Combinatorial Conference (edited by C. St. J. A. Nash-Williams and J. Sheehan) Utilitas Math., Winnipeg, 1976, p. 85-90.

Bondy, J. A., and M. Simonovits, 1974: Cycles of even length in graphs. J. Combin. Theory B **16** (1974), 97-105.

Bondy, J. A., and M. Simonovits, 1980: Longest cycles in 3-connected 3-regular graphs. Canad. J. Math. **32** (1980)4, 987-992.

Brightwell, G. → Bollobás, B.

Broersma, H. J., 1988: Existence of $\Delta_\lambda$-cycles and $\Delta_\lambda$-paths. J. Graph Theory **12** (1988)4, 499-507.

Burštein, M. I., 1978: Kuratowski-Pontrjagin theorem on planar graphs. J. Combin. Theory B **24** (1978)2, 228-232.

Catlin, P. A., 1979: Hajós' graph-coloring conjecture: variations and counterexamples. J. Combin. Theory B **26** (1979)2, 268-274.

Chartrand, G., and H. V. Kronk, 1968: Randomly traceable graphs. SIAM J. Appl. Math. **16** (1968), 696-700.

Chen, R.-Y., 1985: Long cycles in regular 2-connected bipartite graphs (Chinese, English summary). J. Math. (Wuhan) **5** (1985)3, 209-216.

Chen, R.-Y., 1989: Longest cycles in regular 2-connected graphs. J. Fujian Teachers Univ., to appear.

Chetwynd, A., and R. Häggkvist, 1989: Hamilton cycles in regular bipartite graphs. Preprint, Stockholm University.

Chiba, N., and T. Nishizeki, 1989: The Hamiltonian cycle problem is linear-time solvable for 4-connected planar graphs. J. Algorithms **10** (1989)2, 187-211.

Christofides, N. → Ainouche, A.

Chvátal, V., and P. Erdős, 1972: A note on hamiltonian circuits. Discrete Math. **2** (1972), 111-113.

Clark, B. N., Ch. J. Colbourn, and P. Erdős, 1985: A conjecture on dominating cycles. Congr. Numerantium **47** (1985), 189-197.

Clark, L., 1985: Longest cycles in 3-connected planar graphs. Congr. Numerantium **47** (1985), 199-204.

Colbourn, Ch. J., and L. K. Stewart, 1985: Dominating cycles in series-parallel graphs. Ars Combin. **19A** (1985), 107-112.

Colbourn, Ch. J. → Clark, B. N.

Czipszer, J., 1963: Solution to problem 127 (Hungarian). Mat. Lapok **14** (1963), 373-374.

Dahn, B., 1973: $\aleph_0$-kategorische zyklenbeschränkte Graphen. Fund. Math. **80** (1973), 117-131.

Dang, K.-Q., 1989a: Longest cycles in 2-connected bipartite graphs. J. Syst. Sci. and Math. Sci. **9** (1989) 2, 124-127.

Dang, K.-Q., 1989b: A note on R. Häggkvist's conjecture. J. Northeast Univ. Technol. **10** (1989)1, 71-74.

Dang, K.-Q., and L.-Ch. Zhao, 1989: On the circumferences of 2-connected bipartite graphs. Ars Combinatorica **27** (1989), 203-209.

Dawes, R. W., 1983: Constructions of minimally $k$-connected graphs for $k = 1$, 2, and 3. Congr. Numerantium **39** (1983), 273-289.

Dawes, R. W., 1986: Minimally 3-connected graphs. J. Combin. Theory B **40** (1986), 159-168.

Dean, N., and P. Fraisse, 1989: A degree condition for the circumference of a graph. J. Graph Theory **13** (1989)3, 331-334.

Demoucron, G., Y. Malgrange, and R. Pertuiset, 1964: Graphes planaires: reconnaissance et construction de représentations planaires topologiques. Rev. Franç. Rech. Opér. **8** (1964), 33-47.

Descartes, B., 1948: A three colour problem. Eureka, April 1947, Solution March 1948.

Descartes, B., 1954: Solution to advanced problem No. 4526, proposed by P. Ungar. Amer. Math. Monthly **61** (1954), 352.

Diestel, R., 1987: A separation property of planar triangulations. J. Graph Theory **11** (1987)1, 43-52.

Dirac, G. A., 1952a: Some theorems on abstract graphs. Proc. London Math. Soc. **3** (1952)2, 69-81.

Dirac, G. A., 1952b: A property of 4-chromatic graphs and some remarks on critical graphs. J. London Math. Soc. **27** (1952), 85-92.

Dirac, G. A., 1955: Circuits in critical graphs. Monatsh. Math. **59** (1955)3, 178-187.

Dirac, G. A., 1959: Paths and circuits in graphs: Extreme cases. Acta Math. Acad. Sci. Hung. **10** (1959), 357-361.

Dirac, G. A., 1960: In abstrakten Graphen vorhandene vollständige 4-Graphen und ihre Unterteilungen. Math. Nachr. **22** (1960)1/2, 61-85.

Dirac, G. A., 1961: On rigid circuit graphs. Abh. Math. Sem. Univ. Hamburg **25** (1961)1/2, 71-76.

Dirac, G. A., 1963: Some results concerning the structure of graphs. Canad. Math. Bull. **6** (1963), 183-210.

Dirac, G. A., 1964: Homomorphism theorems for graphs. Math. Annalen **153** (1964), 69-80.

Dirac, G. A., 1967: Minimally 2-connected graphs. J. Reine Angew. Math. **228** (1967), 204-216.

Dirac, G. A., 1972: On Hamilton circuits and Hamilton paths. Math. Annalen **197** (1972), 57-70.

Dirac, G. A., 1973: Note on Hamilton circuits and Hamilton paths. Math. Annalen **206** (1973), 139-147.

Dirac, G. A., 1974: The number of edges in critical graphs. J. Reine Angew. Math. **268/269** (1974), 150-164.

Dirac, G. A., 1978: Hamilton circuits and long circuits. Annals of Discrete Math. **3** (1978), 75-92.

Dirac, G. A., and S. Schuster, 1954: A theorem of Kuratowski. Indag. Math. **16** (1954), 343-348.

Dörfler, W., 1972: Bemerkungen zur Ulam-Vermutung. Archiv Math. **23** (1972)4, 442-445.

Dörfler, W., and W. Imrich, 1972: Eine Klasse rekonstruierbarer Graphen. Glasnik Mat. **7** (1972)2, 159-165.

Egawa, Y., 1987: Cycles in $k$-connected graphs whose deletion results in a $(k-2)$-connected graph. J. Combin. Theory B **42** (1987)3, 371-377.

Egawa, Y., R. Glas, and S. C. Locke, 1987: Cycles and paths through specified vertices in $k$-connected graphs. Preprint, 1987.

Egawa, Y., and T. Miyamoto, 1989: The longest cycles in a graph $G$ with minimum degree at least $n/k$. J. Combin. Theory B **46** (1989), 356-362.

Eiselt, F., 1986: Wege mit Sehnen und unabhängige Knotenmengen. Diplomarbeit an der PH Dresden, Sektion Mathematik, 1986.

Ellingham, M. N., D. A. Holton, and C. H. C. Little, 1984: Cycles through ten vertices in 3-connected cubic graphs. Combinatorica **4** (1984)4, 265-273.

Enomoto, H., 1984: Long paths and large cycles in finite graphs. J. Graph Theory **8** (1984)2, 287-301.

Entringer, R. C. → Bondy, J. A.

Erdős, P., 1959: Graph theory and probability. Canad. J. Math. **11** (1959), 34-38.

Erdős, P., 1973: The Art of Counting, Selected Writings. The MIT Press, Cambridge – London, 1973.

Erdős, P., 1977: Problem 3: In: Beiträge zur Graphentheorie und deren Anwendungen. Internat. Koll. Graphentheorie Oberhof (DDR) 10.–16. April 1977, p. 310.

Erdős, P., 1978: Problems and results in graph theory and combinatorial analysis. In: Problèmes Combinatoire et Théorie des Graphes, Colloques Internationaux CNRS 260, CNRS, Paris, 1978, p. 127-129.

Erdős, P., and T. Gallai, 1959: On maximal paths and circuits in graphs. Acta Math. Acad. Sci. Hung. **10** (1959), 337-356.

Erdős, P., and T. Gallai, 1961: On the minimal number of vertices representing the edges of a graph. Publ. Math. Inst. Hung. Acad. Sci. **6** (1961), 181-203.

Erdős, P., and H. Hajnal, 1966: On chromatic number of graphs and set-systems. Acta Math. Acad. Sci. Hung. **17** (1966), 61-99.

Erdős, P., and A. M. Hobbs, 1977: Hamiltonian cycles in regular graphs of moderate degree. J. Combin. Theory B **23** (1977), 139-142.

Erdős, P., and A. M. Hobbs, 1978: A class of Hamiltonian regular graphs. J. Graph Theory **2** (1978), 129-135.

Erdős, P., and L. Pósa, 1965: On independent circuits contained in a graph. Canad. J. Math. **17** (1965), 347-352.

Erdős, P., and A. Rényi, 1963: Asymmetric graphs. Acta Math. Acad. Sci. Hung. **14** (1963), 295-315.

Erdős, P., and H. Sachs, 1963: Reguläre Graphen gegebener Taillenweite mit minimaler Knotenzahl. Wiss. Z. Martin-Luther-Univ. Halle – Wittenberg, Math.-Nat. Reihe, **12** (1963), 251-258.

Erdős, P., → Chvátal, V.; Clark, B. N.

Erdős, Peter L., and E. Győri, 1985: Any four independent edges of a 4-connected graph are contained in a circuit. Acta Math. Hung. **46** (1985)3/4, 311-313.

Even, S., and A. Itai, 1971: Queues, stacks and graphs. In: Theory of Machines and Computations, New York, 1971, p. 71-86.

Fan, G., 1984: New sufficient conditions for cycles in graphs. J. Combin. Theory B **37** (1984)3, 221-227.

Fan, G., 1985: Longest cycles in regular graphs. J. Combin. Theory B **39** (1985)3, 325-345.

Fan, G., 1987: Extremal theorems on paths and cycles in graphs and weighted graphs. Thesis, Waterloo, Ontario, 1987.

Fan, G., 1990a: Long cycles and the codiameter of a graph II. In: Cycles and Rays (edited by G. Hahn, G. Sabidussi, R. E. Woodraw), Kluwer Academic Publishers, Dordrecht, Boston, London, 1990.

Fan, G., 1990b: Long cycles and the codiameter of a graph I. J. Combin. Theory B (to appear).

Fan, G. → Bondy, J. A.

Fassbender, B., 1989: Kriterien vom Ore-Typ für längste Kreise in 2-zusammenhängenden Graphen. Math. Nachr. **142** (1989), 287-296.

Fassbender, B., 1989a: On longest cycles and strongly linking vertex sets. J. Graph Theory **12** (1989)6, 697-702.

Faudree, F. J., and R. H. Schelp, 1975: Path Ramsey numbers in multicolorings. J. Combin. Theory B **19** (1975)2, 150-160.

Feng, T., 1988: A short proof of a theorem about the circumference of a graph. J. Combin. Theory B **45** (1988), 373-375.

Fleischner, H., 1982: Cycle decompositions, 2-coverings, removable cycles and the four-color-desease. In: Progress in Graph Theory (Waterloo, Ont., 1982), Academic Press, Toronto, Ont. 1984, p. 223-246.

Fleischner, H., 1983: Even cycles with prescribed chords in planar cubic graphs. Discrete Math. **44** (1983)3, 275-280.

Fleischner, H., and B. Jackson, 1985: Removable cycles in planar graphs. J. London Math. Soc. **31** (1985)2, 193-199.

Fouquet, J.-L., and H. Thuillier, 1985: Cycles through given vertices in planar 3-connected cubic graphs. Rapport de Recherche N° 234, Univ. de Paris-Sud, Orsay.

Fournier, I., 1985: Longest cycles in 2-connected graphs of independence number $\alpha$. In: Cycles in Graphs (edited by B. R. Alspach and C. D. Godsil), Annals of Discrete Math. **27** (1985), 201-204.

Fournier, I. → Amar, D.

Fournier, I., and P. Fraisse, 1985: On a conjecture of Bondy. J. Combin. Theory B **39** (1985)1, 17-26.

Fournier, J.-C., 1978: Une caractérisation des graphes de cordes. C. R. Acad. Sci. Paris **286** A (1978), 811-813.

Fraisse, P., 1986: Long cycles dans les graphes. Applications aux réseaux de Petri. Thèse (docteur en Sciences), Univ. de Paris-Sud, Centre d'Orsay, 1986.

Fraisse, P. → Dean, N.; Fournier, I.

Fraisse, P., and H. A. Jung, 1989: Longest cycles and independent sets in $k$-connected graphs. In: Recent Studies in Graph Theory (edited by V. R. Kulli), Vishna Internat. Publ. Gulbarga, India, 1989, p. 114-139.

Fraysseix, H. de, 1984: A characterization of circle graphs. Europ. J. Combinatorics **5** (1984), 223-238.

Fuhrmann, R., 1979: Hinreichende Bedingungen für die Existenz von Kreisen vorgeschriebener Länge. Diplomarbeit an der TH Ilmenau, Sektion Mathematik, Rechentechnik und Ökonomische Kybernetik, 1979.

Fulkerson, D. R., 1972: Anti-blocking polyhedra. J. Combin. Theory B **12** (1972), 50-71.

Gabor, C. P., W.-L. Hsu and K. J. Supowit, 1985: Recognizing circle graphs in polynomial time. 26-th Annual IEEE Sympos. FOCS, Portland, Oregon, Oct. 1985, p. 106-116.

Gallai, T., 1962: Graphen mit triangulierbaren ungeraden Vielecken. Magyar Tud. Akad. Mat. Kutadó Int. Közl. **7** (1962), 3-36.

Gallai, T., 1963: Kritische Graphen I. Publ. Math. Inst. Hung. Acad. Sci. **8** A (1963)1/2, 165-191.

Gallai, T., 1964: Kritische Graphen II. Publ. Math. Inst. Hung. Acad. Sci. **8** A (1964), 374-395.

Gallai, T., 1968: Problem 4. In: Theory of Graphs (edited by P. Erdős and G. Katona), Proc. Coll. held at Tihany, Hungary, Sept. 1966, Publ. House Hung. Acad. Sci., Budapest, 1968, p. 362.

Gallai, T., → Erdős, P.

Gavril, F., 1974: The intersection graph of subtrees in trees are exactly the chordal graphs. J. Combin. Theory B **16** (1974), 47-56.

Geoffrey, D. P., and D. P. Sumner, 1978: The edge nucleus of a point-determining graph. J. Combin. Theory B **24** (1978), 189-201.

Gerards, A. M. H., 1988: Homomorphisms of graphs into odd cycles. J. Graph Theory B **12** (1988)1, 73-83.

Germa, A. → Amar, D.

Glas, R., 1987: Längste Wege und Kreise durch vorgegebene Ecken in Graphen. Diplomarbeit, TU Berlin, 1987.

Glas, R. → Egawa, Y.

Golumbic, M. C., 1980: Algorithmic Graph Theory and Perfect Graphs. New York, 1980.

Gordon, L.: Hamiltonian circuits in graphs with many edges. Unpublished report, Sydney Univ., Australia.

Gross, J. L., and T. W. Tucker, 1987: Topological Graph Theory. John Wiley and Sons, New York, 1987.

Grötschel, M., 1977: Graphs with cycles containing given paths. Annals of Discrete Math. **1** (1977), 233-245.

Grötschel, M., 1984: On intersections of longest cycles. In: Graph Theory and Combinatorics, a volume in honour of Paul Erdős (edited by B. Bollobás), Academic Press, London, 1984, p. 171-189.

Grötschel, M., and G. L. Nemhauser, 1984: A polynomial algorithm for the max-cut problem on graphs without long odd cycles. Math. Programming **29** (1984), 28-40.

Grünbaum, B., 1970: Polytopes, graphs, and complexes. Bull. Amer. Math. Soc. **76** (1970), 1131-1201.

Grünbaum, B., 1974: Vertices missed by longest paths and circuits. J. Combin. Theory A **17** (1974), 31-38.
Grünbaum, B. → Barnette, D.W.
Grünbaum, B., and T.S. Motzkin, 1962: Longest simple paths in polyhedral graphs. J. London Math. Soc. **37** (1962), 152-160.
Grünbaum, B., and H. Walther, 1973: Shortness exponents of families of graphs. J. Combin. Theory **14** (1973), 364-385.
Gupta, R.P., J. Kahn, and N. Robertson, 1980: On the maximum number of diagonals of a circuit in a graph. Discrete Math. **32** (1980), 37-43.
Guy, R.K., 1967: The planar and toroidal crossing numbers of $K_n$. In: Beiträge zur Graphentheorie—Internat. Koll. Manebach (GDR) 1967; B.G. Teubner, Leipzig, 1968, p. 37-39.
Guy, R.K., 1971: Latest results on crossing numbers. In: Recent Trends in Graph Theory (Proc. Conf., New York, 1970), Lecture Notes in Math. 186, Springer-Verlag, Berlin-Heidelberg-New York 1971; p. 143-156.
Guy, R.K., 1972: Crossing numbers of graphs. In: Graph Theory and Applications (Proc. Conf., Western Michigan Univ., Kalamazoo, Mich., 1972), Lecture Notes in Math. 303, Springer-Verlag, Berlin-Heidelberg-New York 1972; p. 111-124.
Győri, E. → Erdős, Péter L.
Hadwiger, H., 1943: Über eine Klassifikation der Streckenkomplexe. Vierteljschr. Naturforsch. Ges. Zürich **88** (1943), 133-142.
Hadwiger, H., 1958: Ungelöste Probleme Nr. 26. Elem. Math. **13** (1958), 128.
Häggkvist, R., 1976: Problem. In: Combinatorics, Vol. II, Proc. Fifth Hung. Coll. Keszthely, 1976, Coll. Math. Soc. J. Bolyai 18, North-Holland Publ. Comp., Amsterdam, 1978, p. 1203-1204.
Häggkvist, R., 1979: On $F$-Hamiltonian graphs. In: Graph Theory and Related Topics. Acad. Press, New York and London, 1979, p. 219-231.
Häggkvist, R., 1982: Odd cycles of specified length in nonbipartite graphs. In: Graph Theory (Cambridge, 1981), North-Holland Publ. Comp., Amsterdam, 1982, p. 89-99.
Häggkvist, R., 1985: A note on Hamilton cycles. In: Cycles in Graphs (edited by B.R. Alspach and C.D. Godsil), Annals of Discrete Math. **27** (1985), 233-234.
Häggkvist, R., and B. Jackson, 1985: A note on maximal cycles in 2-connected graphs. In: Cycles in Graphs (edited by B.R. Alspach and C.D. Godsil), Annals of Discrete Math. **27** (1985), 205-208.
Häggkvist, R., and G. G. Nicoghossian, 1981: A remark on Hamiltonian cycles. J. Combin. Theory B **30** (1981)1, 118-120.
Häggkvist, R., and C. Thomassen, 1982: Circuits through specified edges. Discrete Math. **41** (1982)1, 29-34.
Häggkvist, R., → Amar, D.; Chetwynd, A.
Hajnal, A., 1965: A theorem on $k$-saturated graphs. Canad. J. Math. **17** (1965), 720-724.
Hajnal, A., and J. Surányi, 1958: Über die Auflösung von Graphen in vollständige Teilgraphen. Ann. Univ. Sci. Budapest Eötvös Loránd, Sect. Math., **1** (1958), 113-121.
Hajnal, A. → Erdős, P.
Haken, W. → Appel, K.
Halin, R., 1980: Graphentheorie I. Wissenschaftliche Buchgesellschaft, Darmstadt, 1980.
Halin, R., 1981: Graphentheorie II. Wissenschaftliche Buchgesellschaft, Darmstadt, 1981.
Halin, R., 1984: Simplicial decompositions and triangulated graphs. In: Graph Theory and Combinatorics, a volume in honour of Paul Erdős (edited by B. Bollobás), Academic Press, London, 1984, p. 191-196.
Halin, R., 1989: Graphentheorie. Akademie-Verlag, Berlin/Wiss. Buchges. Darmstadt 1989.
Hall, P., 1935: On representatives of subsets. J. London Math. Soc. **10** (1935), 26-30.
Harary, F., and A. Hill, 1962-3: On the number of crossings in a complete graph. Proc. Edinb. Math. Soc. **13** (1962-3), 333-338.
Harary, F., and C. St. J. A. Nash-Williams, 1965: On Eulerian and Hamiltonian graphs and line graphs. Canad. Math. Bull. **8** (1965), 701-709.

Hartman, J. B.-A., 1983: Long cycles generate the cycle space of a graph. Europ. J. Combinatorics **4** (1983), 237-246.

Hartman, J. B.-A. → Bondy, J. A.

Hatzel, W., 1979: Ein planarer hypohamiltonscher Graph mit 57 Knoten. Math. Annalen **243** (1979), 213-216.

Hauschild, K., H. Herre, and W. Rautenberg, 1972: Interpretierbarkeit und Entscheidbarkeit in der Graphentheorie II. Z. Math. Logik Grundl. Math. **18** (1972), 457-480.

Hauschild, K., and W. Rautenberg, 1971: Interpretierbarkeit und Entscheidbarkeit in der Graphentheorie I. Z. Math. Logik Grundl. Math. **17** (1971), 47-55.

Hayes, D. → Schmeichel, E.

Hedetniemi, S., 1971: Characterizations and constructions of minimally 2-connected graphs and minimally strong digraphs. Congr. Numerantium **3** (1971), 257-282.

Hemminger, R. L. → Bondy, J. A.

Herre, H. → Hauschild, K.

Hilbig, F., 1986: Kantenstrukturen in nichthamiltonischen Graphen. Dissertation, TU Berlin, 1986.

Hill, A. → Harary, F.

Hòa, V. D., 1986: Ein Struktursatz für 2-fach zusammenhängende Graphen mit großer Minimalvalenz. Math. Nachr. **128** (1986), 151-160.

Hobbs, A. M., 1979: *J*-components, bridges, and *J*-fragments. In: Graph Theory and Related Topics (edited by J. A. Bondy and U. S. R. Murty), Academic Press, New York, 1979, p. 253-260.

Hobbs, A. M., and J. Mitchem, 1976: The entire graph of a bridgeless connected plane graph is Hamiltonian. Discrete Math. **16** (1976), 233-239.

Hobbs, A. M. → Bollobás, B.; Erdős, P.

Holton, D. A., 1982: Cycles through specified vertices in *k*-connected regular graphs. Ars Combinatorica **13** (1982), 129-143.

Holton, D. A., B. D. McKay, M. D. Plummer, and C. Thomassen, 1982: A nine point theorem for 3-connected graphs. Combinatorica **2** (1982)1, 53-62.

Holton, D. A. → Aldred, R. E. L.; Ellingham, M. N.

Homobono, N., and C. Peyrat, 1989: Graphs such that every two edges are contained in a shortest cycle. Discrete Math **76** (1989), 37-44.

Hopcroft, J. E., and R. E. Tarjan, 1974: Efficient planarity testing. J. Assoc. Comput. Math. **21** (1974), 549-568.

Hsu, W.-L. → Gabor, C. P.

Hübner, St., and K. Ihle, 1987: Ein Satz über die Existenz von Kreisen mit drei topologischen Sehnen. Eulersche Linien und Hamiltonkreise. Diplomarbeit an der PH Dresden, Sektion Mathematik, 1987.

Huneke, Ph. → Archdeacon, D.

Ihle, K. → Hübner, St.

Imrich, W. → Dörfler, W.

Itai, A. → Even, S.

Jackson, B., 1979: Hamilton cycles in regular two-connected graphs. In: Graph Theory and Related Topics (edited by J. A. Bondy and U. S. R. Murty), Academic Press, New York, 1979, p. 261-265.

Jackson, B., 1980a: Hamilton cycles in regular 2-connected graphs. J. Combin. Theory B **29** (1980), 27-46.

Jackson, B., 1980b: Removable cycles in 2-connected graphs of minimum degree at least four. J. London Math. Soc. **21** (1980)3, 385-392.

Jackson, B., 1981: Cycles in bipartite graphs. J. Combin. Theory B **30** (1981)3, 332-342.

Jackson, B., 1985: Long cycles in bipartite graphs. J. Combin. Theory B **38** (1985)2, 118-131.

Jackson, B., 1986: Longest cycles in 3-connected cubic graphs. J. Combin. Theory B **41** (1986)1, 17-26.

Jackson, B., H. Li, and Y.-J. Zhu, 1989: Dominating cycles in regular 3-connected graphs. Submitted.

Jackson, B., and N. C. Wormald, 1990: Cycles containing matchings and pairwise compatible Euler tours. J. Graph Theory 14 (1990)1, 127-138.
Jackson, B. → Ash, P.; Bondy, J. A.; Fleischner, H.; Häggkvist, R.
Jung, H. A., 1965: Anwendung einer Methode von K. Wagner bei Färbungsproblemen für Graphen. Math. Annalen 161 (1965), 325-326.
Jung, H. A., 1978: On maximal circuits in finite graphs. Annals of Discrete Math. 3 (1978), 129-144.
Jung, H. A., 1981: Longest circuits in 3-connected graphs. In: "Finite and Infinite Sets", Vol. I (Eger, 1981, edited by A. Hajnal and V. T. Sós), Coll. Math. Soc. J. Bolyai 37, North-Holland Publ. Comp., Amsterdam, 1984, p. 403-438.
Jung, H. A., 1986: Longest paths joining given vertices in a graph. Abh. Math. Sem. Univ. Hamburg 56 (1986), 127-137.
Jung, H. A., 1989a: Longest ab-paths in regular graphs. In: Graph Theory in Memory of G. A. Dirac, Ann. Discrete Math. 41 (1989)1, 281-298.
Jung, H. A., 1989b: Degree and longest circuits in 3-connected graphs. Manuscript.
Jung, H. A., 1990: Long cycles in graphs with moderate connectivity. In: Topics in Combinatorics and Graph Theory, Essays in Honour of Gerhard Ringel (edited by R. Bodendiek and R. Henn), Physica-Verlag, Heidelberg, 1990, p. 765-778.
Jung, H. A., and C. Nara, 1982: Note on 2-connected graphs with $d(u) + d(v) > n - 4$. Archiv Math. 39 (1982)4, 383-384.
Jung, H. A., → Bauer, D.; Bigalke, A.; Fraisse, P.
Kahn, J. → Gupta, R. P.
Kelly, J. B., and L. M. Kelly, 1954: Paths and circuits in critical graphs. Amer. J. Math. 76 (1954), 786-792.
Kelly, L. M. → Kelly, J. B.
Kelly, P. J., 1957: A congruence theorem for trees. Pacific J. Math. 7 (1957), 961-968.
Kelmans, A. K., 1978a: Graph expansion and reduction. In: Algebraic Methods in Graph Theory, Vol. I (Szeged, 1978, edited by L. Lovász and V. T. Sós), Coll. Math. Soc. J. Bolyai 25, North-Holland Publ. Comp., Amsterdam, 1981, p. 317-343.
Kelmans, A.K., 1978b: The concept of a vertex in a matroid, the non-separating cycles of a graph and a new criterion for graph planarity. In: Algebraic Methods in Graph Theory, Vol. I (Szeged, 1978, edited by L. Lovász and V. T. Sós), Coll. Math. Soc. J. Bolyai 25, North-Holland Publ. Comp., Amsterdam, 1981, 345-388.
Kelmans, A. K., 1980: The concept of a vertex in a matroid and 3-connected graphs. J. Graph Theory 4 (1980)1, 13-19.
Kelmans, A. K., 1981: A new planarity criterion for 3-connected graphs. J. Graph Theory 5 (1981)3, 259-267.
Kelmans, A. K., 1983: A criterion for the existence in a graph of two crossing chain-chords of a given cycle (Russian). In: Algorithmic constructions and their efficiency (Russian), Yaroslav. Gos. Univ., Yaroslavl, 1983, p. 50-60, 139.
Kelmans, A. K., 1984: A strengthening of the Kuratowski planarity criterion for 3-connected graphs. Discrete Math. 51 (1984), 215-220.
Kelmans, A. K., 1986: On 3-connected graphs without essential 3-cuts or triangles. Soviet Math. Dokl. 33 (1986), 698-703.
Kelmans, A. K., and M. V. Lomonosov, 1982a: When $m$ vertices in a $k$-connected graph cannot be walked round along a simple cycle. Discrete Math. 38 (1982)2/3, 317-322.
Kelmans, A. K., and M. V. Lomonosov, 1982b: A cubic 3-connected graph having no cycle through given 10 vertices has the "Petersen form". J. Graph Theory 6 (1982)4, 495.
Kincses, J., 1987: The classification of 3- and 4-Helly dimensional convex bodies. Geom. Dedicata 22 (1987), 283-301.
Kincses, J., 1988: How big can the circuits of a bridge of a maximal circuit be? Combinatorica 8 (1988)2, 201-205.
Koebe, M., 1990: Spider graphs - a new class of intersection graphs (submitted to J. Algorithms).

Koebe, M., 1990a: Colouring of spider graphs. In: Topics in Combinatorics and Graph Theory, Essays in Honour of Gerhard Ringel (edited by R. Bodendiek and R. Henn), Physicy-Verlag, Heidelberg, 1990, p. 435-441.

Köhler, N., 1981a: Maximale Kreise in Graphen. Abh. Math. Sem. Univ. Hamburg **51** (1981), 68-97.

Köhler, N., 1981b: A sufficient condition for a graph to be Hamiltonian. Monatsh. Math. **92** (1981)2, 105-116.

Köhler, N., 1981c: On path-coverings and Hamilton-connectivity of finite graphs. Archiv Math. **36** (1981)5, 470-473.

König, D., 1936: Theorie der endlichen und unendlichen Graphen. Akad. Verlagsgesellschaft, Leipzig, 1936.

Kopylov, G. N., 1977: On maximal paths and cycles in a graph, Soviet Math. Dokl. **18** (1977), 593-596.

Kouider, M. → Bondy, J. A.

Krieger, M. M., 1970: Graphs edge-critical with respect to independence number. Internat. Conf. Combin. Math. (1970), Ann. New York Acad. Sci. **175** (1970), 265-271.

Kronk, H. V. → Chartrand, G.

Krusenstjerna-Hafstrøm, U., and B. Toft, 1980: Special subdivisions of $K_4$ and 4-chromatic graphs. Monatsh. Math. **89** (1980), 101-110.

Kuratowski, K., 1930: Sur le problème des courbes gauches en topologie. Fund. Math. **15** (1930), 271-283.

Lang, R., and H. Walther, 1968: Über längste Kreise in regulären Graphen. In: Beiträge zur Graphentheorie (edited by H. Sachs, H.-J. Voss, and H. Walther), BSB B. G. Teubner Verlagsgesellschaft, Leipzig, 1968, p. 91-98.

Lewin, M., 1975: On maximal circuits in directed graphs. J. Combin. Theory **2** (1967), 383-392.

Li, H., 1990: Edge-hamiltonian property in regular 2-connected graphs. Discrete Math. (to appear)

Li, H. → Zhu, Y.

Larson, J. A., 1979: Some graphs with chromatic number three. J. Combin. Theory B **26** (1979), 317-322.

Lesniak-Foster, L., 1977: Some recent results in Hamiltonian graphs. J. Graph Theory **1** (1977), 27-36.

Li, H. → Jackson, B.; Zhu, Y.

Linial, N., 1976: A lower bound for the circumference of a graph. Discrete Math. **15** (1976), 297-300.

Little, C. H. C. → Ellingham, M. N.

Liu, Z. → Zhu, Y.

Locke, S. C., 1982: Relative lengths of paths and cycles in $k$-connected graphs. J. Combin. Theory B **32** (1982), 206-222.

Locke, S. C., 1985a: A generalization of Dirac's theorem. Combinatorica **5** (1985)2, 149-159.

Locke, S. C., 1985b: A basis for the cycle space of a 3-connected graph. In: "Cycles in Graphs" (edited by B. R. Alspach and C. D. Godsil), Annals of Discrete Math. **27** (1985), 381-398.

Locke, S. C., 1985c: A basis for the cycle space of a 2-connected graph. Europ. J. Combinatorics **6** (1985)3, 215-216.

Locke, S. C., → Bondy, J. A.; Egawa, Y.

Lomonosov, M. V. → Kelmans, A. K.

Lovász, L, 1965: Független köröket nem tartalmazó gráfokról. Mat. Lapok **16** (1965), 289-299.

Lovász, L, 1968: On chromatic number of finite set systems. Acta Math. Acad. Sci. Hung. **19** (1968)1/2, 59-67.

Lovász, L., 1972a: Normal hypergraphs and the perfect graph conjecture. Discrete Math. **2** (1972), 253-267.

Lovász, L., 1972b: A characterization of perfect graphs. J. Combin. Theory B **13** (1972), 95-98.

Lovász, L., 1974: Problem 5. Period. Math. Hungar. **4** (1974), 82.

Lovász, L., 1976: Some finite basis theorems in graph theory. In: Combinatorics, Vol. II (Kesthely, 1976, edited by A. Hajnal and V. T. Sós), Coll. Math. Soc. J. Bolyai 18, North-Holland Publ. Comp., Amsterdam, 1978, p. 717-729.
Lovász, L., 1979: Combinatorial Problems and Exercises. Akadémiai Kiadó, Budapest, 1979.
Lovász, L. → Bondy, J. A.
MacLane, S., 1937: A structural characterization of planar combinatorial graphs. Duke Math. J. 3 (1937), 460-472.
Mader, W., 1967: Homomorphieeigenschaften und mittlere Kantendichte von Graphen. Math. Annalen **17** (1967), 265-268.
Mader, W., 1971: Eine Eigenschaft der Atome endlicher Graphen. Archiv Math. **22** (1971), 333-336.
Mader, W., 1974: Kreuzungsfreie (*a*, *b*)-Wege in endlichen Graphen. Abh. Math. Sem. Univ. Hamburg **42** (1974), 187-204.
Mader, W., 1985: Paths in graphs, reducing the edge-connectivity only by two. Graphs and Combin. **1** (1985)1, 81-89.
Malgrange, Y. → Demoucron, G.
Matthews, M. M., and D. P. Sumner, 1985: Longest paths and cycles in $K_{1,3}$-free graphs. J. Graph Theory **9** (1985), 269-277.
McCuaig, W. D., 1987: Edge reductions in cycliacally *k*-connected cubic graphs. Thesis, Univ. Waterloo, 1987.
McCuaig, W. D., and M. Rosenfeld, 1985: Parity of cycles containing specified edges. In: Cycles in Graphs (edited by B. R. Alspach and C. D. Godsil), Annals of Discrete Math. **27** (1985), 419-431.
McKay, B. D. → Holton, D. A.
Mesner, D. M. → Watkins, M. E.
Meyniel, H., 1976: On the perfect graph conjecture. Discrete Math. **16** (1976), 339-342.
Mitchem, J. → Hobbs, A.
Miyamoto, T. → Egawa, Y.
Moon, J. W., and L. Moser, 1963: Simple paths in polyhedra. Pacific J. Math. **13** (1963), 629-631.
Morgana, A. → Bauer, D.
Moser, L. → Moon, J. W.
Motzkin, T. S. → Grünbaum, B.
Murty, U. S. R. → Bondy, J. A.
Mycielski, J., 1955: Sur le coloriage des graphes. Coll. Math. **3** (1955)2, 161-162.
Nara, C., 1980: On sufficient conditions for a graph to be Hamiltonian. Natur. Sci. Rep. Ochamomizu Univ. **31** (1980), 75-80.
Nara, C. → Jung, H. A.
Nash-Williams, C. St. J. A., 1970: Valency sequences which force graphs to have hamiltonian circuits: interim report. Univ. Waterloo Research Report, 1970.
Nash-Williams, C. St. J. A., 1971a: Edge-disjoint hamiltonian circuits in graphs with vertices of large valency. In: Studies in Pure Mathematics (edited by L. Mirsky), Academic Press, London, 1971, p. 157-188.
Nash-Williams, C. St. J. A., 1971b: Hamilton arcs and circuits. In: Recent Trends in Graph Theory, Springer-Verlag, Berlin – Heidelberg – New York, 1971, p. 197-210.
Nash-Williams, C. St. J. A., 1973: Unexplored and semiexplored territories in graph theory. In: New Directions in Graph Theory (edited by F. Harary), Academic Press, New York, 1973, p. 149-186.
Nash-Williams, C. St. J. A., 1978: The reconstruction problem. In: Selected Topics in Graph Theory (edited by L. W. Beineke and R. J. Wilson), Academic Press, London – New York – San Francisco, 1978, p. 205-236.
Nash-Williams, C. St. J. A. → Harary, F.
Nemhauser, G. L. → Grötschel, M.
Nešetřil, J., 1966: *k*-chromatic graphs without cycles of length at most 7 (Czech). Comment. Math. Univ. Carolinae **7** (1966), 373-376.

Nešetřil, J., and V. Rödl, 1979: A short proof of the existence of highly chromatic hypergraphs without short cycles. J. Combin. Theory B **27** (1979)2, 225-227.
Nicoghossian, G. G. → Häggkvist, R.
Nishizeki, T. → Chiba, N.
Olaru, E., 1969: Über die Überdeckung von Graphen mit Cliquen. Wiss. Z. TH Ilmenau **15** (1969), 115-121.
Ore, O., 1960: Note on Hamiltonian circuits. Amer. Math. Monthly **67** (1960), 55.
Ore, O., 1963: Hamilton connected graphs. J. Math. Pures Appl. **42** (1963), 21-27.
Ore, O., 1967a: The four-color problem. Academic Press, New York – London, 1967.
Ore, O., 1967b: On a graph theorem by Dirac. J. Combin. Theory **2** (1967), 383-392.
Pertuiset, R. → Demoucron, G.
Petersen, J., 1891: Die Theorie der regulären Graphen. Acta Math. **15** (1891), 193-220.
Peyrat, C. → Homobono, N.
Plummer, M. D., 1967: On a family of line-critical graphs. Monatsh. Math. **71** (1967), 40-48.
Plummer, M. D., 1968: On minimal blocks. Trans. Amer. Math. Soc. **134** (1968), 85-94.
Plummer, M. D., 1973: Problem. In: Infinite and Finite Sets, Vol. III (Keszthely, 1973, edited by A. Hajnal, R. Rado, and V. T. Sós), Coll. Math. Soc. J. Bolyai 10, North-Holland Publ. Comp., Amsterdam, 1975, p. 1549-1550.
Plummer, M. D. → Bondy, J. A.; Holton, D. A.
Pósa, L., 1961: Problem Nr. 127 (Hungarian). Mat. Lapok **12** (1961), 254.
Pósa, L, 1963: On the circuits of finite graphs. Publ. Math. Inst. Hung. Acad. Sci. **8** (1963), 355-361.
Pósa, L. → Erdős, P.
Pöschel, R. → Wessel, W.
Rautenberg, W. → Hauschild, K.
Read, R. C., 1957: Maximal circuits in critical graphs. J. London Math. Soc. **32** (1957), 456-462.
Rényi, A. → Erdős, P.
Robertson, N., 1984: Minimal cyclic-4-connected graphs. Trans. Amer. Math. Soc. **284** (1984)2, 665-687.
Robertson, N., and P. Seymour, 1984: Generalizing Kuratowski's theorem. Congr. Numerantium **45** (1984), 129-138.
Robertson, N., and P. D. Seymour, 1990: Graph minors. VIII. A Kuratowski theorem for general surface. J. Combin. Theory B **48** (1990)2, 255-288.
Robertson, N. → Gupta, R. P.
Rosenfeld, M. → Mc Cuaig, W. D.
Sachs, H., 1968: Problem 50. In: Theory of Graphs (edited by P. Erdős and G. Katona), Proc. Coll. held at Tihany, Hungary, Sept. 1966, Publ. House Hung. Acad. Sci. Budapest, 1968, p. 366.
Sachs, H., 1970a: Einführung in die Theorie der endlichen Graphen. Teil I, B. G. Teubner, Leipzig, 1970.
Sachs, H., 1970b: On the Berge conjecture concerning perfect graphs. In: Combinatorial Structures and their Applications, Gordon and Breach, New York, 1970, p. 377-384.
Sachs, H., 1972: Einführung in die Theorie der endlichen Graphen, Teil II. B. G. Teubner, Leipzig, 1972.
Sachs, H. → Erdős, P.
Saito, A., 1989: Long cycles through specified vertices in *k*-connected graphs. J. Combin. Theory B **47** (1989), 220-230.
Sauer, N., 1967: Extremaleigenschaften regulärer Graphen gegebener Taillenweite, I, II. Sitzungsber. Österr. Akad. Wiss., Math.-Nat. Kl., Abt. II, **176** (1967), 9-43.
Schaar, G., M. Sonntag, and H.-M. Teichert, 1988: Hamiltonian properties of products of graphs and digraphs. Teubner–-Texte zur Mathematik 108, B. G. Teubner, Leipzig, 1988.
Schäuble, M., 1966: Problem 51. In: Theory of Graphs. Proc. Coll. Tihany, Hung., Sept. 1966, Publ. House Hung. Acad. Sci. 1968, p. 368.
Schelp, R. H. → Faudree, F. J.

Schmeichel, E., and D. Hayes, 1984: Some extensions of Ore's theorem. In: Graph Theory with Application to Algorithms and Computer Science, 5. Quadrennial Intern. Conf. Western Michigan Univ. Kalamazoo, June 4-8, 1984.
Schmeichel, E. → Bauer, D.
Schmitz, W., 1975: Über längste Wege und Kreise in Graphen, Rend. Sem. Mat. Univ. Padova **53** (1975), 97-103.
Schuster, S. → Dirac, G. A.
Seymour, P. D., 1981: On nowhere zero 6-flows. J. Combin. Theory B **30** (1981), 130-135.
Seymour, P. D., and R. W. Weaver, 1984: A generalization of chordal graphs. J. Graph Theory **8** (1984)2, 241-251.
Seymour, P. D. → Robertson, N.
Simonovits, M., 1967: A new proof and generalizations of a theorem of Erdős and Pósa on graphs without $k+1$ independent circuits. Acta Math. Acad. Sci. Hung. **18** (1967), 191-206.
Simonovits, M., 1974: Extremal graph theorems with symmetrical graphs. Additional chromatic conditions. Discrete Math. **7** (1974), 349-376.
Simonovits, M., 1983: Extremal graph theory. In: Selected Topics in Graph Theory, Vol. II (edited by L. W. Beineke and R. J. Wilson), Academic Press, London – New York 1983, p. 161-200.
Simonovits, M. → Bondy, J. A.
Širáň, J., 1983: Edges and Kuratowski subgraphs of non-planar graphs. Math. Nachr. **113** (1983), 187-190.
Širáň, J., R. Bodendiek, and K. Wagner, 1990: Planarität - ein einführender Überblick. Kapitel 9 in "Graphentheorie, Band 2: Weitere Methoden, Masse-Graphen, Planarität und minimale Graphen" von K. Wagner und R. Bodendiek, Bibliogr. Inst., Wiss. Verlag, Mannheim, Wien, Zürich, 1990, p. 99-171.
Skupień, Z., 1985: An improvement of Jung's conditions for Hamiltonicity. In: 30. Internat. Wiss. Koll. TH Ilmenau 1985. Vortragsreihe "Graphen und Netzwerke – Theorie und Anwendung"; p. 111-113.
Slater, P. J., 1966: A classification of 4-connected graphs. J. Combin. Theory **1** (1966), 306-336.
Sonntag, M. → Schaar, G.
Stewart, L. K. → Colbourn, Ch. J.
Sumner, D. P. → Geoffrey, D. P.; Matthews, M. M.
Supowit, K. J. → Gabor, C. P.
Surányi, J. → Hajnal, A.
Surányi, L., 1968: The covering of graphs by cliques. Studia Sci. Math. Hungar. **3** (1968), 345-349.
Surányi, L., 1973: On line-critical graphs. In: Infinite and finite sets, Vol. III (Keszthely, 1973, edited by A. Hajnal, R. Rado and V. T. Sós), Coll. Math. Soc. J. Bolyai 10, North-Holland Publ. Comp., Amsterdam, 1975, p. 1411-1444.
Surányi, L., 1976: A note on a conjecture of Gallai concerning $\alpha$-critical graphs. In: Combinatorics, Vol. II (Keszthely, 1976, edited by A. Hajnal and V. T. Sós), Coll. Math. Soc. J. Bolyai 18, Nort-Holland Publ. Comp., Amsterdam, 1978, p. 1065-1074.
Tait, P. G., 1880: Remarks on the colouring of maps. Proceedings Roy. Soc. Edinburgh **10** (1880), 501-503.
Tarjan, R. E. → Hopcroft, J. E.
Teichert, H.-M. → Schaar, G.
Thomassen, C., 1974a: A minimal condition implying a special $K_4$-subdivision in a graph. Archiv Math. **25** (1974), 210-215.
Thomassen, C., 1974b: On hypohamiltonian graphs. Discrete Math. **10** (1974), 383-390.
Thomassen, C., 1976: Hypohamiltonian graphs and digraphs. In: Theory and Applications of Graphs, Proc., Michigan 1976 (edited by Y. Alavi and D. R. Lick). Lecture Notes in Mathematics 642, Springer-Verlag, Berlin – Heidelberg – New York, 1976, p. 557-571.
Thomassen, C., 1978: On separating cycles in graphs. Discrete Math. **22** (1978), 57-73.

Thomassen, C., 1980: Planarity and duality of finite and infinite graphs. J. Combin. Theory B **29** (1980), 244-271.

Thomassen, C., 1981a: Nonseparating cycles in $k$-connected graphs. J. Graph Theory 5 (1981)4, 351-354.

Thomassen, C., 1981b: Kuratowski's theorem. J. Graph Theory **5** (1981)3, 225-241.

Thomassen, C., 1983a: Girth in graphs. J. Combin. Theory B **35** (1983), 129-141.

Thomassen, C., 1983b: A theorem on paths in planar graphs. J. Graph Theory **7** (1983)2, 169-176.

Thomassen, C., 1983c: Graph decomposition with applications to divisions and path systems modulo $k$. J. Graph Theory **7** (1983), 261-271.

Thomassen, C., 1984: A refinement of Kuratowski's theorem. J. Combin. Theory B **37** (1984)3, 245-253.

Thomassen, C., 1986: Configurations in graphs of large minimum degree, connectivity or chromatic number. Manuscript.

Thomassen, C., 1988: Embeddings and minors. In: Handbook of Combinatorics (edited by R. L. Graham, M. Grötschel and L. Lovász), North-Holland Publ. Comp., Amsterdam, 1988.

Thomassen, C., and B. Toft, 1981: Non-separating induced cycles in graphs. J. Combin. Theory B **31** (1981), 199-224.

Thomassen, C. → Aldred, R. E. L.; Bermond, J. C.; Häggkvist, R.; Holton, D. A.

Thullier, H. → Fouquet, J.-L.

Tian, Y., and L. Zhao, 1988: On the circumference of 1-though graphs. Kexue Tonghao **35** (1988)7, 538-541.

Titov, V. K., 1967: Some operations over graphs and some classes of graphs induced by these operations (Russian). In: 8th All-union Coll. on General Algebra, Resumé of reports, Univ. of Latvia, 1967.

Titov, V. K., 1975: A constructive description of some classes of graphs. Doctoral dissertation, Moscov, 1975.

Toft, B., 1986: Graph colouring theory. Mat. Inst., Odense Univ. Preprint 1986, No. 2.

Toft, B. → Krusenstjerna-Hafstrøm, U.; Thomassen, C.

Tomescu, I., 1983: On Hamilton-connected regular graphs. J. Graph Theory **7** (1983), 385-390.

Topics on Perfect Graphs, 1984 (edited by C. Berge and V. Chvátal), North-Holland Publ. Comp., Amsterdam, 1984.

Tucker, A. C., 1977: Critical perfect graphs and perfect 3-chromatic graphs. J. Combin. Theory B **23** (1977), 143-149.

Tucker, T. W. → Gross, J. L.

Turán, P., 1941: Egy gráfelméleti szélsöétékfeladratról. Mat. Fiz. Lapok **48** (1941), 436-452.

Turán, P., 1954: On the theory of graphs. Coll. Math. **3** (1954)1, 19-30.

Tutte, W. T., 1946: On Hamiltonian circuits. J. London Math. Soc. **21** (1946), 98-101.

Tutte, W. T., 1947: A family of cubical graphs. Proc. Cambridge Phil. Soc. **43** (1947), 459-474.

Tutte, W. T., 1956: A theorem on planar graphs. Trans. Amer. Math. Soc. **82** (1956), 99-116.

Tutte, W. T., 1958: Matroids and graphs. Trans. Amer. Math. Soc. **88** (1958), 144-174.

Tutte, W. T., 1960: Convex representations of graphs. Proc. London Math. Soc. **10** (1960), 304-320.

Tutte, W. T., 1961: A theory of 3-connected graphs. Indag. Math. **29** (1961), 441-455.

Tutte, W. T., 1963: How to draw a graph. Proc. London Math. Soc. **3** (1963)13, 743-768.

Tutte, W. T., 1965: Lectures on matroids. J. Res. Nat. Bur. Standards **69** (1965), 1-47.

Tutte, W. T., 1966: Connectivity in graphs. Univ. Toronto Press, 1966.

Tutte, W. T., 1971: Introduction to the Theory of Matroids. Amer. Elsevier Publ. Comp., New York, 1971.

Tutte, W. T., 1975: Separation of vertices by a circuit. Discrete Math. **12** (1975), 173-184.

Tutte, W. T., 1977: Bridges and Hamiltonian circuits in planar graphs. Aequat. Math. **15** (1971), 1-33.

Tutte, W. T., 1984: Graph theory. Encyclopedia of Mathematics and its Applications, 21. Adison-Wesley Publ. Co., Reading, Mass., 1984.

Tverberg, H., 1989: A proof of Kuratowski's theorem. In: Graph Theory in Memory of G. A. Dirac, Annals of Discrete Math. **41** (1989)2, 417–419.

Veldman, H. J., 1983a: Existence of dominating cycles and paths. Discrete Math. **43** (1983)2/3, 281–296.

Veldman, H. J., 1983b: Existence of $D_\lambda$-cycles and $D_\lambda$-paths. Discrete Math. **44** (1983)3, 309–316.

Veldman, H. J., 1986: Existence of spanning and dominating trails and circuits. J. Graph Theory **10** (1986)1, 23–31.

Veldman, H. J. → Bauer, D.

Voss, H.-J., 1968: Some properties of graphs containing $k$ independent circuits. In: Theory of Graphs, Proc. Coll. Tihany, Hung., Sept. 1966, Publ. House Hung. Acad. Sci., Budapest, 1968, p. 321–334.

Voss, H.-J., 1973: Eigenschaften zweifach zusammenhängender Graphen mit vorgeschriebener Maximalkreislänge. Mitteilungen der MGdDDR, Heft 2/3 (1973), 139–151.

Voss, H.-J., 1975: Reguläre Graphen gegebener Maximalkreislänge mit maximaler Knotenzahl. Elektron. Informationsverarb. u. Kybernetik **11** (1975)10/11/12, 580–581.

Voss, H.-J., 1976a: Maximal circuits and paths in graphs. Extreme cases. In: Combinatorics, Vol. II (Keszthely, 1976, edited by A. Hajnal and V. T. Sós), Coll. Math. Soc. J. Bolyai 18, North-Holland Publ. Comp., Amsterdam, 1978, p. 1099–1122.

Voss, H.-J., 1976b: Open problem. In: Combinatorics, Vol. II (Keszthely, 1976, edited by A. Hajnal and V. T. Sós), Coll. Math. Soc. J. Bolyai 18, North-Holland Publ. Comp., Amsterdam, 1978, p. 1218.

Voss, H.-J., 1977a: Bridges of longest circuits and of longest paths in graphs. In: Beiträge zur Graphentheorie und deren Anwendungen. Internat. Koll. Graphentheorie Oberhof (DDR) 10.–16. April 1977, p. 275–286.

Voss, H.-J., 1977b: Kommentar zu Professor Erdős's Problem 3. In: Beiträge zur Graphentheorie und deren Anwendungen. Internat. Koll. Graphentheorie Oberhof (DDR) 10.–16. April 1977, p. 310–311.

Voss, H.-J., 1977c: Graphs with prescribed maximal subgraphs and critical chromatic graphs. Comment. Math. Univ. Carolinae **18** (1977)1, 129–142.

Voss, H.-J., 1979: Über einen 3-Farbensatz der Graphentheorie. In: Tagungsbericht des Koll. über Geometrie und Kombinatorik, Karl-Marx-Stadt am 6. 4. 1979, p. 68–75.

Voss, H.-J., 1980: Graphs having odd circuits with at least $k$ chords. Elektron. Informationsverarb. u. Kybernetik **16** (1980)1/3, 77–86.

Voss, H.-J., 1981: Problem 3. In: Einige offene Probleme der Graphentheorie (edited by R. Paninski and J. Harandt), Wiss. Z. TH Ilmenau **27** (1981)5, 83–91.

Voss, H.-J., 1982a: Graphs having circuits with at least two chords. J. Combin. Theory B **82** (1982), 264–285.

Voss, H.-J., 1982b: Bridges of longest circuits and path coverings of labelled trees. Periodica Math. Hung. **13** (1982)3, 173–189.

Voss, H.-J., 1982c: Properties of 2-connected cubic graphs. In: 27. Internat. Wiss. Koll. TH Ilmenau 1982. Vortragsreihe "Graphen und Netzwerke – Theorie und Anwendung", p. 41–44.

Voss, H.-J., 1982d: Chords in cycles of graphs having girth ≥5 (Russian). In: Graphs, Hypergraphs and Discrete Optimal Problems (edited by L. F. German), Mat. Issled. No. 66 (1982), "Shtiintsa", Kishinev, 1982, p. 160–172, 197.

Voss, H.-J., 1985a: Chords of circuits in graphs with given minimum degree. Elektron. Informationsverarb. u. Kybernetik **21** (1985)4/5, 201–208.

Voss, H.-J., 1985b: Problem 23, 24. In: Graphs, Hypergraphs and Applications, Proc. Conf. Graph Theory, held at Eyba, Oct. 1984 (edited by H. Sachs), Teubner-Texte zur Mathematik 73, B. G. Teubner, Leipzig, 1985, p. 218.

Voss, H.-J., 1985c: Bridges of circuits through specified edges. In: Graphs, Hypergraphs and Applications, Proc. Conf. Graph Theory, held at Eyba, Oct. 1984 (edited by H. Sachs), Teubner-Texte zur Mathematik 73, B. G. Teubner, Leipzig, 1985, p. 198–202.

Voss, H.-J., 1985d: Large cycles with many diagonals. In: Algebra und Graphentheorie. Beiträge Jahrestagung "Algebra und Grenzgebiete" in Siebenlehn (DDR), 28. Oct.–1. Nov. 1985 (edited by J. Machner and G. Schaar), Bergakademie Freiberg, 1985, p. 119–122.

Voss, H.-J., 1985e: Symmetric subgraphs and the reconstruction conjecture. In: 30. Internat. Wiss. Koll. TH Ilmenau 1985. Vortragsreihe "Graphen und Netzwerke – Theorie und Anwendung", p. 123–126.

Voss, H.-J., 1988a: On longest cycles in graphs with given minimum degree. Dresdner Reihe zur Forschung **9** (1988), 65–68.

Voss, H.-J., 1988b: Maximale gerade und ungerade Kreise in Graphen III. Wiss. Z. PH "K. F. W. Wander" Dresden, Math.-Nat. Reihe, **22** (1988), 41–52.

Voss, H.-J., 1989a: Maximale gerade und ungerade Kreise in Graphen II. Wiss. Z. TH Ilmenau **35** (1989)3, 55–64.

Voss, H.-J., 1989b: Bridges of longest circuits applied to the circumference of regular graphs. In: Graph Theory in Memory of G. A. Dirac, Annals of Discrete Math. **41** (1989)2, 437–452.

Voss, H.-J., 1990a: On 2-connected 3-regular graphs with given circumference having maximum order (Russian). Submitted to Mat. Issled., Shtiintsa, Kishinev.

Voss, H.-J., 1990b: Cycles with many diagonals in cyclically $k$-vertex-connected graphs. In: Contemporary Methods in Graph Theory, Essays in Honour of Klaus Wagner (edited by R. Bodendiek), BI Wiss. Verlag, Mannheim, Wien, Zürich, 1990.

Voss, H.-J., and C. Zuluaga, 1977: Maximale gerade und ungerade Kreise in Graphen I. Wiss. Z. TH Ilmenau **23** (1977)4, 57–70.

Voss, H.-J. → Walther, H.

Wagner, K., 1937: Über eine Eigenschaft der ebenen Komplexe. Math. Annalen **114** (1937), 570–590.

Wagner, K., 1964: Beweis einer Abschwächung der Hadwigervermutung. Math. Annalen **153** (1964), 139–141.

Wagner, K. → Bodendiek, R.; Širan, J.

Walther, H., 1965a: Über reguläre Graphen gegebener Taillenweite und minimaler Knotenzahl. Wiss. Z. HfE Ilmenau **11** (1965), 93–96.

Walther, H., 1965b: Eigenschaften von regulären Graphen, gegebener Taillenweite und minimaler Knotenzahl. Wiss. Z. HfE Ilmenau **11** (1965), 167–168.

Walther, H., 1967a: Über die Länge eines längsten Kreises in regulären Graphen beliebigen Zusammenhanges. Wiss. Z. TH Ilmenau **13** (1967), 427–429.

Walther, H., 1967b: Über die Anzahl der Knotenpunkte eines längsten Kreises in planaren, kubischen, dreifach zusammenhängenden Graphen. Studia Sci. Math. Hungar. **2** (1967), 391–398.

Walther, H., 1969: Über die Nichtexistenz eines Knotenpunktes, durch den alle längsten Wege eines Graphen gehen. J. Combin. Theory **6** (1969), 1–6.

Walther, H., 1969a: Über das Problem der Existenz von Hamiltonkreisen in planaren regulären Graphen. Math. Nachr. **39** (1969), 277–296.

Walther, H., 1970: Über die Nichtexistenz zweier Knotenpunkte eines Graphen, die alle längsten Kreise fassen. J. Combin. Theory **8** (1970), 330–333.

Walther, H., 1979: Anwendungen der Graphentheorie. VEB Deutscher Verlag der Wissenschaften, Berlin/Friedr. Vieweg & Sohn, Braunschweig – Wiesbaden, 1979.

Walther, H., 1984: Ten Applications of Graph Theory. D. Reidel Publ. Comp., Dordrecht – Boston – Lancaster, 1984 (Revised Translation of Walther, H., 1979).

Walther, H. → Grünbaum, B.; Lang, R.

Walther, H., and H.-J. Voss, 1974: Über Kreise in Graphen. VEB Deutscher Verlag der Wissenschaften, Berlin, 1974.

Watkins, M. E., and D. M. Mesner, 1967: Cycles and connectivity in graphs. Canad. J. Math. **19** (1967), 1319–1328.

Weaver, R. W. → Seymour, P. D.

Wessel, W., 1975: A first family of edge-critical wheels, Periodica Math. Hungar. **6** (1975), 229–233.

Wessel, W., 1976: A second family of edge-critical wheels. In: Combinatorics, Vol. II (Keszthely, 1976, edited by A. Hajnal and V. T. Sós), Coll. Math. Soc. J. Bolyai 18, North-Holland Publ. Comp., Amsterdam, 1978, p. 1123-1145.

Wessel, W., 1981: Eigenschaften, Operationen, Kritizität von Graphen, Dissertation B, Fak. Math.-Nat. TH Ilmenau, 1981.

Wessel, W., 1985: Kreissehnengraphen – ein neues Hilfsmittel zur Darstellung von Graphen. In: 30. Internat. Wiss. Koll. TH Ilmenau 1985. Vortragsreihe "Graphen und Netzwerke – Theorie und Anwendung", p. 131-134.

Wessel, W., 1987: Chord graphs—new means for representing graphs. Applicationes Mathematicae (Zastowania Matematyki) **19** (1987)3/4, 619-627.

Wessel, W., and R. Pöschel, 1985: On circle graphs. In: Graphs, Hypergraphs and Applications, Proc. Conf. Graph Theory, held at Eyba, Oct. 1984 (edited by H. Sachs), Teubner-Texte zur Mathematik 73, B. G. Teubner, Leipzig, 1985, p. 198-202.

White, A. T., 1973: Graphs, Groups and Surfaces. North-Holland Publ. Comp., Amsterdam, 1973.

White, A. T., and L. W. Beineke, 1978: Topological graph theory. In: Selected Topics in Graph Theory (edited by L. W. Beineke and R. J. Wilson), Academic Press, London, 1978, p. 15-49.

Whitney, H., 1931: A theorem on graphs. Annals of Math. **32** (1931), 378-390.

Whitney, H., 1932a: Non-separable and planar graphs. Trans. Amer. Math. Soc. **34** (1932), 339-362.

Whitney, H., 1932b: Congruent graphs and the connectivity of graphs. Amer. J. Math. **54** (1932), 150-168.

Whitney, H., 1933a: 2-isomorphic graphs. Amer. J. Math. **55** (1933), 245-254.

Whitney, H., 1933b: A set of topological invariants for graphs. Amer. J. Math. **55** (1933), 231-235.

Williamson, S. G., 1980: Embedding graphs in the plane—Algorithmic aspects. Annals of Discrete Math. **6** (1980), 349-384.

Wojda, A. P., 1988: Hamiltonian cycles through matchings. Demonstratio Math **22** (1988), 547-553.

Wojda, A. P. → Benhocine, A.

Woodall, D. R., 1972: Sufficient conditions for circuits in graphs. Proc. London Math. Soc. **24** (1972), 739-755.

Woodall, D. R., 1973: The binding number of a graph and its Anderson number. J. Combin. Theory B **15** (1973), 225-255.

Woodall, D. R., 1975: Maximal circuits of graphs II. Studia Sci. Math. Hungar. **10** (1975), 103-109.

Woodall, D. R., 1976: Maximal circuits of graphs I. Acta Math. Acad. Sci. Hung. **28** (1976), 77-80.

Woodall, D. R., 1977: Circuits containing specified edges. J. Combin. Theory B **22** (1977), 274-278.

Woodall, D. R., 1978: A sufficient condition for hamiltonian circuits. J. Combin. Theory B **25** (1978), 184-186.

Wormald, N. C. → Jackson, B.

Yu, Zh., and Y. Zhu, 1990: Longest cycles in regular 3-connected graphs. Preprint 1990.

Yu, Z. → Zhu, Y.

Zamfirescu, T., 1972: A two-connected planar graph without concurrent longest paths. J. Combin. Theory B **13** (1972), 116-121.

Zamfirescu, T., 1974: L'histoire et l'état présent des bornes connues pour $P_k^j$, $C_k^j$, $\bar{P}_k^j$ et $\bar{C}_k^j$. Colloque sur la Théorie des Graphes (Paris 1974), Cahiers Centre Etude Recherche Opér. **17** (1975)2/3/4, 427-439.

Zamfirescu, T., 1975: Graphen, in welchen je zwei Eckpunkte von einem längsten Weg vermieden werden. Ann. Univ. Ferrara, Ser. VII (N. S.), **21** (1975), 17-24.

Zamfirescu, T., 1976: On longest paths and circuits in graphs. Math. Scand. **38** (1976)2, 211-239.

Zarankiewicz, K., 1953: The solution of a certain problem on graphs of P. Turán. Bull. Acad. Polon. Sci., Cl. III, **1** (1953)5, 167-168.

Zarankiewicz, K., 1954: On a problem of P. Turán concerning graphs. Fund. Math. **41** (1954)1, 137-145.

Zeidl, B., 1958: Über 4- und 5-chrome Graphen. Monatsh. Math. **62** (1958), 212-218.

Zhang, C.-Q., 1987: Longest cycles and their chords. J. Graph Theory **11** (1987)4, 521-529.

Zhang, C.-Q., 1988: Hamilton cycles in claw-free graphs. J. Graph Theory **12** (1988)2, 209-216.

Zhang, C.-Q., 1989a: Bridges of longest cycles. Discrete Math **78** (1989), 195-211.

Zhang, C.-Q., 1989b: Circumference and girth. J. Graph Theory **13** (1989)4, 485-490.

Zhao, L.-Ch. → Dang, K.-Q.; Tian, Y.

Zhu, R. Y., 1983: On maximal circuits in 2-connected graphs (Chinese). Qufu Shiyuan Xuebao **4** (1983), 8-9.

Zhu, Y., and H. Li, 1990: Hamilton cycles in regular 3-connected graphs. Preprint 1990.

Zhu, Y. → Yu, Zh.

Zhu, Y., Z. Liu and Z. Yu, 1985: An improvement of Jackson's result on Hamilton cycles in 2-connected regular graphs. In: Cycles in Graphs (edited by B. R. Alspach and C. D. Godsil), Annals of Discrete Math. **27** (1985), 237-248.

Zhu, Y., Z. Liu and Z. Yu, 1986: 2-connected $k$-regular graphs on at most $3k+3$ vertices to be Hamiltonian. J. Syst. Sci. and Math. Sci. **6** (1986)1, 34-49 and **6** (1986)2, 136-145.

Zhu, Y.-J. → Jackson, B.; Yu, Zh.

Zuluaga, C., 1977: Längste Kreise in Graphen. Diplomarbeit an der TH Ilmenau, Sektion Math., Rechentechnik u. Ökonomische Kybernetik, 1977.

Zuluaga, C. → Voss, H.-J.

Zykov, A. A., 1949: On some properties of linear complexes (Russian). Mat. Sbornik, N. S., **24** (1949), 163-188 (Amer. Math. Soc. Transl. **79** (1952)).

# Symbol Index

# Author Index

# Subject Index

The letter $G$ always denotes a graph.

9 780792 308997